Proceedings for the 2015 International Workshop on

MATERIALS, MANUFACTURING TECHNOLOGY, ELECTRONICS AND INFORMATION SCIENCE (MMTEI2015)

Proceedings for the 2015 International Workshop on

MATERIALS, MANUFACTURING TECHNOLOGY, ELECTRONICS AND INFORMATION SCIENCE (MMTEI2015)

Wuhan, China, 9 – 11 October 2015

Editor

Xiao Long Li

Indiana State University, USA

World Scientific

NEW JERSEY • LONDON • SINGAPORE • BEIJING • SHANGHAI • HONG KONG • TAIPEI • CHENNAI • TOKYO

Published by

World Scientific Publishing Co. Pte. Ltd.
5 Toh Tuck Link, Singapore 596224
USA office: 27 Warren Street, Suite 401-402, Hackensack, NJ 07601
UK office: 57 Shelton Street, Covent Garden, London WC2H 9HE

British Library Cataloguing-in-Publication Data
A catalogue record for this book is available from the British Library.

MATERIALS, MANUFACTURING TECHNOLOGY, ELECTRONICS AND INFORMATION SCIENCE (MMTEI2015)
Proceedings of the 2015 International Workshop on Materials, Manufacturing Technology, Electronics and Information Science

ISBN 978-981-3109-37-7

Preface

This proceedings published by World Scientific consists of 47 selected papers presented at the 2015 International Workshop on Materials, Manufacturing Technology, Electronics and Information Science (MMTEI2015), which was successfully held in Wuhan, China during October 9-11, 2015. The MMTEI2015 was sponsored by China Three Georges University, Chengdu University and Advance Science, Technology and Industry Research Center (ASTIRC) and published by World Scientific.

The MMTEI2015 covers a wide range of fundamental studies, technical innovations and industrial applications in Material Science and Application, Mechanical Engineering and Mechatronics, Electronics Engineering and Microelectronics, and Information Science. The MMTEI2015 provides a forum for scientists, scholars, engineers and students from universities all around the world and the industry to present ongoing research activities, and hence to foster research relations between the universities and the industry. This conference provides opportunities for the delegates to exchange new ideas and application experiences face to face, to establish business or research relations and to find global partners for future collaboration.

This proceedings tends to collect the up-to-date, comprehensive and worldwide state-of-art knowledge on Material Science and Application, Mechanical Engineering and Mechatronics, Electronics Engineering and Microelectronics, and Information Science. All of accepted papers were subjected to strict peer-reviewing by 2-3 expert referees. The papers have been selected for this proceedings based on originality, significance, and clarity for the purpose of the conference. The selected papers and additional late-breaking contributions made an exciting technical program on conference. The conference program was extremely rich, featuring invited keynote speeches, panel discussions, high-impact oral and poster presentations.

Finally, on behalf of the MMTEI2015 organizing committee, we would like to take this opportunity to express my appreciation to all the authors who submitted their manuscripts to this book. And, we would like to express my deepest and sincere gratitude to Dr. Yan H. Ng from World Scientific for his elaborate support dedicated to the publication of this proceedings.

Contents

CHAPTER 1

Material Science and Application

Welding defects of SUPER304H steel and their countermeasures*

Xiao-Ming Liu†, Yun-Peng Gao, Zhi-Gang Wei
Inner Mongolia Electric Power Science and Research Institute, Hohhot, 010020, China
†E-mail: 31535029@qq.com

Hou-Xia Yan
JINSHAN Thermal Power Plant of Inner Mongolia Energy Power Investment LTD. Hohhot, 010106, China
E-mail: yanhouxia@163.com

In order to guarantee welding quality and prevent welding defects, welding procedure and welding defects of SUPER304H were studied. Welding procedure was judged by qualification tests in accordance with W*elding Procedure Qualification Regulation.* Welding defects were discussed using observations of the metallurgical structure as well as photographs. The results show that normal procedure is very important for gaining excellent welding quality. Weld defects such as lack of penetration, lack of fusion, overlap, penetration, gas pores and overheat easily appear when welding under abnormal weld procedures, which include insufficient or excessive weld input, much higher weld interlayer temperature, and no inert gas protection in the interior. The authors therefore suggest carrying out normal welding technology strictly during welding operation to avoid weld defects effectively and to obtain high quality weld joins.

Keywords: SUPER 304 H steel; welding crack; welding strength; heat input.

1. Introduction

The SUPER304H steel is mainly employed to prepare the superheater tube and reheater tube with supercritical and ultra-supercritical units. As to its chemical composition, strengthening elements include Cu, Nb and N added in TP304H steel. Among which, functions of the elements including Cu, Nb, and N in the alloy can be divided into two aspects: on one hand, they can prevent the Austenite grain from growing up, so as to endow the SUPER304H steel with a

* Work partially supported by 2011 second batch of science and technology project of Inner Mongolia electric (Group) CO., LTD.

grain size finer than TP304H steel; on the other hand, the carbon nitride compound of Nb plays an strengthening role. In addition, the intermetallic compound generated in the aging precipitation of Cu can also improve strength of the SUPER304H steel [1, 2]. SUPER304H is a new steel which is ideally characterized with high temperature strength, and rupture ductility and erosion resistance under high temperature. Therefore, the study of this new material in the power system has been research focus.

In the practical engineering application of the SUPER304H steel, the welding quality can determine the overall engineering quality essentially. From the perspective of the material prosperity, SUPER304H steel is an austenitic stainless steel, which has poor thermal conductivity and great linear expansion coefficient: its resistivity is four times lower than that of the low carbon steel, and its linear expansion coefficient is two times greater than that of the low carbon steel, which leads to great welding stress during the welding process. In the view of the technology execution, although the welding procedure qualification performed on the SUPER304H steel is able to ensure the welding quality, there exist problems such as the inefficient execution in the procedure and so on, in its practical application. Consequently, the wielding stress gives rise to defects such as welding crack, and the defects such as porosity, incomplete penetration, incomplete fusion, undercut, overlap and so on result from the execution during wielding process. Therefore, analyzing the wielding defects of the SUPER304H steel and countermeasures of wielding defects are of great guiding significance for preventing wielding defects and obtaining perfect weld joint.

2. Experimental Materials and Methods

The SUPER304H austenitic stainless steel was used as the Base metal in the experiment. Its chemical compositions and mechanical properties are shown in Table 1 and Table 2[3] respectively. The Gas Tungsten Arc Welding (GTAW) technology was adopted as the wielding method. In this research, the wielding procedure qualification was conducted on the Super 304H steel according to the *DL/T868-2004 Welding Procedure Qualification Regulation.*

Finally, standard welding parameters of the steel Super 304H were determined included as follow: when the welding position is 2 G, the parameters consisted of : current range from 75 to 80A, the voltage of 11 to12V, welding speed of 50~70 mm/min, argon flow for 10~12 L/min, and back argon flow of 10~18L/min; in case the welding position is 5 G, the current of 75~78A, while the voltage is in the range from 11 to 12V. Other parameters included welding speed of 50~60 mm/min, argon flow of 10~12 L/min, and back argon flow 8~18

L/min. The inlayer temperature of the welding required being smaller than100°C [4]. In the welding process, despite the great welding stress of the Austenitic stainless steel SUPER304H, non-standard welding technology also led to welding defects.

Table 1. Chemical Composition of SUPER304H

Steel Type	C	Si	Mn	P	S	Ni	Cr	Cu	Nb	N
SUPER304H	0.08	0.030	0.82	0.035	0.001	8.60	18.50	2.91	0.41	0.35

Table 2. Mechanical properties of SUPER304H

Steel Type	Re(MPa)	Rm(MPa)	A(%)
SUPER304H	>235	>590	>30

3. Welding Defects and Its Countermeasures

3.1. *Welding defects caused by material property*

Metallographic structure of the SUPER304H steel is shown as the single austenitic structure. The austenitic steel showed no cold crack tendency, had a preferably overall welding property. However, the concentration of alloying elements including Cr, and Ni was relatively high in the SUPER304H steel, which leading to an obvious hot crack tendency. While welding steel like the SUPER304H is expected to generate hot crack in the weld if the heat input was not well controlled or the interlayer temperature was inappropriate, as illustrated in Figure 1 and Figure 2.

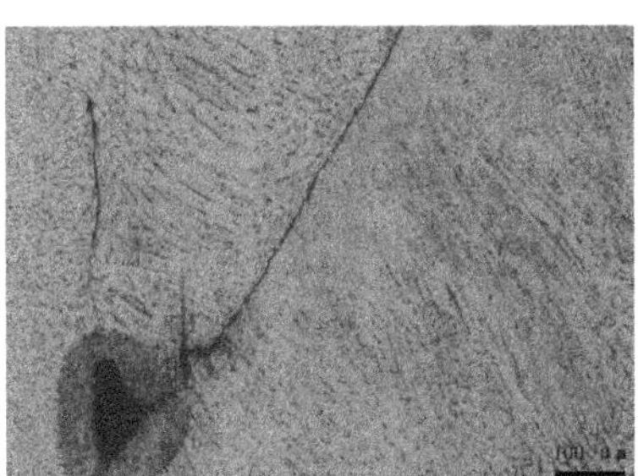

Fig. 1. Hot crack of vertical position weld

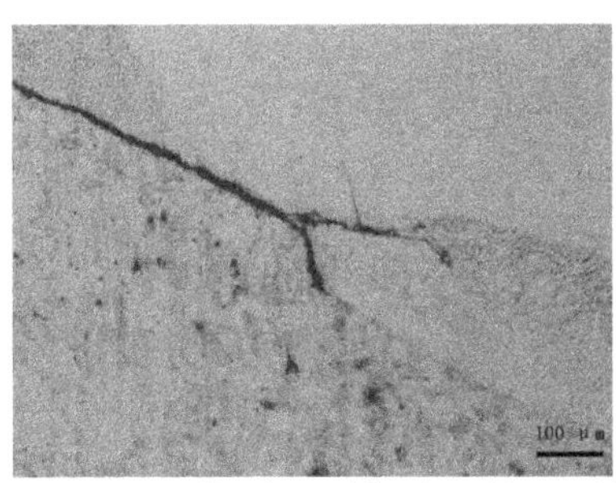

Fig. 2.Hot crack of horizontal fixed position weld

The welding hot crack of the SUPER304H steel mainly consists of three types: solidification crack, high temperature liquidation crack and high temperature brittle crack. In terms of the formulation process, it can be divided into three parts: firstly, in the solidifying crystallization process of the deposited metal fused in the welding line of the SUPER304H steel, some residual liquid films in the solidification grains might be ripped by the shrinkage stress, and thus resulted in the crack. However, the SUPER304H steel contained the alloying elements with high content, and the deposited metal had a relatively poor fluidity, which was not able to provide enough liquid metal to fill the crack in time. As a result, the solidification crack was formed. For second part, the base metal SUPER304H steel was melt partially when heated at high temperature in the over-heated part of the welding heat affected zone. And then, in the solidification-crystallization process, the partially melted base metal in the grain boundary also resulted in such phenomenon, where the liquid film was ripped and could not be filled immediately. Finally, the high temperature liquidation crack was formed in the overheated part of the welding heat affected zone, in which the base metal SUPER304H steel lay in. In the third part, when the base metal SUPER304H steel did not melt as heated by high temperature in the over-heated part of the welding heat affected zone, grains of the SUPER304H steel accelerated its growing speed and coarsen terribly, which was expected to reduce its ductility and toughness. As the ductility and toughness reduced to a certain level, the cracks were generated under the influence of the stress, and finally developed into high temperature brittleness crack. Considering the countermeasure of SUPER304H steel, , the welding heat needed to be as low as possible in addition to strict control over the interlayer temperature[5–9].

3.2. *Welding defects caused by the inappropriate welding technology*

A strict execution on the welding technology can guarantee perfect welded joint, there is no exception to all steels. However, in the engineering practice of welding SUPER304H steel, due to being limited by factors such as welder and objective conditions of welding scene, the welding technology could not be always conducted strictly. Phenomena such as insufficient heat input, overhigh heat input, the interlayer temperature greater than set value, non Argon welding inside and utilizing a method of filling wire outside, etc, are likely to appear, which gives rise to the welding defects.

(1) Insufficient heat input.

Insufficient heat input mainly caused by two factors: a small welding current and a fast welding speed. Both factors tend to result in defects such as the incomplete penetration (Figure 3) and incomplete fusion (Figure 4) in the welding process.

Fig. 3. Incomplete penetration of weld

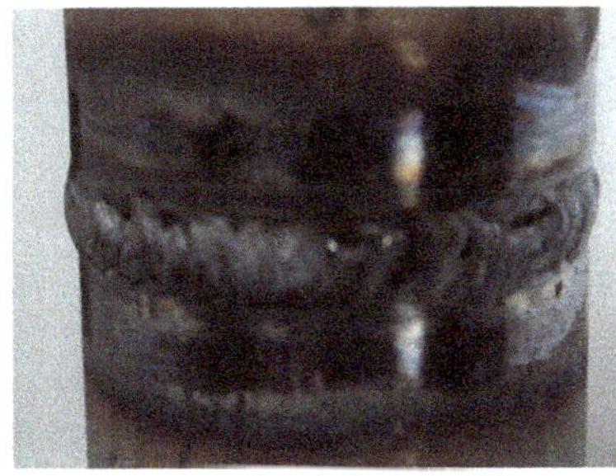

Fig. 4. Incomplete fusion of weld

(2) High heat input.

Similarly to insufficient heat input, high heat input usually sources from big welding current and slow welding speed. If the heat input is too high, overlap defects in the welding root, and undercut defects in the surface of the weldment are produced, as shown in Figures 5–6.

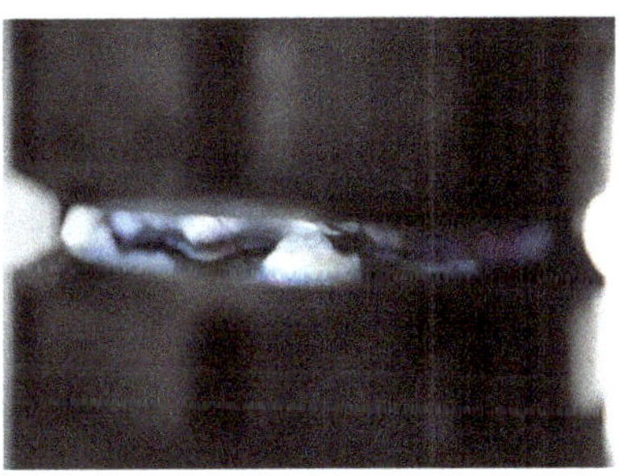

Fig. 5. Overlap of in the welding root

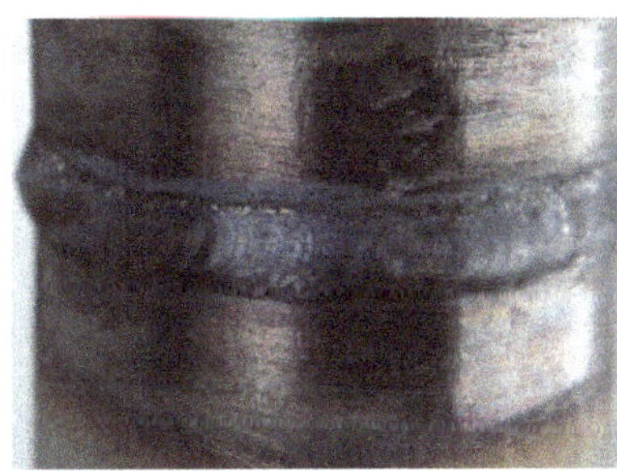

Fig. 6. Undercut in the surface of the weldment

(3) Excess of the welding interlayer temperature.

The interlayer temperature for welding steel SUPER304H was expected to be lower than 100°C. If the interlayer temperature was too high, fluidity of the molten steel reduced apparently, and the molten even stopped flowing. Under such circumstance, if the welding was still going on, the interlayer temperature of the weld was expected to increase, the welding operation become more difficult; while thickness of the weld increased also. As a result, the porosities were likely to be generated in the welding line (Figure 7–8), which further affecting mechanical property of the welding line. Therefore, the interlayer temperature had to be strictly controlled in the welding process.

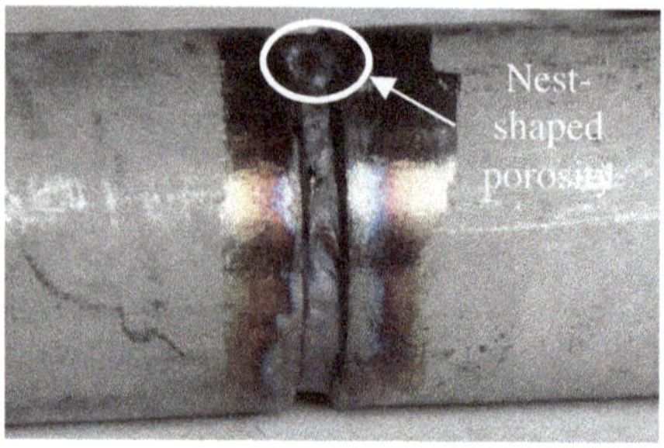

Fig. 7. Nest-sharped porosity of weld

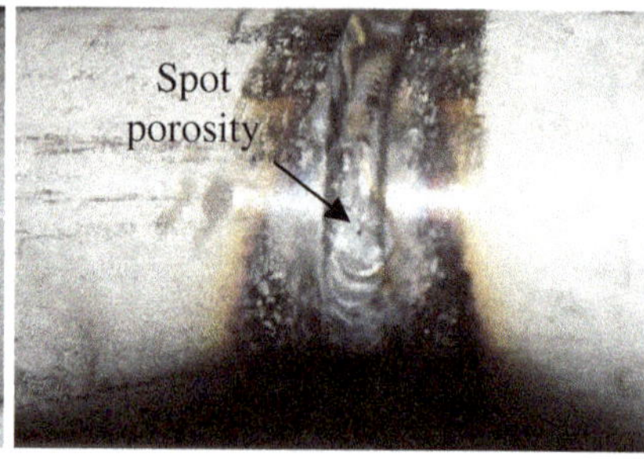

Fig. 8. Spot porosity of weld

(4) Lack of argon filling inside.

In order to prevent the alloying elements in the high temperature zone from oxidation, the argon filling protection needed to be carried on the reverse side in the whole welding process, which could provide favorable protection for the root of the backing weld. The overburning phenomenon (Figure 9) was found on the weld root and base metal when there was no argon protection in the root. Moreover, welding on the root was proved to be disqualified and influenced mechanical properties of the root[10].

(5)Welding defects caused by utilizing the method of filling wire outside.

Owing to the molten iron of the steel SUPER304H was sticky; the conventional wire filling method could not ensure the molding of weld root. Besides, it was likely to produce defects like incomplete fusion in the weld interlayer and incomplete penetration in the weld root and so on in welding process, as shown in Figure 10. Therefore, the method of filling wire inside had to be adopted while conducting the backing weld.

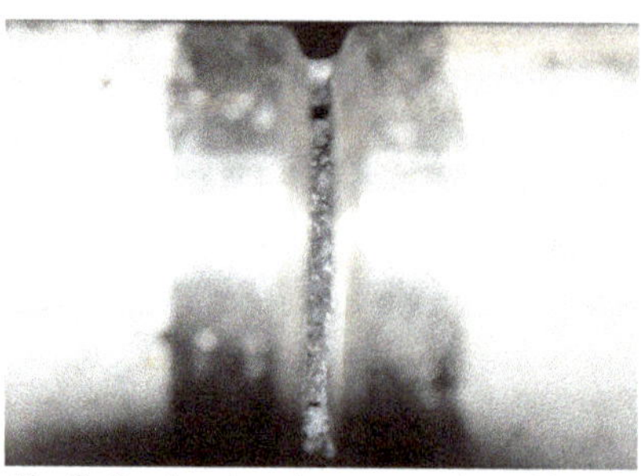

Fig. 9. Overburning phenomenon in weld root

Fig.10. incomplete penetration in weld root

Above all, the welding defects of the SUPER304H steel were caused not only by its own material properties, but also by improper selection of the non-standard welding technology parameters. Under the mutual influences of the fine grain strengthening and dispersion strengthening, the property of the

SUPER304H steel was enhanced. Meanwhile, the crack problem for welding the steels like the SUPER304H needed to be solved. Welding methods used to other Cr-Ni Austenitic steels were not suitable here. Because the welding methods for common Cr-Ni Austenitic steels usually involve an increasing content of ferrite former in order to generate sufficient δ phase. Then, in welding process of SUPER304H steel, the risks arising from δ phase brittleness tend to occur in the weld using common method. To avoid high temperature crack in the weld, the appropriate technology parameters (as shown in Table 3) need to be adopted, namely, a method of reducing heat input of welding and interlayer temperature with strict operation.

Inappropriate welding technology parameter is expected to result in a series of welding defects in welding the SUPER304H steel. For example, defects like incomplete penetration, incomplete fusion and overlap, undercut are apt to occur as welding heat input is low or high; while high interlayer temperature and the non Argon welding inside are likely to give rise to porosity and burnt phenomenon on the root; the method of filling wire outside leads to defect like incomplete penetration. In summary, a strict execution over the standard welding technology parameter is required in the experiment, so that it is an important guarantee for obtaining a perfect welded joint.

4. Conclusion

(1) Standard welding parameters of the steel Super 304H were determined included as follow: when the welding position is 2 G, the parameters consisted of : current range from 75 to 80A, the voltage of 11 to12V, welding speed of 50~70 mm/min, argon flow for 10~12 L/min, and back argon flow of 10~18L/min; in case the welding position is 5 G, the current of 75~78A, while the voltage is in the range from 11 to 12V. Other parameters included welding speed of 50~60 mm/min, argon flow of 10~12 L/min, and back argon flow8~18 L/min. The inlayer temperature of the welding required being smaller than100°C.

(2) A commonly seen defect shown in welding steel SUPER304H is welding crack, is mainly caused by the non-standard heat input and relatively high interlayer temperature except for technology influence. Besides, poor weldability of such material itself is the essential reason leading to the welding crack.

(3) Similarly to other steels, owing to insufficient heat input, overhigh heat input of welding, excess of the interlayer temperature, non Argon welding inside and method of filling wire outside, the defects such as, incomplete penetration, incomplete fusion, undercut, overlap, porosity and over burning are produced.

This research suggests to strictly adopting standard technology parameters in welding operation so as to avoid welding defects.

References

1. Li Dong-Sheng, Dai Qi-Xun, Wang Guo-Jian, et al. High-temperature oxidation resistance of austenitic stainless steel Super304H[J]. Journal of Jiangsu University (Natural Science Edition). 2012, 33(5) pp. 581–585.
2. Li Xin-Mei, Zou Yong, Zhang Zhong-Wen, et al. Microstructure and Property of Novel Heat-resistance Steel Super304H After Being Aged at High Temperature[J]. Journal of Materials Engineering. 2009, (5) pp. 38–42.
3. Peng Fang-Fang, Zhu Guo-Liang, Song Jian-Xin. Analysis of Key Manufacturing Process in Domestic Producing for Super304H Steel Tube Applied in Ultra-supercritical Unit[J]. Journal of Power Engineering. 2008, 28(5) pp. 803–806.
4. Li Xin-Mei, Zou Yong, Zhang Zhong-Wen, et al. Microstructure and Property of Novel Heat-resistance Steel Super304H After Being Aged at High Temperature[J]. Transactions of Materials and Heat Treatment. 2009, 30(6) pp. 51–56.
5. Zhang Chunlei, Xiong Xiahua, Zheng Zhijun et al. Research Progress on Intergranular Corrosion Resistance of Super304H Austenitic Stainless Steel [J]. Special Steel. 2013, 34(3) pp. 34–38.
6. Li Xin-Mei, Zhang Zhongwen, Zou Yong, et al. Lattice Parameter Evolution of Super304H Steel γ Phase Aged at High Temperature[J]. Materials Review. 2014, 28(12) pp. 102–104.
7. Bai Jia-Wei, Zhan Qian, Chi Cheng-Yu, et al. Strengthening effect of precipitate phase in Super304H austenitic heat-resistant steel after long-term aging[J]. The Chinese Journal of Nonferrous Metals. 2014, (12) pp. 3042–3050.
8. Li Xin-Mei, Zhang Zhongwen, Zou Yong, et al. Study on Susceptibility to Intergranular Corrosion of Super304H and TP304H Stainless Steel[J]. Hot Working Technology. 2011, 40(6) pp. 5–7.
9. Li Ping, Li An-Na, Pang Sheng-Jiao, et al. Corrosion Behavior of Super304H with Alkali Resistance Metal Sulfate[J]. Journal of Materials Engineering. 2015, 43(1) pp. 54–58.
10. Ou Ping, Sun Jian, Cui Zheng-qiang, et al. Microstructure of Super304H Austenitic Heat-Resistant Steel After Long-term Aging[J]. Transactions of Materials and Heat Treatment. 2014, 35(5) pp. 86–91.

A study of screen/stencil printing on LTCC substrate*

Fei Zhao†, Yuan-Lan Dang

The 54th Research Institute of CETC,
Shijiazhuang Hebei 050081, China
†E-mail: fly.zf@163.com

Lines and vias for connecting in LTCC (Low Temperature Co-fired Ceramic) modules herein are both separately formed with screen and stencil by printing, whose performance is strongly dependent on various factors. In this paper, the key factors, including speed, snap-off distance and some special techniques, which influence the printing quality of line printing and via filling are discussed. Important details for the whole printing process are also emphasized.

Keywords: LTCC, line printing, via filling, key factors.

1. Introduction

With the rapid development of communication, electronic devices and systems are demanded to have the features of miniaturization, light-weight, high-speed, high-frequency, high-reliability and low cost, especially for the apparatus utilized in harsh environment. LTCC technology[1], featured by low loss materials, low firing temperature and precise procedures during multilayer lamination, is one of the effective methods for fabricating electronic systems with great integration[2] and mechanical systems. Printing, an ancient technique which originated from cloth structuring, is considered to be one of the most important procedures in LTCC process[3]. Figure 1 shows a sketch of screen printing, where squeegee with certain pressure is used to sweep the paste from one side to the other on a screen with tension, thus the paste penetrating the opening of screen is deposited on a substrate which is fixed firmly by vacuum on the printing table.

* Work partially supported by Processing Foundation of CETC54 (No. N1124012).

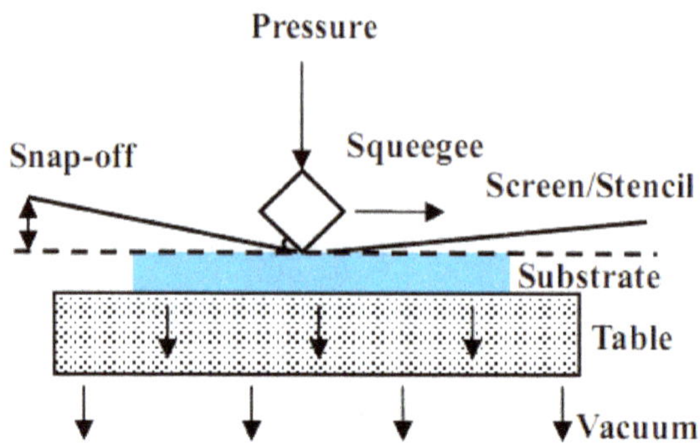

Fig. 1. Sketch of printing

It can be summarized from the reports[4] that printing is affected by three main factors, in which paste characterization including viscosity and rheology, shape and hardness of squeegee, roughness of substrate, screen parameters, such as mesh count, wire diameter, thickness of emulsion, are considered to be fundamental but important "material factor", illustrated in Figure 2. "Printing parameter factor", including printing pressure, printing speed and snap-off distance (the distance between screen and substrate), plays a key role during printing process and can be adjusted at certain "environment", such as temperature, humidity and cleanliness, which are contained in the third type of factor. As mentioned above, there are so many variables affecting the printed pattern, especially for the thickness of circuit lines and structure of filled vias. In order to get better printed profile, a study should be implemented to determine how the variables change. In this article, line printing and via filling are fulfilled to analyze some key influences to the states of printing.

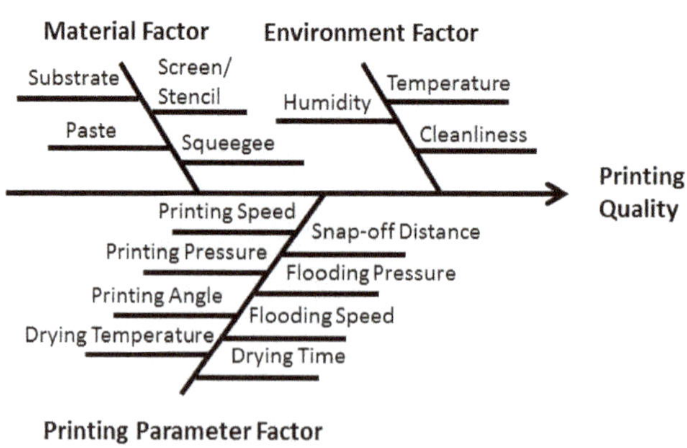

Fig. 2. Influence factors to printing quality

2. Experiment

Printing quality of fine line is related to the thickness of the paste, which would influence the frequency characterization of the whole module. Furthermore, the height of filled vias could also affect the flatness and connection of the lines, which are printed on the vias.

To monitor the different affection of the factors listed above to the printing quality, line printing and via filling are separately done with DuPont's conductor paste and via filling paste and then inspected by optical profiler, microscope and X-Ray inspection.

3. Results and Discussion

3.1. *Line printing*

As concerned above, one of the most important material factor during printing is the status of the paste, in which viscosity plays a key role in the whole run. As is shown in Figure 3, the thickness of the printed line varies while the viscosity of the paste changes. With the same printing parameter, except for viscosity at 290 Pa·s, the thickness of the paste is around 11.8 μm, while at 350 Pa·s, the thickness goes to 12.5 μm, and finally the thickness increases to 15.4 μm at 374 Pa·s.

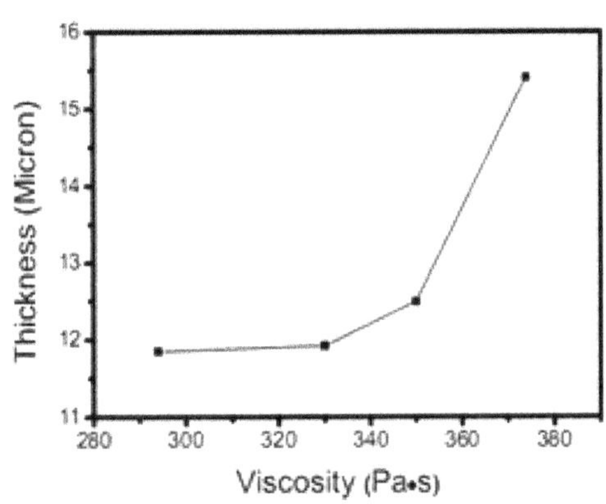

Fig. 3. Relationship between thicknesses to paste viscosity

The original account for the relationship between thickness and paste viscosity could be explained by the fluidity of the paste after printed[5]. The higher the viscosity is, the worse the spreading behavior of the paste acts, especially when paste is printed on rougher substrates. At higher viscosity, the spreading of the line printed is blocked so as to outstanding figure, while a slumping height at decreasing viscosity.

Except for the viscosity of paste utilized in printing process, snap-off distance is another key factor in line printing, which would directly affect the print quality and thickness of the figure. Figure 4 shows the relationship between the thickness and the snap-off distance, from which it can be seen that at the beginning, thickness increases as distance rising. While to some extent, it falls to less than 10 μm at the distance of 2 mm.

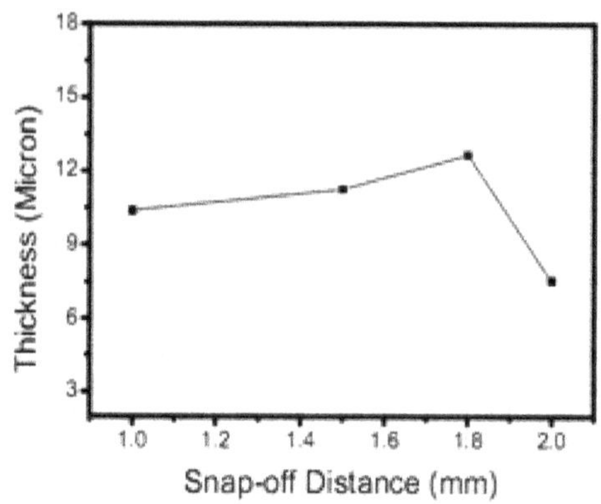

Fig. 4. Relationship between thickness changes to snap-off distance

Snap-off distance should be set properly during the process, because with too small distance, the screen will be stuck on the substrate, resulted in indistinct image. With the increase of the distance, paste has much more space to fall through the screen after pressed by squeegee and the thickness of the paste may get higher. But on the contrary, if the screen leaves too far to touch the substrate during printing, the thickness will definitely decrease.

In some cases, there are not only one but various types of paste with different functions, such as conductor line, resistor, solderable and bondable paste, needed to be printed on the same substrate. Herein, printing order should be considered in order to get better printing quality. Figure 5 illustrated the difference of printing effect for conductor paste and resistor paste, the latter of which is black. In Figure 5(a), resistor is printed firstly and after some drying time, conductor paste is printed on the top of resistor with overlap, while printing is executed oppositely in Figure 5(b), where it can be observed that when resistor paste is printed on the existed conductor paste, the area of overlap (shown by circle in the figure) may have leakage of paste spreading along with the edge of the conductor line. This situation would not happen when the resistor has been printed firstly and followed by conductor printing. Actually, the influence of printing order not only affects the appearance of printing, but also the value of sheet resistance Rs, defined in formulation (1), where R is the resistance of resistor with length of l and width of w.

$$R = R_S \frac{l}{w} \tag{1}$$

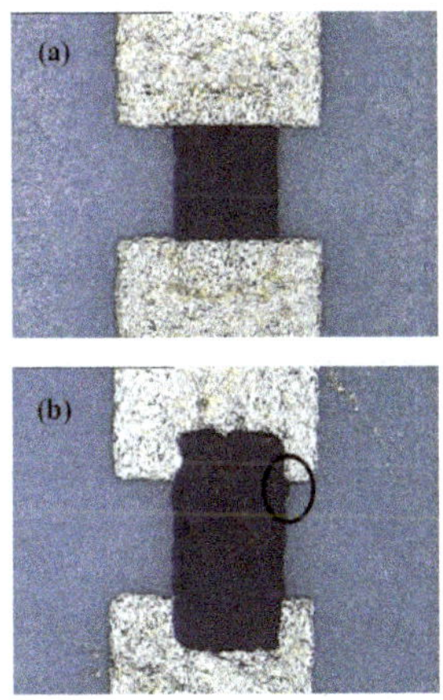

Fig. 5. (a) Conductor overlaps resistor (b) Resistor overlaps conductor. Printing order difference

The appearance distinction, identified as the leakage of resistor paste, shown in Figure 5(b) is resulted from the lower viscosity of resistor paste, which distributes less than 100 Pa·s. Thus, the paste will spread successfully after printed, especially for space provided by existed conductor lines. Similarly, the lower resistance is also related to the wider and irregular shape of the square.

According to experience, fine line printing is also related to squeegee and its angle. With 60 shore's hardness and 45° angle squeegee, fine line up to 75 μm could be printed successfully.

3.2. *Via filling*

Via filling in LTCC is accomplished by stencil to achieve the connection between multi-layers. In order to get better connection, the vias should be fully filled but not excessively, which may lead to protuberance and broken line on them.

Printing speed during via filling is a sensitive factor for the filled status of vias, especially for 100 μm ones. Table 1 shows the results of filled vias with different speed. As measured by optical profiler, at lower speed of 5 mm/s, vias could be fully filled with the height of 17.4 μm, while 14.1 μm with 10 mm/s speed. When the speed increases to 15 mm/s, the vias are failed to be filled, whose height is tested to be -13.5 μm, which means that the paste is lower than the substrate, indicating to unfilled ones.

Paste left in the vias is largely dependent on speed of squeegee that lower speed at certain pressure makes much paste to go through the stencil, and while the squeegee moves too fast, the paste would have little time to pass through.

Table 1 Via-filled results with different speed

No.	Speed (mm/s)	Filled Result	Height (μm)
1	5	√	17.4
2	10	√	14.1
3	15	×	-13.5

In practice, vias filled by printing often go too high because of the usage of 50 to 100 μm thick stencil, leading to broken lines printed on the vias. In order to solve this problem, a flattening technique is used to make the height of vias low enough to be printed on. Table 2 shows the comparison between flattened vias and non-flattened ones. After flattening, the height of vias with dimension of 150 μm changes from 36.2 μm to 12.5 μm, which is much lower and would not block the lines printed on them, as is shown in Figure 6. One more thing should be considered that after flattened, the vias could also gain about 10 to 30 μm larger in diameter, which should be offset in the design layout in accurate manufacturing.

To compensate in volume of paste, changing the stencil opening size is still an effective way[3], which could also improve the misalignment during via filling.

Fig. 6. Picture of lines printed on top of flattened via

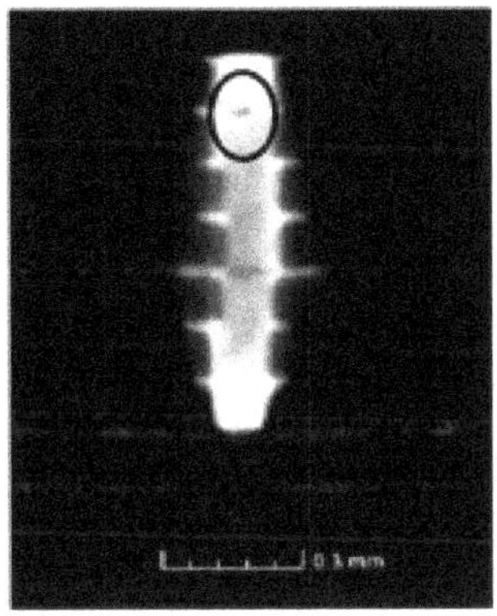

Fig. 7. X-ray inspection of filled vias

Table 2 Non-flattened and flattened vias

Diameter (μm)	Non-flattened Height (μm)	Flattened Height (μm)
152.9	33.9	10.1
148.1	33.3	12.4
162.0	41.4	14.9
Average	36.2	12.5

In certain condition, vias around 100 μm could be fulfilled with proper paste, speed and pressure. AOI[6] provides a convenient and highly effective way to inspect the defects during each printing step. Still, X-ray inspection to substrates is an effective way to catch the defects, shown in figure 7.

4. Conclusion

According to the analysis to the factors affecting the printing quality, fine line printing and via filling are successfully achieved by controlling paste viscosity, printing speed, snap-off distance, printing order and special technique, such as via flattening and suitable opening. For further development of LTCC process, these experiences could help adjusting quality quickly to eliminate unexpected defects.

Acknowledgements

This research is sponsored by Processing Foundation of CETC54 (No. N1124012).

References

1. W. A. Vitriol, J. L. Steinberg. Development of a low temperature co-fired ceramic technology. Intern. J. Microelectron and Electron. Pkg., 593–598, 1983.
2. Ke-Li W, Yong H, LTCC technology and its applications in high frequency front end modules, Antennas, Propagation and EM Theory, 730–734, 2003.
3. Vasudivan S., Arulvanan P., Chua K. M., Wong C. K, Process issues and characterization of LTCC substrates, Electronic Components and Technology, 1933–1937, 2004.
4. Jerzy A. O., Frank L. H., A study of the off-contact screen printing process. I. Model of the printing process and some results derived from experiments, IEEE Trans. on Components, Packaging, and Manufacturing Technology, 358–367, 1990.
5. Michael N., Weiming Z., Dong Z., Annette K., Screen-printing simulation study on solar cell front side AG paste, IEEE Photovolatic Specialists Conference, 2008.
6. Zhiyi T., The technique status and trend of LTCC, Equipment for Electronic Products Manufacturing, 166(11): 1–9, 2008.

Heavy metal pollution prediction based on hybrid ARIMA and grey model*

Ji-Yuan Zhang† and De-Xia Li
School of Computer Science and Engineering,
GuiLin University of Electronic and Technology,
GuiLin, Guang Xi 541004, China
†E-mail: 27141432@QQ.com
www.guet.edu.cn

Ting-Lei Huang
GuiLin University of Electronic and Technology,
GuiLin, GuangXi 541004, China

Accidents involving heavy metal pollution of water and the environment cause huge casualties and loss of property, making the monitoring of heavy metal pollution of great importance. In order to overcome the disadvantages of the traditional model algorithm, such as low predictive accuracy and poor fusion ability, a new combined forecasting model was developed based on the autoregressive integrated moving average model and grey theory. The autoregressive integrated moving average model time series forecasting methods were used to make a preliminary forecast for heavy metal historical data. Then a non-interval GM(1,1) model was built. The model residual of initial forecasts were put as the input of GM(1,1) to be modified. The non-interval GM(1,1) model is established regarding the residual value of the ARIMA. Finally, the initial forecasts of ARIMA model and the residual prediction value of GM(1, 1) fusion and combination forecast GM-ARIMA model were constructed. The heavy metals monitoring data of a river in Hebei province in 1978–2007 were taken as the sample to be analyzed. The predictive results of the autoregressive integrated moving average model, GM(1,1) model and the GM-ARIMA model are compared. The results show that the GM-ARIMA model has better predictive performance and fusion ability.

Keywords: Pollution forecast; ARIMA model; grey model; combined forecasting.

With the industry development, a large amount of heavy metal is discharged into rivers, lakes and oceans and the heavy metal pollution is increasingly becoming serious[1]. The pollution accidents have damaged people's health and caused social crisis. Domestic and foreign scholars have carried out many researches

* This work is supported by National High Technology Research and Development Program 863 (NO. 2013AA065502).

such as heavy metal test methods, morphological content characteristics and influence on vegetation covers[2-4].

In this paper, we use a new combined forecasting model which base on the autoregressive integrated moving average model and grey theory to establish the prediction model of heavy metal pollution in water environment, and to probe the change law of heavy metal concentration in water.

1. ARIMA Model

ARIMA is the abbreviation of Autoregressive Integrated Moving Average Model which is a famous time series forecasting method, proposed by Boks and Jenkins in early 1970s[5]. The method is widely used in the fields of meteorology, astronomy, economics, management, machinery, biology, medicine, water conservancy, chemical industry, agriculture, forestry, etc., and has achieved good results[6].

The model is suitable for the non-stationary time series, which uses d times difference to transform the non-stationary time series into stationary time series[7], then determine order and estimate parameters to get value of parameters p and q. After that the model ARIMA(p,d,q) can be used to predict and analyze time series. Essentially, the ARIMA model is processed by the ARMA model after the time series is stationary[8].

ARIMA (P, D, q)'s basic idea:

The forecast object's data series which followed by time are processed into non-white noise sequence. The unitary root test of data series is done with ADF(Augmented Dickey Fuller) test method to determine whether the series are stable or not. According to the properties of the sample's correlation coefficient and partial correlation coefficient, the samples are fitted with the ARIMA (D, Q, P) model by appropriate order. The future tendency can be forecasted from the past and present values of the time series with the fitting model.

ARIMA ARIMA(p,d,q) are denoted as:

$$\begin{aligned} y_t &= \varphi_1 y_{t-1} + \cdots + \varphi_p y_{t-p} \\ &\quad - \theta_1 \varepsilon_{t-1} - \cdots - \theta_q \varepsilon_{t-q} + \varepsilon_t \end{aligned}, \tag{1}$$

Where T represents the time serial number, ϕ_i and θ_i are the model's parameters, ε_t is a white-noise process whose mean is 0 and variance σ_t^2. With the backward shift operator B, the upper formula can be written as:

$$\varphi(B) y_t = \theta(B) y_t \,. \tag{2}$$

The expressions of $\varphi(B)$ and $\theta(B)$ are like this :

$$\begin{cases}\varphi(B)=1-\varphi_1 B-\varphi_2 B^2-\ldots-\varphi_p B^p \\ \theta(B)=1-\theta_1 B-\theta_2 B^2-\ldots-\theta_p B^p\end{cases}. \tag{3}$$

2. GM (1, 1) Model

The gray system theory, which is proposed by Professor Deng Julong in 1982, is a method to study the grey system analysis, modeling, forecasting, decision-making and control[10]. In the grey system theory, the model is formed from generating-transformation of less or less accurate original data sequence which represent the gray system's behavior character. It is used to describe the continuous process of gray system, so which is called Gray Model, abbreviated GM (m, n) model. GM (m, n) is a n-order differential equation with m variables. GM (1, 1) model is usually used.

The basic idea of GM (1,1) model is that Sequence 0 is formed from the original data and Sequence 1 is formed from Cumulative generation method of original data, then establish the differential equation model of the GM model. The GM(1,1) model is a one order and one parameter differential equation model.

3. Combined Model

ARIMA and GM(1,1) work steps are as follows:

Step 1: Determining the value of parameter d.

Given a non-stationary time series of data {Yt} which equation is written like this:

$$\Delta y_t = \mu+\tau t+\phi y_{t-1}+\gamma_1\Delta y_{t-1}+\cdots+\gamma_k\Delta y_{t-k}+\varepsilon_t. \tag{4}$$

Where Δy is the first order differential processing of the studied variables, the Statistical significance of parameter can be gotten from Critical value table which is set by Dickey and Fulle.

{Yt} is processed by d times differential transformation, then it should be suitable for the stationary condition of AR(p) Model and MA(q) Model[11]. That is:

$$\begin{cases}Y_t=\varepsilon_1-\phi_1\varepsilon_{t-1}-\phi_2\varepsilon_{t-2}-\cdots-\phi_q\varepsilon_{t-q} \\ Y_t=\varphi_1 Y_{t-1}+\varphi_2 Y_{t-2}+\cdots+\varphi_p Y_{t-p}+\varepsilon\end{cases}, \tag{5}$$

suitable for

$$\begin{cases}E(Y_t)=E(\varepsilon_1)-\phi_1 E(\varepsilon_{t-1})-\phi_2 E(\varepsilon_{t-2})-\cdots-\phi_q E(\varepsilon_q)=0 \\ |\varphi_1|+|\varphi_2|+\cdots+|\varphi_p|<1\end{cases}, \tag{6}$$

Step 2: Determining the value of parameter p and q.

According to the self-correlation functions truncation and partial self-correlation function's tailing of sample sequence, the model type is preliminarily to determined, then the optimal value of p and q are selected with AIC criterion step by step.

The expression of AIC criterion:

$$AIC(n,m) = \ln L^2 + 2(n+m)/T. \tag{7}$$

In the equation, L is the maximum likelihood function, T is sample size, (n+m) represent the number of model's parameter. If there are constant terms in the ARIMA model, the number of parameters need add 1. The principle to determine the order in AIC criterion is based on:

$$AIC(p,q) = \min_{0\le n,m\le 1} AIC(n,m), \tag{8}$$

After ARIMA(p, d, q) Model is fitted, ARIMA(p,d,q) Model is got.

Step 3: Establishing the Residual model GM(1,1).

Get the residual series from ARIMA model.

$$\varepsilon_0 = \{\varepsilon\ (2), \varepsilon\ (3), \cdots, \varepsilon\ (n)\}\ ,\ \varepsilon(t) = X(t) - \hat{X}(t)\}\ , \tag{9}$$

ε_0 is the residual series of ARIMA, X(t) is actual value and $\hat{X}(t)$ is the predictive value at t moment.

ε_0 is separated into positive and negative residual series to model respectively, and the negative sequence is modeled with the absolute value.

The formula of residual model GM(1,1) is:

$$b = \frac{\Delta\varepsilon_1^+(t_k)}{\Delta t_k} + az_1^+(t_k). \tag{10}$$

Where $\varepsilon_1^+(k) = \sum_{i=1}^{k} \varepsilon_0^+(i)$ is the series generated from accumulative value of ε_0's positive residual series. $z_1^+(k) = \frac{1}{2}(\varepsilon_1^+(k) + \varepsilon_1^+(k-1))$ is the mean series of ε_0's positive residual series.

Step4: GM(1,1) forecasting

Get the fitting calculation formula of Residual prediction series [12]:

$$\hat{\varepsilon}_0^+(\mathrm{t}_k) = \frac{b\Delta t_k - a\Delta t_k \varepsilon_1^+(t_{k-1})}{1+0.5a\Delta(t_k)}. \tag{11}$$

The value of parameters are:

$$a=\frac{\Delta a}{\Delta}=\frac{\mathrm{cd}-(n-1)e}{(n-1)f-c^2},\quad b=\frac{\Delta b}{\Delta}=\frac{\mathrm{df}-ce}{(n-1)f-c^2},\quad c=\sum_{k=2}^{n} z_1^+(t_k)\Delta t_k,$$

$$d=\sum_{k=2}^{n}\varepsilon_0^+(t_k),\quad e=\sum_{k=2}^{n} z_1^+(t_k)\Delta t_k\varepsilon_0^+(t_k),\quad f=\sum_{k=2}^{n}(z_1^+(t_k)\Delta t_k)^2.$$

Step 5: Calculating the forecasting value of GM_ARIMA combined model

The sequence of positive and negative sequence is arranged in the original sequence, and the residual prediction sequence is obtained.

The prediction of the ARIMA model is Superimposed on residual prediction of GM (1, 1), which is the forecasting value exactly.

4. Model Application

The Hg concentration data of a river in Hebei Province from 2007 to 2010 are take as samples in our hybrid model. The data are the average of concentration of every month, which are collected at a fixed timed. From time series line of chart 1, it can be seen that the emission of Hg has a significant increase in the trend over years.

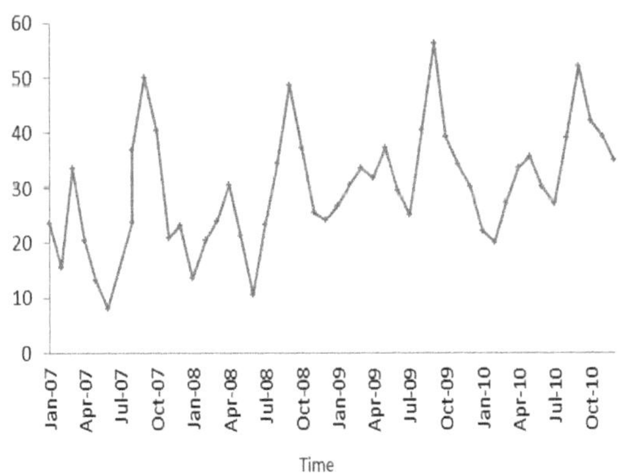

Fig. 1 The concentration data of Hg from 2007 to 2010 years

4.1. *Determining the parameter d of ARIMA model*

According to the ADF Unitary root test method, the stability of the original series is tested. As shown in the table below, the ADF test statistic is -1.346, which is greater than the p value(-2.341) of the t test whose significant level is 1% .So Unitary root exists, the original hypothesis is accepted, the series is non-stationary. The differential process is necessary.

Table 1 Series stationary test

ADF	-1.346
1%	-2.341
5%	-1.263
10%	-3.698

After 1-order difference, the ADF test statistics is less than the p test whose significant level is 1%,5%,20%. That's mean the sample trend is eliminated, there's no unitary root any more, the concentration series is stationary. Because of the series' stationarity, ARIMA (p,d,q) model's order, d=1.

4.2. *Determining the parameter p and q of ARIMA model*

Through analyzing the self-correlation function and partial self-correlation function of stationary series, the tailing of new series' self-correlation function and partial self-correlation function is discovered, which is suitable for the character of ARMA(p, q). Then the optimal value of p and q are selected with AIC criterion step by step, meanwhile the other parameters are referenced.

Table 2 The autocorrelation and partial autocorrelation function

	self-correlation	Partial self-correlation	Q Statistics	P value
1	0.336	0.336	14.225	0.001
2	0.112	0.003	15.457	0.013
3	0.047	0.021	17.553	0.016
4	-0.014	-0.007	17.119	0.008
5	-0.025	-0.013	24.336	0.021
6	-0.065	-0.021	28.336	0.048
7	-0.102	-0.085	33.589	0.032
8	-0.145	-0.112	48.556	0.52

According to the rule of model identification, the 1-4 order of the self-correlation function of the series change significantly, and the p value is assumed as 4; the value of the partial self-correlation function in first 3 order vary significantly, and the q value is assumed as 3. So here is the preliminary model shown below.

Table 3 The data of AIC for different ARIMA models

ARIMA(p, d, q)	AIC Value	ARIMA(p, d, q)	AIC Value
ARIMA(1,1,1)	4.225	ARIMA(3,1,2)	4.231
ARIMA(2,1,1)	4.147	ARIMA(4,1,2)	4.336
ARIMA(3,1,1)	4.012	ARIMA(1,1,3)	4.321
ARIMA(4,1,1)	4.056	ARIMA(2,1,3)	4.241
ARIMA(1,1,2)	4.256	ARIMA(3,1,3)	4.187
ARIMA(2,1,2)	4.008	ARIMA(4,1,3)	4.028

According to table 3, the AIC value of the model ARMA(2,2) is the minimum. So ARIMA(2,1,2) has the best fitting effect in 12 models, which is selected as our model. The unknown parameters estimation is shown in Table 4.

Table 4 Parameter estimation of ARIMA (2,1,2) model

Reference Model	Coefficient	Standard error	T Statistics	Probability
AR2	0.586	0.146	0.632	0.000
MA1	-0.125	0.108	-0.652	0.021
MA2	0.354	0.324	7.254	0.013

According to table 4, the test values of the parameters are less than 0.1 under the confidence level of 10%. So ARIM(2,1,2) is the final model. The denotation is:

$$(1-0.586)(1-B)\ Y_t = (1+125B+0.354B^2)a_t\ , \tag{12}$$

4.3. *Analyzing experimental result*

(1) Forecasting in 2011.

According to the Hg concentration of 2007-2010, the ARIMA model, GM model and GM-ARIMA model are used to forecast the value in 2011. The result is shown in figure 2, compared with predicted and actual values.

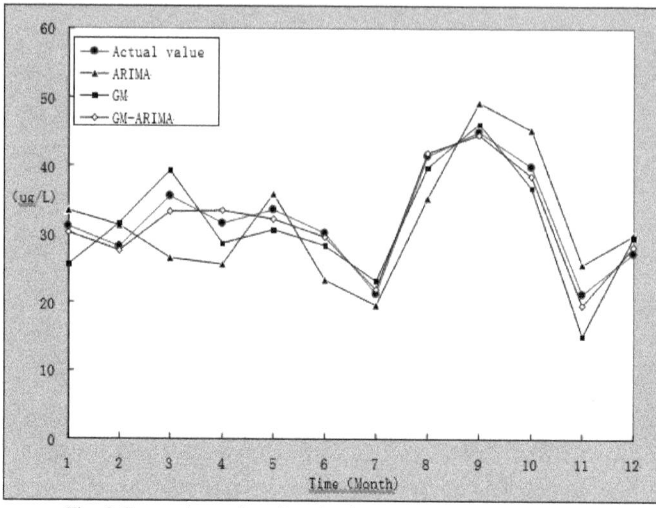

Fig. 2 Comparison of predicted values and actual values in 2011

As can be seen from Figure 2, the predicted and actual values alternate roughly simultaneously and most of the error is not large. The predicted result is satisfying.

(2) Forecasting in 2012 and 2013.

The further experiment are to forecast the heavy metal concentrations in 2012 and 2013. The results are shown in Figure 3.

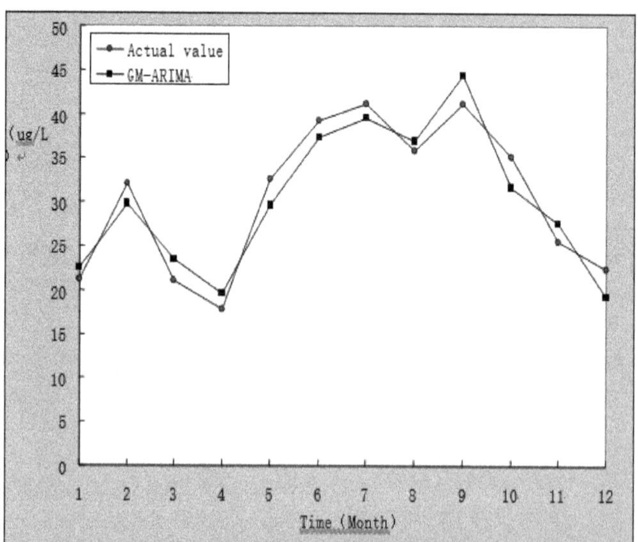

Fig. 3 Predictive value 2012.

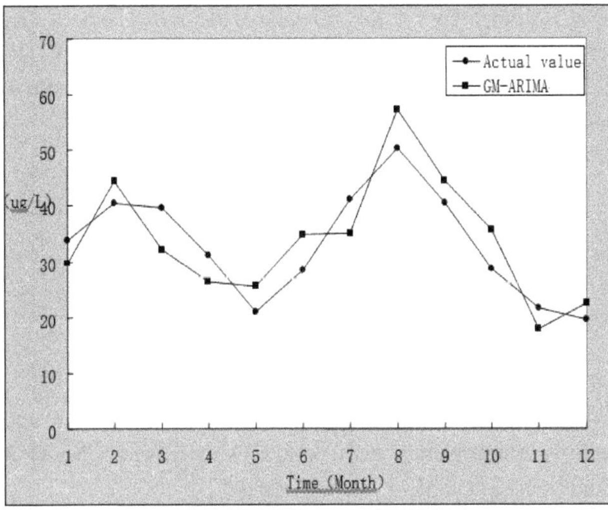

Fig.4 Predictive value in 2013

According to the analysis of Figure 3 and Figure 4, the error of the predicted and the actual values becomes large and the accuracy become lower.

(4) Analyzing the accuracy.

There are 4 accuracy evaluation indexes for predictive model: Relative error, Absolute error, Mean square error, Mean absolute error.

According to the experiment of three models, the results and errors of the heavy metal Hg concentration in 2011 are shown in the Table 5.

Table 5 The relative error of the prediction model

Time	Actual value	ARIMA(2,1,2)		GM(1,1)		GM-ARIMA	
		Predictive value	Relative error	Predictive value	Relative error	Predictive value	Relative error
1	31.1	33.4	-7.34	29.6	4.83	30.1	3.21
2	28.1	31.2	-1.10	31.6	12.45	27.5	2.13
3	35.6	26.5	2.55	39.2	-10.11	33.2	6.74
4	31.6	25.6	1.89	28.6	9.46	33.4	5.69
5	33.6	35.9	6.84	30.6	8.92	32.1	4.46
6	30.1	23.2	2.29	28.3	3.53	29.6	1.66
7	21.2	19.5	8.01	23.1	-9.04	21.9	-1.88

By the absolute error of the formulas, the mean square error, the mean absolute error and the mean absolute percentage error of the model are calculated. The comparison of accuracy evaluation indexes for GM-ARIMA, GM and ARIMA is shown in Table 6.

Table 6 Comparison of accuracy evaluation index

Accuracy	GM-ARIMA	ARIMA	GM(1,1)
MSE	3.59	5.01	8.23
MAE	2.89	4.23	7.45
MAPE	3.21	4.56	7.69

By Table 5 and 6, the relative error of GM-ARIMA model is obviously reduced, The Maxim of RE is 6.74%. The accuracy evaluation index of hybrid GM-ARIMA is much better than others. The hybrid GM-ARIMA is more effective than the single GM or ARIMA model.

5. Conclusion

Using the measured data of heavy metal pollution, t GM and ARIMA model are combined to forecast the future data. The GM(1,1) is implemented to reduce residual error of ARIMA, so the accuracy of ARIMA is improved greatly. The hybrid model of GM and ARIMA takes advantage of the merits of both ones, achieves higher accuracy and gives us a new approach to forecast the heavy metal pollution.

References

1. Fan Xiao-Shan, Luo Hong. Spatial and industrial distribution pattern of heavy metals emission in industrial waste water[J]. China Environmental Science, 2013, 33(4): 655–662 (in Chinese).
2. Hu Li, Zhao Nan-Jing, Liu Wen-Qing. Research on Parameter Optimization of LIBS Used for Multi-Element Measurements in Water[J]. Spectroscopy and Spectral Analysis , 2014, 34(4): 869–873.
3. Nie Fa-Hui, Li Tian, Wang Shuo. The Form and Distribution of Heavy Metals from Highway Runoff in Shanghai [J]. Journal of Hunan University (Natural Sciences), 2009, 36(3): 76–80.
4. Lado M, Hur B M, Treated domestic sewage irrigation effects on soil hydraulic properties in arid and semiarid zones: A review[J]. Soil & Tillage Research, 2009, 106(1): 152–163.

5. Box G. E P and Jenkins G M. Time Series Analysis: Forecasting and Control [M]. San Francisco: Holdenday.1970.
6. Xiao Yi, Xiao Jin, Liu John, Wang Shouyang. A Multiscale Modeling Approach Incorporating ARIMA and Anns For Financial Market Volatility Forecasting[J]. Journal of Systems Science & Complexity, 2014 27(1): 225–236.
7. Zhang G P. Time series forecasting using a hybrid ARIMA and neural network model[J] . Neuro computing, 2003, (50): 159–175.
8. Zhu Bang-Zhu, Lin Jian. A Novel Nonlinear Ensemble Forecasting Model Incorporating ARIMA and LSSVM[J]. Mathematics in Practice and Theory. 2009, 39(12): 34–40.
9. Xiao Xinping, Deng Julong. Parameter Characteristics of GM (0, N) Model Under Multiple Transformation[J]. Systems Engineering and Electronics. 2000. 22(10): 1–3.
10. Hu Zhaoqing Mao Chengxiong, Lu Jiming. Excitation controller based on grey prediction[J]. Proceedings of the CSU-EPSA, 2002, 14(1): 30–33.
11. Wang Zhen-Lei, Tang Ku, Wang Xin. A Multi-Model soft sensing method based on D-S and ARIMA model [J]. Control and Decision. 2014. 29(7): 1160–1166.
12. Peng Huiting. Research of Power Generation Forecast Based on the GM-ARIMA Model[D].Gansu: Lanzhou University, 2014: 9–19.

The behavior of the additive Yb_2O_3 doped in the anodes during electrolysis

Han-Bing He[†] and Yuan Wang
School of Metallurgy and Environment, Central South University, Changsha, Hunan, China, 410083
[†]*Email address: hehanbinghhb@163.com*

Investigation of the behavior of additive Yb_2O_3 in the anode during electrolysis is an important method for improving the corrosion resistance of the grain boundary to high-temperature molten salt electrolyte. The author examined the Yb content of cermets after both sintering and electrolysis and the experimental results showed that a dense layer of $NiFe_2O_4$-$NiAl_2O_4$-$FeAl_2O_4$ ceramic favorably formed on the surface of the anodes only if the electrolysis time was greater than 10 hours. Moreover, $NiFe_2O_4$, NiA_2O_4 and $FeAl_2O_4$ were produced and dissolved continuously as a result of the chemical and electrochemical corrosion that took place after the formation of the dense layer. As the electrolysis time was extended, Yb_2O_3 or $YbFeO_3$ gradually dissolved into the electrolyte and primary aluminum as an impurity substance.

Keywords: Behavior; grain boundary; dense layer; Yb_2O_3; $YbFeO_3$.

1. Introduction

$NiFe_2O_4$-NiO-based cermets are one of the most promising inert anode materials for aluminum electrolysis because of their good corrosion resistance and high electrical conductivity in high-temperature molten salt [1–6]. However, due to the differences in corrosion between the metal phase and ceramic phase in cermets, the preferential corrosion of the metal phase causes some problems such as electrolyte permeability, swelling and cracking [7–10]. Thus, it is necessary to take effective measures to improve the corrosion resistance of the metal phase to high-temperature molten salt electrolyte. At present, adding additives to strengthen the grain boundary and forming a dense ceramic spinel layer on the surface of $NiFe_2O_4$-based cermets inert anodes are the important way to accomplish this.

[†] This work was supported by the National Natural Science Foundation of China (NO. 51404298), the key Project of Science and Technology of Hunan Province (2014FJ2002), the Project of post-doctoral foundation of Central South University, the China Scholarship and the Free Exploration Plan of Central South University (201510533203).

Xi et al.[11] pointed out that the addition of MnO_2 increases the relative density of nickel ferrite, refines the grain and improves the thermal shock resistance of samples containing MnO_2. Xi[12] also reported that V_2O_5 can improve corrosion resistance, and distribute Ni_2FeVO_6 along the grain boundary, which can control the chemical dissolution of the ceramic anode. The reinforced grain boundary can then control the grain boundary corrosion rate.

Tao et al.[13] reported that the grains coarsen after doping with Yb_2O_3, and the reaction product phase is distributed along the $NiFe_2O_4$ grain boundary in the way of point and line. Moreover, the reaction product of Y_2O_3 and the ceramic phase is distributed in the NiO and along the $NiFe_2O_4$ grain boundary.

However, little work has been reported on the behavior of additives on the grain boundary during electrolysis. To the best of the author's knowledge, the additive Yb_2O_3 is advantageous for the relative density, corrosion resistance and electrical conductivity of cermets[14], but very few studies have evaluated the behavior of Yb_2O_3 on the grain boundary during electrolysis, especially during the formation of a $NiFe_2O_4$-$NiAl_2O_4$-$FeAl_2O_4$ dense layer on the surface of inert anodes.

2. Experimental

2.1 *Preparation of cermets*

xYb_2O_3-15(20Ni-Cu)/(85-x)($NiFe_2O_4$-10NiO) (x=0, 0.25, 0.5, 0.75 and 1.0) cermets were prepared from reagent-grade nickel powder (99.9 wt%, Tianjing, China), copper powder (99.9 wt%, Tianjing, China), nickel oxide (99 wt%, Jinchuan, China) and iron oxide (99 wt%, Qidong, China) using an isostatic pressing-sintering process. The raw materials were Fe_2O_3 and NiO, ball-milled in distilled water for 2 hours, dried for 12 hours at 90°C and then calcinated in air at 1200°C for 4 hours. Yb_2O_3 and 20Ni-Cu powder were added to the calcinated powder, and the mixture was uniaxially compacted to form cylindrical blocks (about 20 mm in diameter and 40 mm thick) at 200 MPa and sintered at 1300°C for 4 hours in a nitrogen atmosphere with an original oxygen partial pressure of 100 ppm.

2.2 *Characterization*

The electrolyte used consisted of Na_3AlF_6 (99 wt%, Shanghai, China), AlF_3 (98 wt%, Shanghai, China), CaF_2 (98.5 wt%, Tianjing, China) and Al_2O_3 (98 wt%,

Tianjing, China). The composition was 5% CaF_2, 7.43% Al_2O_3 and the balance cryolite (NaF/AlF_3=2.3). A 3-cm distance was used between anode and cathode, and the current density was 1.0 A/cm^2. Al_2O_3 was added at 15-min intervals based on the electrolytic consumption rate at 80% cathode current efficiency.

The anode was analyzed with XRD and SEM/EDS. The phase composition of the electrolyte and cermet was identified by X-ray diffraction analysis using a Philips PW1390 X-ray diffractometer with Cu-Kα radiation. The microstructure was analyzed with a JSM25600LV scanning electron microscope and XJP26A metallographic microscope.

3. Results and Discussion

Figure 1 shows the metallographic images of the surfaces of the $0.5Yb_2O_3$-15(20Ni-Cu)/(10NiO-$NiFe_2O_4$) anodes after sintering at 1300°C for 4 hours, and Figure 2 presents the metallographic images of the $0.5Yb_2O_3$-15(20Ni-Cu)/(10NiO-$NiFe_2O_4$) polished anodes after electrolysis for 10 hours. The images show a very clear interface between the anode components and the relative density is high. A metal-depleted dense layer formed on the surface of the cermets after electrolysis for 10 hours in high-temperature molten salt regardless of whether the Yb_2O_3 was doped or undoped in the cermets, and the thickness of the metal-depleted dense layer was about 20-50 μm. This suggests that a dense layer of $NiFe_2O_4$-$NiAl_2O_4$-$FeAl_2O_4$ ceramic favorably formed on the surface of the anodes during electrolysis.

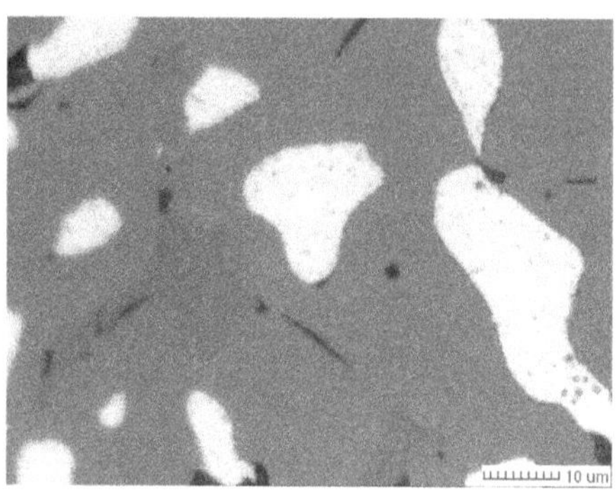

Fig. 1 The metallographic images of the surfaces of the $0.5Yb_2O_3$-15(20Ni-Cu)/(10NiO-$NiFe_2O_4$) anodes after sintering at 1300°C for 4 hours

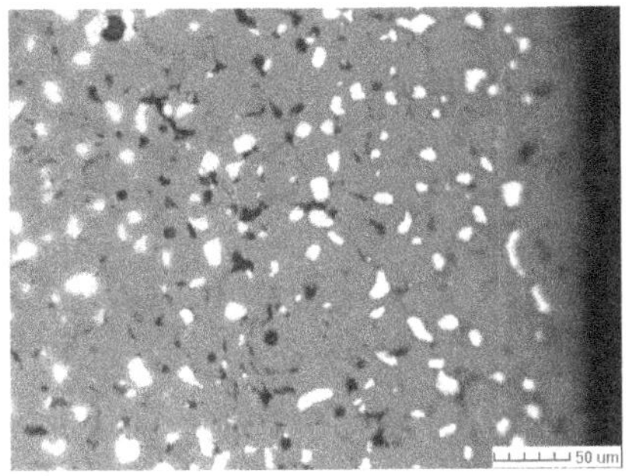

Fig. 2 The metallographic images of the $0.5Yb_2O_3$-15(20Ni-Cu)/(10NiO-$NiFe_2O_4$) polished anodes after electrolysis for 10 hours

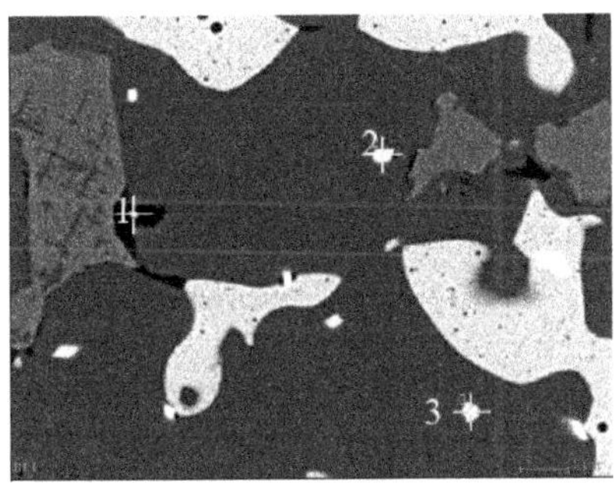

Fig. 3 The SEM image and EDX analysis of the surface $0.5Yb_2O_3$-15(20Ni-Cu)/(10NiO-$NiFe_2O_4$) polished anodes after sintering at 1300°C for 4 hours

Table 1. Element content distribution at each point

Position	Content of Ni (wt%)	Content of Fe (wt%)	Content of Cu (wt%)	Content of Yb (wt%)	Content of O (wt%)
1	2.39	8.82	0.39	10.74	77.66
2	1.53	15.48	0.39	16.70	65.90
3	2.27	18.98	0.38	14.82	63.55

Figure 3 presents the SEM images of the surface $0.5Yb_2O_3$-15(20Ni-Cu)/(10NiO-$NiFe_2O_4$) polished anodes after sintering at 1300°C for 4 hours, and Figure 4 shows the SEM images of the $0.5Yb_2O_3$-15(20Ni-Cu)/(10NiO-$NiFe_2O_4$) polished anodes after electrolysis for 20 hours. Every interface is very clear, and the results of the EDX analysis show that the white particles in Figure 3 are Yb_2O_3, which indicate that the Yb content was 10.74.at%, 16.70. at% and 14.82. at%, respectively.

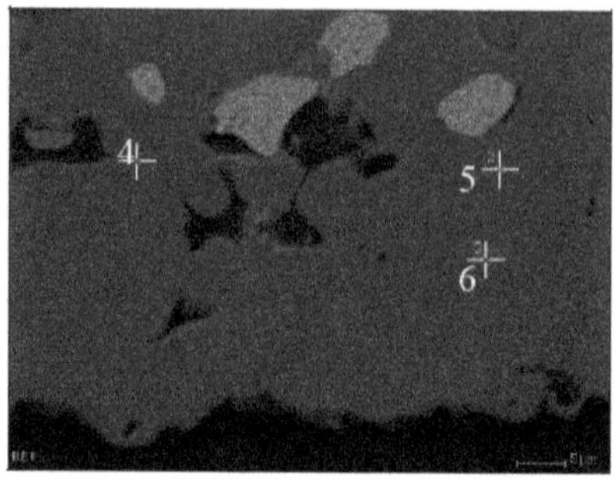

Fig. 4 the SEM images and EDX analysis of the $0.5Yb_2O_3$-15(20Ni-Cu)/(10NiO-$NiFe_2O_4$) polished anodes after electrolysis for 20 hours

Table 2. Element content distribution at each point

Position	Content of Ni (wt%)	Content of Fe (wt%)	Content of Cu (wt%)	Content of Yb (wt%)	Content of O (wt%)
4	4.76	18.45	0.42	12.13	64.24
5	9.28	8.83	0.43	15.40	66.07
6	6.59	16.22	0.22	8.43	68.54

A metal-depleted dense layer was produced on the surface of the $0.5Yb_2O_3$-15(20Ni-Cu)/(10NiO-$NiFe_2O_4$) anode after electrolysis for 20 hours, and its thickness was 20 μm. According to the EDX analysis, the Yb content was 12.13.at%, 15.40.at% and 8.43.at%, respectively, which was a little lower than the sample after sintering. These results indicate that the Yb_2O_3 or $YbFeO_3$ was corroded by the high-temperature molten salt electrolyte during electrolysis, and the Yb element gradually entered the electrolyte and the primary aluminum.

The formation law of the dense $NiFe_2O_4$-NiA_2O_4-$FeAl_2O_4$ ceramic layer has been investigated previously[15]. The progression of mutual diffusion, volume expansion and interaction took place during the formation of the dense layer, indicating that Yb_2O_3 or $YbFeO_3$ was involved in this process and also took part in the mutual reaction, which is a part of the dense $NiFe_2O_4$-NiA_2O_4-$FeAl_2O_4$ ceramic layer.

To survey the differences in corrosion resistance between Yb_2O_3 or $YbFeO_3$ and $NiFe_2O_4$-NiA_2O_4-$FeAl_2O_4$, electrolysis was carried out over an extended time in the lab-scale electrolytic tank. Figures 5, 6, 7 and 8, respectively, show the SEM images of the bottom of the anodes and EDX analysis of the 15(20Ni-Cu)/(10NiO-$NiFe_2O_4$), $0.25Yb_2O_3$-15(20Ni-Cu)/(10NiO-$NiFe_2O_4$), $0.5Yb_2O_3$-15(20Ni-Cu)/(10NiO-$NiFe_2O_4$) and $0.75Yb_2O_3$-15(20Ni-Cu)/(10NiO-$NiFe_2O_4$) cermets after electrolysis for 100 hours with an alumina concentration of 7.43%. The compact layer was about 50 μm, 30 μm, 30 μm and 100 μm thick at the bottom of the anode, respectively. Unfortunately, There are some fractures on the $1.0Yb_2O_3$-15(20Ni-Cu)/ (10NiO-$NiFe_2O_4$) anode during electrolysis so the thickness of the compact layer was not measurable, revealing that the excess Yb_2O_3 was disadvantageous to the corrosion resistance of the anode. In addition, on the basis of the EDX analysis in Figs. 5–8, the Yb element was not found, indicating that the Ni, Fe and Cu elements were corroded during the dynamic corrosion balance of the dense layer on the surface of the anodes. At the same time, either corrosion of the Yb element or preferential corrosion of the Yb element had taken place.

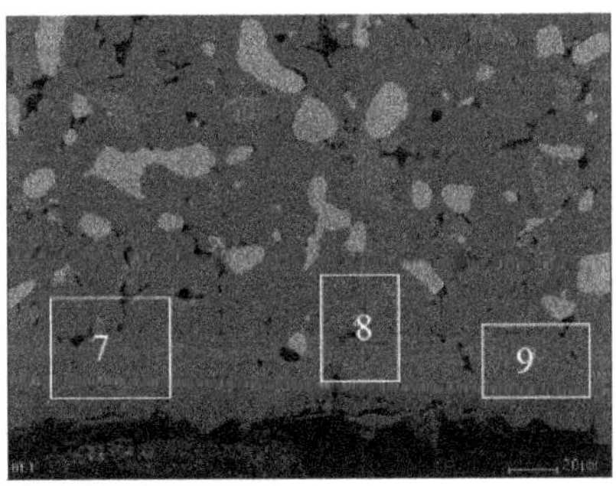

Fig. 5 The SEM images of the bottom of the anodes and EDX analysis of the 15(20Ni-Cu)/(10NiO-$NiFe_2O_4$) cermets after electrolysis for 100 hours with an alumina concentration of 7.43%

Table 3. Element content distribution at each point

Position	Content of Ni (wt%)	Content of Fe (wt%)	Content of Cu (wt%)	Content of Al (wt%)	Content of O (wt%)
7	32.37	63.54	0.82	3.27	-
8	14.40	29.91	0.45	1.31	53.93
9	14.17	30.46	0.59	0.80	53.98

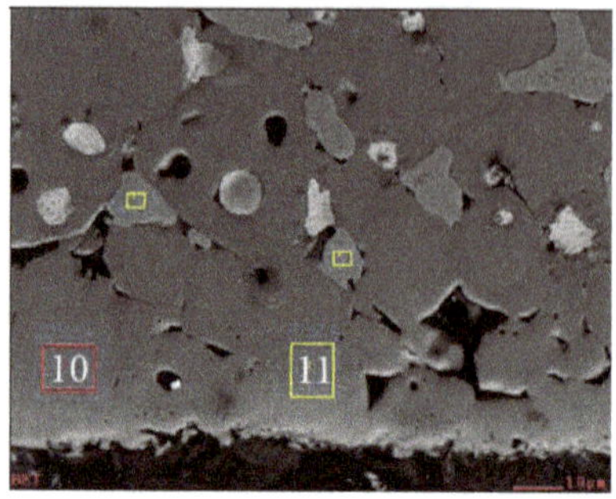

Fig. 6 The SEM images of the bottom of the anodes and EDX analysis of the $0.25Yb_2O_3$-15(20Ni-Cu)/(10NiO-$NiFe_2O_4$) cermets after electrolysis for 100 hours with an alumina concentration of 7.43%

Table 4. Element content distribution at each point

Position	Content of Ni (wt%)	Content of Fe (wt%)	Content of Cu (wt%)	Content of O (wt%)
10	14.22	35.83	0.32	49.63
11	14.70	36.02	0.43	48.85

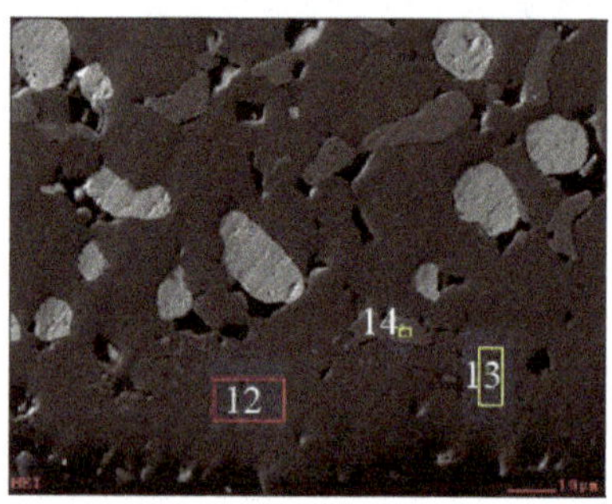

Fig.7 The SEM images of the bottom of the anodes and EDX analysis of the $0.5Yb_2O_3$-15(20Ni-Cu)/(10NiO-$NiFe_2O_4$) cermets after electrolysis for 100 hours with an alumina concentration of 7.43%

The dense ceramic layer can prevent the molten salt electrolyte from penetrating into the anode due to its high density and excellent corrosion resistance. Therefore, during the dynamic corrosion balance of the dense layer, several reactions may occur, as shown in equations 1 to 9. The $NiFe_2O_4$, NiA_2O_4 and $FeAl_2O_4$ formed and dissolved continuously during electrolysis because of chemical and electrochemical corrosion. Meanwhile, the Yb_2O_3 or $YbFeO_3$ gradually dissolved into the electrolyte and the primary aluminum as the impurity substance.

Table 5. Element content distribution at each point

Position	Content of Ni (wt%)	Content of Fe (wt%)	Content of Cu (wt%)	Content of O (wt%)
12	12.63	31.19	0.52	55.67
13	12.68	31.22	0.38	55.72
14	47.01	5.86	0.49	46.64

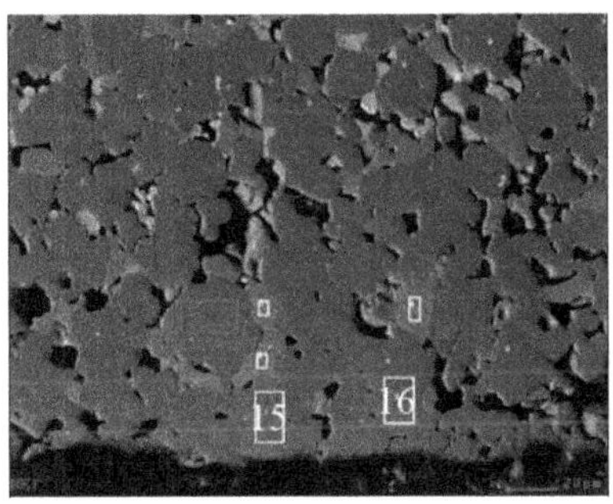

Fig. 8 The SEM images of the bottom of the anodes and EDX analysis of the $0.75Yb_2O_3$-15 (20Ni-Cu)/(10NiO-$NiFe_2O_4$) cermets after electrolysis for 100 hours with an alumina concentration of 7.43%

Table 6. Element content distribution at each point

Position	Content of Ni (wt%)	Content of Fe (wt%)	Content of Cu (wt%)	Content of O (wt%)
15	12.60	33.12	0.36	53.91
16	13.78	31.46	0.43	54.33

$$Al_2O_3 + NiO = NiAl_2O_4 \tag{1}$$

$$FeO + Al_2O_3 = FeAl_2O_4 \tag{2}$$

$$FeO + NiO + O_2 = NiFe_2O_4 \tag{3}$$

$$3NiAl_2O_4 + 2AlF_3 = 3NiF_2 + 4Al_2O_3 \tag{4}$$

$$3FeAl_2O_4 + 2AlF_3 = 3FeF_2 + 4Al_2O_3 \tag{5}$$

$$3NiFe_2O_4 + 12AlF_3 = 3NiF_2 + 6FeF_3 + 4Al_2O_3 \tag{6}$$

$$Yb_2O_3 + 2AlF_3 = 2YbF_3 + Al_2O_3 \tag{7}$$

$$YbFeO_3 + 2AlF_3 = YbF_3 + FeF_3 + Al_2O_3 \tag{8}$$

$$2NiO + 4FeO + O_2 = 2NiFe_2O_4 \tag{9}$$

4. Conclusions

(1) A dense layer of $NiFe_2O_4$-$NiAl_2O_4$-$FeAl_2O_4$ ceramic is favorably formed on the surface of anodes only if the electrolysis time is greater than 10 hours.

(2) The Yb_2O_3 or $YbFeO_3$ is gradually corroded by the high-temperature molten salt electrolyte during electrolysis.

(3) The $NiFe_2O_4$, NiA_2O_4 and $FeAl_2O_4$ are formed and dissolved continuously during electrolysis because of chemical and electrochemical corrosion. Meanwhile, Yb_2O_3 or $YbFeO_3$ is gradually dissolved into the electrolyte and the primary aluminum as the impurity substance.

References

1. Z. L. Tian, Y. Q. Lai, Z. Y. Li, J. Li, K. C. Zhou and Y. X. Liu, Cup-Shaped Functionally Gradient $NiFe_2O_4$-based Cermet Inert Anodes for Aluminium Reduction, *JOM*, ,61,34 (2009).
2. Li, Jie, Zhang, Gang, Lai, Yan-Qing, Zhang Yong and Tian, Zhong-Liang. Densification and Sintering Dynamics of 10NiO-$NiFe_2O_4$ Composites Doped with CaO. Journal of Central South University of Technology, 14, 629(2007).
3. H.B. He, K.C. Zhou, Z.Y. Li and B.Y. Huang: Effect of BaO addition on electric conductivity of xCu/$NiFe_2O_4$–10NiO cermets, Trans. Nonferr. Met. Soc. China, 18 (5), 1134(2008).
4. Yuqiang Tao, Zhiyou Li, Kechao Zhou. Effects of debinding atmosphere on the microstructure and sintering densification of nickel ferrite. Ceramics International, 39 (1), 865(2013).
5. Yuqiang Tao, Zhiyou Li, Dou Zhang, Huiwen Xiong, Kechao Zhou. Microstructural Evolution of A Nickel Ferrite–Copper Alloy Cermet During Sintering and High-Temperature Oxidation. Journal of the American Ceramic Society, 95 (1), 3031(2012).
6. Yuqiang Tao, Zhiyou Li, Lei Zhang, Xueping Gan, Kechao Zhou. Preparation of Nickel Ferrite-Copper Composite by Pressureless Infiltration. Journal of the American Ceramic Society, 95 (12), 3739 (2012).

7. Han-Bing He, Zhao-Hui Chen. Corrosion behavior of $NiFe_2O_4$-10NiO based cermet inert anode surface for aluminum electrolysis, Metalurgia International, 18 (10), 58(2013).
8. Liu, Baogang, Zhang, Lei, Zhou, Kechao, Li, Zhiyou and Wang, Hao. Electrical conductivity and molten salt corrosion behavior of spinel nickel ferrite, Solid State13, 1483(2011).
9. Han-Bing He, Yuan Wang, Jia-Ju Long, Zhao-Hui Chen. Corrosion of $NiFe_2O_4$-10NiO based cermet inert anode for aluminium electrolysis, Transactions of Nonferrous Metals Society of China, 23(12), 3816(2013).
10. Liu Jian-Yuan, Li Zhi-You, Tao Yu-Qiang, Zhang Dou and Zhou Ke-Chao, Phase evolution of 17(Cu-10Ni)-($NiFe_2O_4$-10NiO) cermet inert anode during aluminum electrolysis. Transactions of Nonferrous Metals Society of China, 21 (3), 566(2011).
11. Xi Jin-Hui, Liu Yi-Han, Yao Guang-Chun. Effect of MnO_2 on properties of inert anode of $NiFe_2O_4$ spinel. Journal of Functional Materials, 36 (3), 374 (2005). (in Chinese)
12. Xi Jin-Hui, Xie Ying-Jie, Yao Guang-Chun, Liu Yi-Han. Effect of additive on corrosion resistance of $NiFe_2O_4$ ceramics as inert anodes. Transactions of Nonferrous Metals Society of China, 18 (1), 356(2008).
13. Tao Yu-Qiang, Liu Jian-Yuan, Li Zhi-You, Zhou Ke-Chao, Tan Zhan-Qiu. Conductivity and corrosion resistance of 10Cu-($NiFe_2O_4$-10NiO) inert anode doped Yb_2O_3 or Y_2O_3. The Chinese Journal of Nonferrous Metals, 21(5), 1137(2011). (in Chinese)
14. He Han-Bing, Xiao Han-Ning. Effect of Yb_2O_3 doping on performances of 15(20Ni-Cu) / ($NiFe_2O_4$-10NiO) cermets, The Chinese Journal of Nonferrous Metals, 21 (3), 611 (2011). (in Chinese).
15. H. He. The formation law of dense $NiFe_2O_4$-$NiAl_2O_4$-$FeAl_2O_4$ ceramic layer on the surface of $NiFe_2O_4$ based ceramic inert anode for aluminium electrolysis, Materials Research Innovations, 18(2), 138(2014).

Effect of hot working processes on the electrical conductivity and mechanical properties of Cu-Te-Se alloys*

Lin Jiao, Da-Chuan Zhu† and Tao Zeng

School of Materials Science and Engineering,
Sichuan University, Chengdu, Sichuan, China
†*E-mail: zhudachuan@scu.edu.cn*
www.scu.edu.cn

Samples of Cu-Te-Se alloys were smelted in a vacuum induction furnace, and the effects of different hot working processes on their electrical conductivity and mechanical properties were studied. An eddy current conductivity meter and a Micro Vickers hardness tester were used to detect the electric conductivity and Vickers hardness of alloys, respectively. The morphologies of precipitated phase after aging were observed with a Scanning Electron Microscope (SEM). Combined with an Energy Dispersive Spectrometer (EDS), the elemental constituents of precipitated phase were analyzed. The phase composition after aging was measured by means of X-ray Diffraction (XRD), while the alloys' tensile strengths were measured via an electronic universal testing machine. The alloys' microstructures were observed using a metallographic microscope. The results showed that the relative electric conductivity of Cu-Te-Se alloys, after selenium was used to replace tellurium, could be kept at more than 95.7% IACS. Their mechanical properties were also slightly improved. The Vickers hardness and tensile strength of alloys showed undulating patterns, achieving 155.3HV and 462MPa, respectively.

Keywords: Cu-Te-Se alloys; electric conductivity; mechanical properties; hot working processes.

1. Introduction

Because of their excellent electric conductivity, thermal conductivity, corrosion resistance and processing performance, copper and copper alloys are widely used in electrical engineering, light industry, machinery manufacturing, building industry, national defense industry and many other fields[1–5]. However, due to the low strength and hardness of pure copper, even the high strength of pure copper through cold deformation is easily lost in the process of annealing[6], the

* This work is supported by the Science & Technology Pillar Program of Sichuan Province, China. (0674-144PTCFXJ082-F)

study on high strength and high conductivity copper alloys are highly necessary[7]. At present, the study of copper alloys with high strength and high conductivity are mainly concentrated on Cu-Cr, Cu-Zr, Cu-Mg, and Cu-Fe series. But these kinds of copper alloys are not perfect, the disadvantages such as higher cost, adding difficult, low conductivity, and so on still exist, so it is urgent to research and develop new type copper alloys with high strength and high conductivity[8-16].The previous researches showed that tensile strength of copper alloys could be improved by adding tellurium into pure copper, thus the tensile strength could reach 500MPa while the relative electric conductivity could keep 94%~98% IACS by adopting different processing technology [17].

Although Cu-Te alloy has excellent electrical and mechanical properties, the application is limited because tellurium is one of the scarce resources and belongs to strategic resource. At the same time, the market price of tellurium is much higher than that of selenium. While selenium is in the same group with tellurium in the Periodic Table, thus they may have similar physical and chemical property. Considered the reserves and cost difference between tellurium and selenium, selenium was used to replace tellurium in order to study and develop the related performance of Cu-Te-Se alloys. In this paper, the samples of Cu-Te-Se alloys were smelted in vacuum induction furnace, and the electric conductivity and mechanical properties under different states of casting, forging, solid solution and aging, have been studied respectively.

2. Experiments

2.1. *Materials*

The copper alloy ingots were smelted in ZG-25 Vacuum Induction Furnace by adding different amount of tellurium and selenium into industrial pure copper. The content of tellurium and selenium of ingots was detected by ADVAN-XRF and shown in Table 1.

Table 1. The designed and measured content of Te and Se

No. of Samples	1#	2#
Designed content of Se	0.15	0.25
Designed content of Te	0.15	0.25
Measured content of Se	0.109	0.149
Measured content of Te	0.125	0.243

2.2. *Methods*

Firstly, the copper sheets of ϕ30mm×3 mm and cubes of 10mm×10mm×10 mm were cut from the casting ingots, respectively. Then, the conductivity and hardness were detected and microstructures were observed. As-cast samples were divided into the first group. Then, after homogenized at 850°C for 4 h, the ingots were forged until the thickness of the copper plate reached 18mm. After forged, the samples were divided into three groups. The second group was no longer processed; the third and the fourth group were reheated to 850°C, keeping at this temperature for 2 h and quenched in cold water. Then the fourth group was aged at 420°C for 4 hours.

In order to study the electric conductivity and mechanical properties of Cu-Te-Se alloys. Firstly, the samples were burnished with sand paper to 2000# and then the electric conductivity was detected by use of 7051Eddy current conductivity meter. The Vickers hardness of alloys was detected by use of DHV-1000Z Micro Vickers hardness tester. And then the cubic samples were polished and etched to observe metallographic microstructure. The precipitated phase composition was determined with X'Pert Pro MPD type X-ray diffraction manufactured by Philips Company. JSM-5910LV type SEM was used to observe the morphology of precipitated phase, and EDS were used to measure the constituents of the precipitated phases.

3. Results

3.1. *Microstructure*

The microstructure of the alloys was shown in Figure 1. As shown in the Figure 1.a and 1.b, the microstructure was equiaxial crystal grain, of which size was from 30μm to 100μm. There was an amount of second phase existing in the as-casting alloys, and the second phase distributed mainly along the grain boundary. The microstructure of the forged alloys was shown in the figure 1.c and 1.d. The grains got deformation and refinement compared with the cast, of which size was less than 25μm.After the samples were solid dissolved at 850°Cfor 2 h and quenched in cold water, the crystal grain was equiaxial and the size was in the range of 25-50μm, which can be seen evidently from figure 1.e and 1.f. The microstructure of samples after aging treatment was shown in figure 1.g and 1.h. As we can see that grains became more uniform, while the twins and dispersive phases existed in the grains.

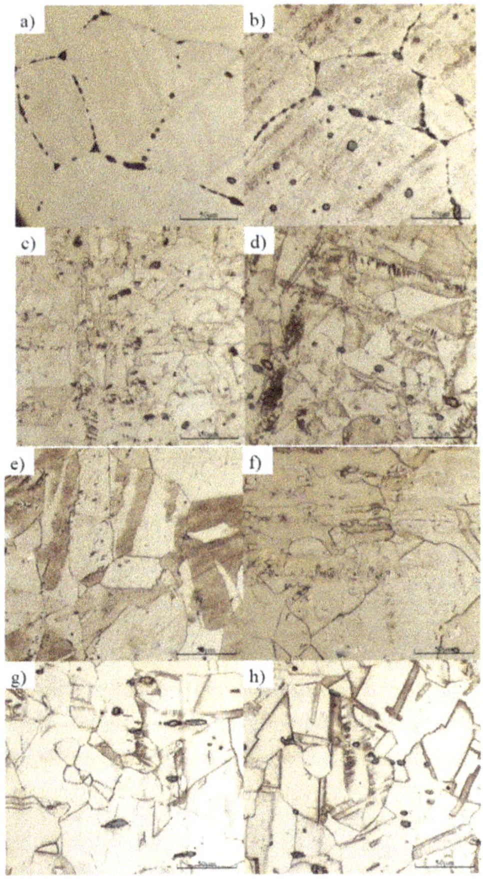

Fig. 1. The microstructure of Cu-Te-Se alloys under different process.

a) Casting of 1# alloy; b) Casting of 2# alloy; c) Forging of 1# alloy; d) Forging of 2# alloy; e) Solid solution of 1# alloy; f) Solid solution of 2# alloy; g) Aging of 1# alloy; h) Aging of 2# alloy

3.2. *Electric conductivity*

The relative electric conductivity of alloys was shown in Figure 2. As we can see from the Figure 2, under different states of casting, forging, solid solution and aging, the relative electric conductivity of alloys kept rising, higher than 95.7% IACS. For example 1#, the relative electric conductivity reached 96.5% IACS at casting state, after aging treatment, it reached a maximum of 99.8% IACS. Obviously, the relative electric conductivity of 1# alloy was higher than that of 2# alloy at the same state.

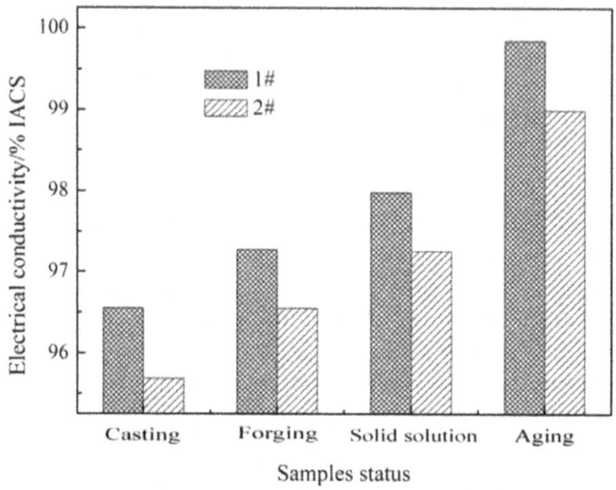

Fig. 2. The relative electric conductivity of Cu-Te-Se alloys.

3.3. *Vickers hardness*

As can be seen from figure 3, the Vickers hardness of alloys was low at casting state, but it increased after the sample was forged, reaching 155.3HV for the forged alloy 2#. However, the vickers hardness of alloys declined after solid dissolution, while appeared increasing slightly after aging. In a word, at the same state, the Vickers hardness of 2# alloy was higher than that of 1# alloy.

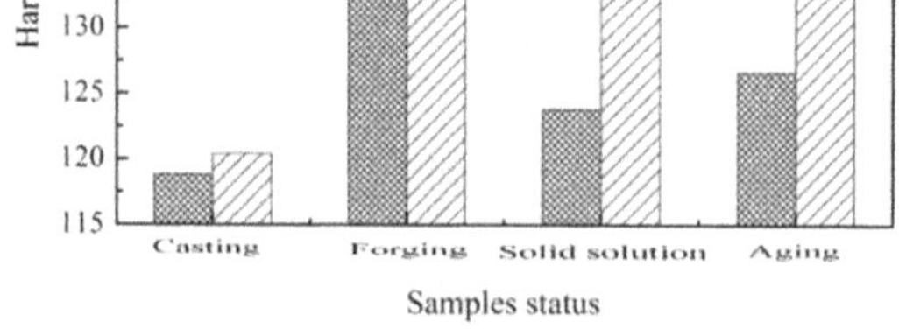

Fig. 3. The Vickers hardness of Cu-Te-Se alloys.

3.4. *Tensile strength*

The tensile strength of alloys was shown in Figure 4. Figure 4 demonstrated the tensile strength of alloys was low at casting state, but it increased for the forged alloy, which of 2# alloy after forged reached 462MPa. Although it decreased after solid solution, it appeared a rising after aging treatment. That is to say, with different microstructures of samples at casting, forging, solid solution or aging state, the tensile strength of alloys has experienced the process of rising→reducing→rising. For the microstructure at thc same state, the tensile strength of 2# alloy was higher than that of 1# alloy.

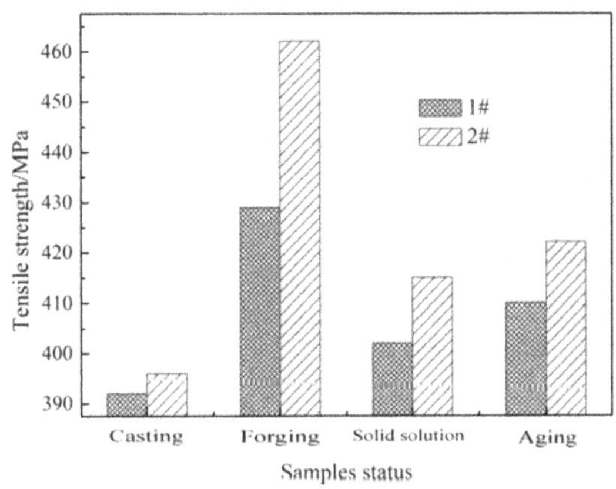

Fig. 4. The tensile strength of Cu-Te-Se alloys.

3.5. *Precipitated phase analysis*

The scanning morphology and EDS test results of precipitated phase after aging were shown in Figure 5. The precipitation phase distributed along the grain boundary and within grains, and because of the different content of selenium and tellurium, the quantity of precipitated phase of sample 2# was more than that of sample 1#, which can be seen evidently from Figure 5. The elemental constituents of the precipitated phase marked as plot A, B were detected by EDS and the tested results were shown in Figure 5 and Table 2. According to the mass ratio and atomic ratio of the precipitated phase as shown in Table 2, the precipitated phase could be inferred as $Cu_2(Te, Se)$ preliminarily.

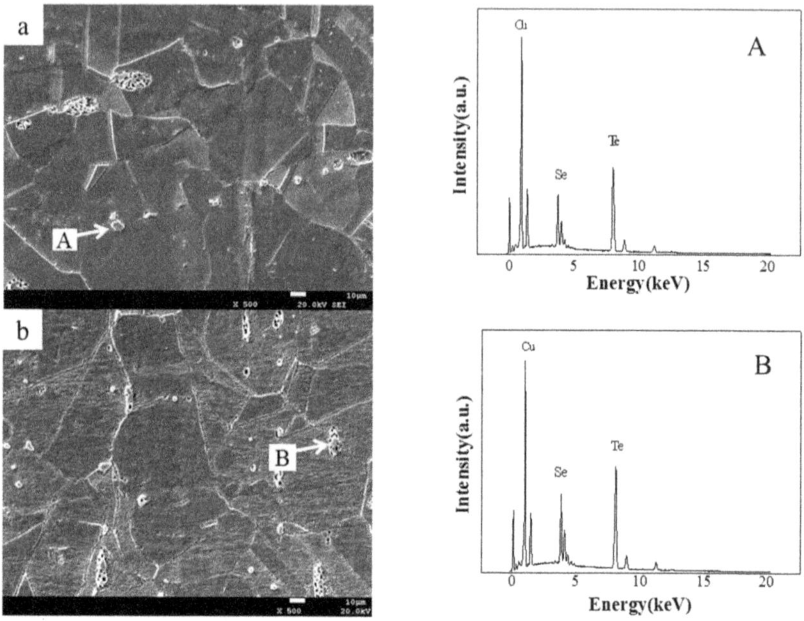

Fig. 5. The scanning morphology and EDS test results of precipitated phase.
a) Aging of 1# alloy; b) Aging of 2# alloy

Table 2. The EDS test results of Precipitated phase

	A		B	
Elements	ω(%)	at%	ω(%)	at%
Cu K	56.27	67.40	58.31	69.32
Se L	17.74	17.10	16.44	15.73
Te L	26.00	15.51	25.25	14.95.7
Total			100.00	

4. Analysis and Discussion

Based on all these above experimental results, it showed that material composition and working processes of the Cu-Te-Se alloy have an effect on its strength and electric conductivity. Compared with the Cu-Te alloys[18], the tensile strength and the electric conductivity of Cu-Te-Se alloys were studied. As well known, the conductivity of alloys is mainly affected by combined scattering effects to electron resulting from pores, inclusions, solute atoms, impurities, vacancies, dislocation and grain boundary [19-20]. After the series treatments of casting, forging, solid solution and aging, the electric conductivity of alloys kept rising (Fig. 2). As-cast, the electric conductivity of alloys was lower than that of subsequent states. The possible reasons were that the casting defects produced in the process of solidification were relatively high, which would intensify the electronic scattering, thus, leading to a lower electric conductivity. However, samples were homogenized at the temperature of 850°C for 4 h before being forged, coupled with forging process, the casting defects of alloys were reduced, thus the electric conductivity should be improved. But after forging, the grains became smaller and deformed (Fig. 1), thus grain boundaries increased, leading to a reducing of electric conductivity. As a comprehensive result, the electric conductivity of alloys was improved after being forged (Fig.2), it is possible because the influence of the eliminating of defects on the electric conductivity takes the dominant position compared to deformation during the process. After solid solution, the grains became coarser and more uniform than that after being forged (Fig. 1e and f), leading to the improvement of the electric conductivity. Although trace Te and Se can dissolve into Cu matrix, the solid solubility is relatively low according to Cu-Te and Cu-Se phase diagram[21], thus the factor for improving the electric conductivity may play a dominant role during solid solution process. After aging, the second phase precipitated and distributed along grain boundary and within intracrystalline (Fig.1g and h, Fig. 5). Furthermore, the remaining crystal defects in the alloy would be further eliminated, so the electric conductivity of alloys was improved again (Fig. 2), which of 1# and 2# alloy was as high as 99.8% IACS and 98.9% IACS, respectively.

Meanwhile, it can be seen from Fig.3 and Fig.4, under different states of casting, forging, solid solution and aging, the vickers hardness and tensile strength of alloys showed undulate. Because the casting defects existed in the ingots were relatively high and the crystalline grains were relatively coarse, the hardness and tensile strength appeared low. After being forged, the grains of sample became smaller than those of as-casting with no obvious directivity (Fig.1c and d). At the same time, more dislocations, grain boundaries and subgrains existed after forged, thus the tensile strength and hardness increased. After solid solution, the elements of Te and Se dissolved into Cu matrix, as a result, the effect of solution strength should be intensified. However, the solid solubility of Te and Se in copper is very low and has little difference from room temperature to 850°C, the effect of solution strength was not obvious, and the deformed structure disappeared essentially and the grains grew up, so the tensile strength and hardness decreased notably. During the process of aging treatment, the second phase precipitated as small $Cu_2(Te,Se)$ with dispersive distribution (Fig.5 and Table 2). Under the interaction between precipitation and dislocation, the second phase would assemble around the dislocation and prevent dislocation from moving[19-20,22-24], thus the tensile strength and hardness should be increased after aging treatment. However, the solid solubility of Te and Se in copper is very low, the effect of precipitation strength was not obvious, so the tensile strength and hardness increased slightly compared with that after solid solution (Fig.3 and Fig.4).

5. Conclusions

This research showed that under different states of casting, forging, solid solution and aging, substitution of Se for Te in copper can make the relative electric conductivity of Cu-Te-Se alloys keep in the range of 95.7~99.8% IACS, equivalent to that of Cu-Te alloy, while the tensile strength and hardness were slightly improved, reaching the maximum of 462MPa and 155.3HV, respectively, thus the application of this kind of alloys is expected to develop.

Acknowledgments

The authors thank the Project (0674-144PTCFXJ082-F) supported by the Science & Technology Pillar Program of Sichuan Province, China.

References

1. Z X Xie, H Y Gao, J Wang, et al. Effect of homogenization treatment on microstructure and properties for Cu–Fe–Ag in situ composites. *Mat. Sci. Eng. A***529**: 388–392 (2011).
2. S H Hong, H T Jeong, et al. Deformation and recrystallization textures of surface layer of copper sheet. Mat. Sci. Eng. A229,174–181 (1997).
3. X W Li, D T Zhang, et al. Microstructure and mechanical properties of dissimilarpure copper/1350 aluminum alloy butt joints by friction stir welding.T.Noferr. Metal. Soc.22, 1298–1306 (2012).
4. Raygan Sh, Ehsanian Mofrad H, Pourabduli M, Ahadi FK. Effect of rolling and annealing processes on the hardness and electrical conductivity values of Cu-13.5% Mn-4%Ni alloy. J. Mater. Process. Tech. 211, 1810–1816 (2011).
5. A Kauffmann, J Freudenberger. Severe deformation twinning in pure copper by cryogenic wire drawing. Acta. Mater.59, 7816–7823 (2011).
6. T Konkova, S Mironov, etal. Microstructural response of pure copper to cryogenic rolling. Acta. Mater.58, 5262–5273 (2010).
7. Ruoshan Lei, Shiqing Xu, et al. Microstructure and properties of nanocrystalline copper–niobium alloy with high strength and high conductivity. Mat. Sci. Eng. A586, 367–373 (2013).

8. S Gorsse, B Ouvrard, et al. Microstructural design of new high conductivity-high strength Cu-based alloy. J. Alloy Compd.633, 42–47 (2015).
9. K Oikawa, H Ido, H Mitsui. Development of Machinable High-Strength Copper-Based Alloys by Sulfide Dispersion. Mater. Trans.44, 2088–2093 (2003).
10. CD Xia, W Zhang, et al. High strength and high electrical conductivity Cu-Cr system alloys manufactured by hot rolling-quenching process and thermomechanical treatments. Mat. Sci. Eng. A538, 295–301 (2012).
11. S L Zhang, J M E Harper, F M Heurle. High Conductivity Copper-Boron Alloys Obtained by Low Temperature Annealing. J. Electron. Mater.30, 11–13 (2001).
12. M Li, J K Heuer, J F Stubbins, D J Edwards. Fracture Behavior of High-strength, High-conductivity Copper Alloys. J. Nucl. Mater.283, 977–981 (2000).
13. M Lopez, J A Jimenez, et al. Precipitation strengthened high strength-conductivity copper alloys containing ZrC ceramics. Composites Part A: Applied Science and Manufacturing 38, 272–279(2007).
14. Loffler F, Siewert C, Ascher C. Manganic thin film sensor for force sensing. SurfCoat Technol.174–175, 1287–1292 (2003).
15. Ye, Y., et al. Enhanced strength and electric conductivity of Cu-Zr-B alloy by double deformation–aging process. J. Alloy Compd. 249–254 (2014).
16. Maki, K., et al. Solid-solution copper alloys with high strength and high electric conductivity. Scripta. Mater.68, 777–780 (2013).
17. D C Zhu, M Z Song, D M Yang. Study on the oxidation resistance of Cu2Te alloys with high electrical conductivity. Funct. Mater.13, 713–715 (2004).
18. D C Zhu, K Tang, M Z Song, et al. Effects of annealing process on electrical conductivity and mechanical property of Cu-Te alloys. T. Nonferr. Metal. Soc.16, 459–462 (2006).
19. Mofrad, H.E., et al. Effect of cold-working and aging processes on the microstructure mechanical properties and electric conductivity of Cu-13.5%Mn-4%Ni-1.2%Ti alloy. Mater. Design182–191 (2012).
20. Wang, F., et al. Cu-Ti-C alloy with high strength and high electric conductivity prepared by two-step ball-milling processes. Mater. Design. 70–74 (2014).
21. J Q Yu. Binary alloy phase diagrams. Shang Hai: Science and technology press, 1987.

22. A Kauffmann, J Freudenberger, et al. Efficiency of the refinement by deformation twinning in wire drawn single phase copper alloys. Mat. Sci. Eng. A624, 71–78 (2015).
23. Shijie, Z., et al. Application of Lanthanum in High Strength and High Conductivity Copper Alloys. J. Rare Earth24, 385–388 (2006).
24. Gao, N., et al. Influence of prior deformation on the age hardening of a phosphorus-containing Cu-0.61wt.%Cr alloy. Mat. Sci. Eng. A342, 270–278 (2003).

Batch processing method for sheet metal parts of space fitting curve contour*

Wen-Suo Ma†, Lei Ding and Kai Chen

School of Mechatronics Engineering, Henan University of Science and Technology, Luoyang, China
†E-mail: mawensuo@126.com
www.haust.edu

Sheet metal parts are widely used in the modern industrial manufacturing process. In the production of sheet metal parts with complex spatial profiles, traditional approximate fabrication methods are not precise enough, while automatic programming methods have a low production efficiency and require the handling of large amounts of data. To analyse the unique forming characteristics of sheet metal parts, 3D software was first used to establish the 3D shell model containing the sheet metal part's space-fitting curve contour (SFCC). Next, to obtain the 3D model of the SFCC's projection on a plane, the SFCC of the sheet metal part was projected onto the plane and stretched in the specified direction. Following this, a Boolean addition was performed on the shell model and sheet metal part's plane projection model, thereby obtaining a solid model of the sheet metal part. The accuracy of the unfolding plane part is then derived using commercial 3D software. The NC program of the unfolding plane part's contour is automaticaly programmed by the software and can be used in production via wire-cutting or processing of stamping die. Finally, the unfolding plane metal part is bent and formed to meet the design requirements. This method can improve the processing efficiency and guarantee the quality of the products, thus making it suitable for batch production of the sheet metal parts of SFCC.

Keywords: Space fitting curve contour; Unfolding of sheet metal; Automatic programming; Batch processing.

1. Introduction

The characteristics of the sheet metal parts are light weight and low manufacturing cost, and the sheet metal parts can be adapted to the mass production. They are widely used in medical devices, communication, electronic

* This work is supported by grant 14A460012 of the Education Department Henan Province.
† Work partially supported by Advanced Machinery and Equipment Manufacturing Collaborative Innovation Center in Henan Province.

and electrical appliances, automotive and other industries.The traditional processing method cannot meet the requirements of the precision machining when processed the complex sheet metal parts of space fitting curve. Most of the sheet metal parts of the raw materials are usually thin, stiffness and easy deformation, so they cannot be used for a one-time cutting process.

The shape of the sheet metal parts of SFCC is complex and has not determined analytic expression. Even sometimes we can meet a part of a curve or surface which is expressed in a similar transcendental function in the engineering design process. These complex curves or surfaces cause parts are formed very difficult. In order to obtain the feasible processing method, the discrete points can be fitted by a specific algorithm. The commonly used algorithm fitting process is complex and easy to make mistakes, so it is very difficult to achieve automatic processing.

The projection method, cutting plane method and surface triangular discretization method are often used in machining complex curve and curved surface parts contour to obtain all kinds of approximate machining cutter path. Projection method is difficult to control the machining accuracy due to the influence of the two-dimensional curve by upfront design guide cutter motion. For some complex and changeable curved surfaces, the accuracy of the cutter path which is generated by the guide curve on the surface to be projected is not guaranteed[1]. Liu et al put forward a new method of path generation to approximate realized the combination of surface residues such as highly processed on the basis of projection method, namely the space ring equidistant cutting processing , but it needs a large amount of calculation and the algorithm operation is not stable[2]. Compared with the projection method, the cutting plane method can save converting from cutter contact to the cutter site, but when dealing with cross line interference algorithm is too complex and unstable. For some more complex intersecting line, this method is still not a good solution[3]. Zhang proposed a method of generating non interference tool path based on cutting plane method. This method can be used in the numerical control processing of the composite surfaces which are made of NURBS surfaces and analytic surface elements, but this algorithm also needs to deal with large amounts of data[4]. The discrete process can be omitted from the cross section line to the cutter contact by the surface triangular discretization method. The algorithm is simple, but the defect is that it needs a lot of iterative calculation procedure and the application range is relatively narrow. Hwang et al presented a new tool path generation algorithm, which is based on the surface triangular discretization model of a flat end mill and a circular milling cutter, and expands the application range of the surface triangulation method processing combination[5]. By means of combining triangular mesh method and projection method, Xie et al put forward an algorithm of NC tool path generation based on adaptive projection method for freeform surfaces[6]. This method can effectively solve the combination between surface processing programming of surface crack, discontinuous and partial overlap phenomenon, thus improves the machining efficiency and precision.

In this paper, a new method for the sheet metal parts of SFCC in a batch production is proposed to improve the machining precision of the parts.

2. The Structure Characteristics of the Sheet Metal Parts of SFCC

The curve modeling technology of sheet metal parts is more and more widely used in modern industrial product design and manufacturing. In order to meet the needs of the equipment running track, some special space curves are designed. If the space curve has a formula for determining equation, the machining path can be written in a simple programming part of NC machining program. For some uncertain analytic expression of the space discrete points, it can constitute a closed boundary by the method of fitting curve. But the numerical iterative calculation is large in the fitting process. When faced with complex curve form, even no convergence, no solution, the space curve fitting trajectory which has no detailed expression cannot achieve automatically programming processing.

3. The Forming Method of the Sheet Metal Parts of SFCC

Through the analysis of the structure characteristics of the sheet metal parts of SFCC, the precise machining of the parts can be obtained by using the sheet metal unfolding function module of the 3D software. In order to realize automatic programming processing, the process of the plane unfolding sheet metal is aimed at the curve or multi segment fitting curve which based on the complex space curved surface and can avoid the iterative calculation in the processing of solving these problems.

The flat sheet metal part can be automatical produced with the programme which is achieved by 3D software. The problems of process engineering which are met in processing the sheet metal parts of SFCC, such as making the stamping die and designing the bending die, can be also solved effectively by 3D software.Through the analysis of the structure characteristics of the sheet metal parts, the forming method of the sheet metal parts is shown in Figure 1.

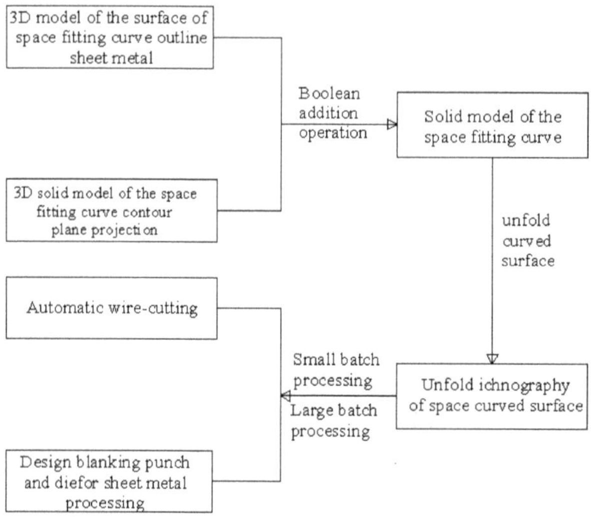

Fig. 1 The processing method on sheet metal parts of SFCC

3.1. *3D model contained sheet metal parts of SFCC*

Through the 3D modeling software, the surface model of the space sheet metal fitting curve contour can be established. In this paper, the CAM plate part which is used to change the shape of the fabric in the high speed braiding machine is described as an example. The shape of the sheet metal parts is complex and cannot be processed by conventional processing method. In the curved surface of the CAM plate surface is an ellipse paraboloid, and its axial profile curve equation is Eq. (1).

$$\begin{cases} f(y,z)=0 \\ x=0 \end{cases} \tag{1}$$

A 3D shell model is established through rotating with the thickness of the sheet metal parts as the radial dimension. The model is shown in Figure 2.

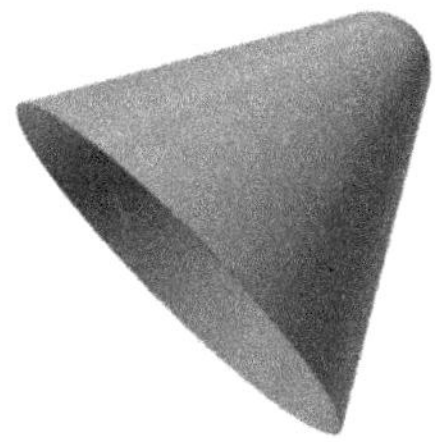

Fig. 2 3D shell model contained the CAM plate

3.2. *3D solid model of the plane projection parts of SFCC*

The contour shape boundary curve which belongs to the sheet metal part of SFCC is designed in a corresponding position of the three dimensional surface of the shell model, then the sheet metal part of SFCC is projected onto the plane, finally the 3D solid model of the plane projection part of SFCC can be formed through stretched in the specified direction. A 3D solid model of the CAM plate boundary space fitting curve contour projection plane is shown in Figure 3.

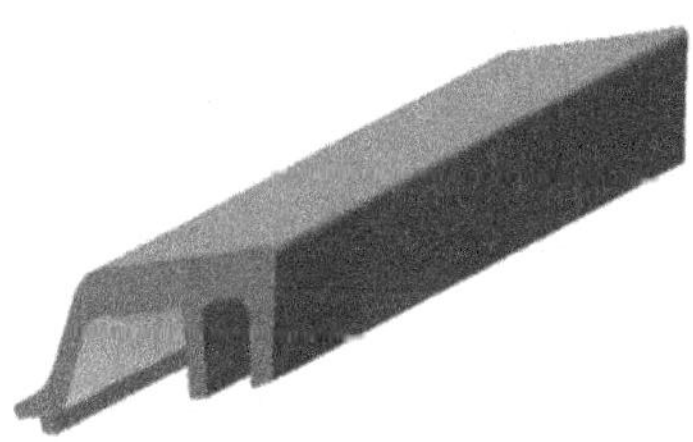

Fig. 3 3D solid model of the plane projection CAM plate

3.3. *Boolean operation of solid models*

The 3D solid model of the plane projection part and 3D the shell model contained sheet metal par are carried out by Boolean addition operation, and the 3D entity model of the space fitting curve is obtained. The model of a three dimensional model which is used to carry out the model of Boolean addition operation by the elliptic parabolic of the CAM plate and plane projection of the boundary space fitting curve contour. The model is shown in Figure 4.

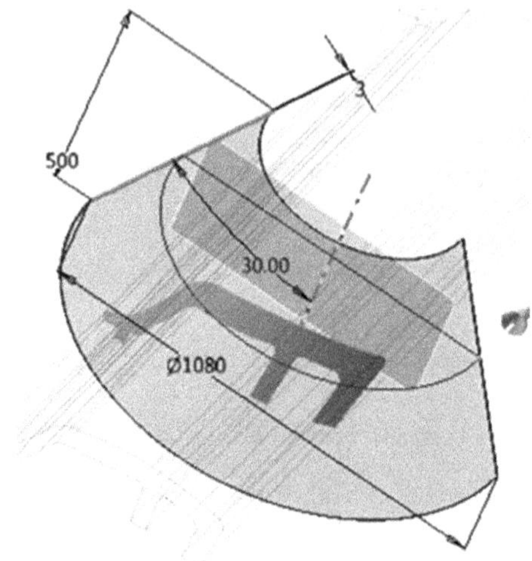

Fig. 4 3D entity model of Cam plate of space curved surface

3.4. *Unfolding of the sheet metal parts of SFCC*

Sheet metal parts are made of uniform thickness of metal thin plate. The unfolding method and processing involve many problems, such as intersecting line of complex curved surface, computing the curved surface unfolding, welding process and cutting process. They cannot be carried out in accordance with the general method of space curved surface. Choi et al studied the sheet metal CAD/CAM technology which based on the knowledge, and generated the corresponding NC code for the sheet metal unfolding or punching operation[7]. Although the function of the unfolding module is mentioned in theory, the details are still missing. Shpitalni et al proposed the using of artificial intelligence and fuzzy technology to unfold the box type of 3D parts roughly, but the results show that the conceptual design accuracy of the drawing unfolded sheet metal parts is lower[8]. Liu et al studied the thickness of plate, groove angle and weld on the influence of the sheet metal unfolding, then established mathematical model and achieved automatic drawing parts of the sheet metal through the secondary development of AutoCAD in the case of feasible processing technology[9]. However, it is not related to the mathematical modeling of the complicated curved surface sheet metal parts process.

The structure and size of the sheet metal parts will be the important parameters in the layout designing, sheet cutting and stamping forming after they are unfolded. If the workblank size of the unfold sheet metal parts is accurate and reasonable, it can reduce the subsequent machining allowance, save the material and reduce the cost of production; otherwise, it will lead to a lot of waste and even cause the workblank scrap. With the development of computer technology, CAD/CAM technology has been widely used in the mechanical industry. The unfolding principle and algorithm of complex curved surface sheet metal parts based on modern computer technology have become the developing trend of the sheet metal industry. The unfold ichnography is shown in Figure 5 which comes from the CAM plate space curved surface parts by the 3D software module function of the sheet metal unfold.

Fig. 5 The unfold space curved surface of the Cam plate

3.5. *Forming of the sheet metal parts of SFCC*

According to the different production quantities, the automatic numerical control line cutting process is adopted to manufacture the single piece and small batch sheet metal parts. When large quantities of sheet metal parts are produced, the production efficiency can be improved by designing the blanking punch and die for sheet metal stamping. The size value of the punching should be calculated before design the blanking punch and die. In the 3D software modeling environment, after designing and editing of the planar sketch which has been unfolded, the blanking punch and die can be obtained. Respectively, the blanking punch and die for the cam plate are shown in Figure 6 and Figure 7.

Fig. 6 Blanking punch for the CAM plate

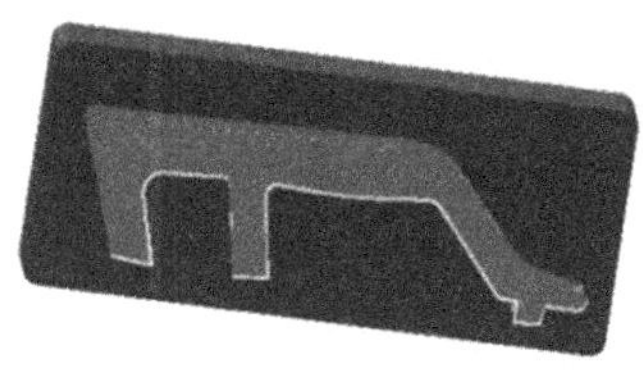

Fig. 7 Blanking die for the CAM plate

According to the unfolding of the plane contour shape design the bending mold of the surface sheet metal parts and then bend them to forming. In order to simplify the mould, reducing the springback angle and improving the processing accuracy of the bending parts, the bending die of the CAM plate is chosen as the polyurethane rubber mould. After 3D solid model is designed, the model of the workblank parts is built by the post-processing function of 3D software. Then combined with the performance of the selected machine tool, a reasonable cutter path and machining parameters are set up for automatic programming, which can realize the fast processing of blanking punch, die and bending die.

4. Conclusions

In this paper, a feasible method for the processing of sheet metal part with a complex space fitting curve is studied. The 3D model of the sheet metal parts of SFCC is established and the model is unfolded into a planar graph accurately through using the CAD software, then the plane part is used to carry out the automatic programm. The program can be used for wire-cutting when processed small quantities of parts or machining the blanking punch and die for the flat sheet metal parts. This method not only can realize the fast processing of the sheet metal parts of complex SFCC, but also has an high production efficiency. Additionally, the product size is stable. It can meet the requirement of the batch production of parts.

Acknowledgments

This project was supported by a grant from the Education Department Henan Province (Project serial number: 14A460012). The authors also thank Xinchang Benfa Electromechica Co. Ltd for the help in setting up the experiments and for the use of facilities.

References

1. Ruijiang Ran, Practical research and development of NC programming system, Master's thesis, Beijing University of Aeronautics and Astronautics (1994), In Chinese.
2. Huaiguang Liu, Guangrong Yan and Yanqiu Chen, Journal of Engineering Graphics, (2008) No. 01, pp. 1–7, In Chinese.
3. Changming Liu, Research and implementation of NC Machining Technology, Master's thesis, Beijing University of Aeronautics and Astronautics (2001), In Chinese.

4. Haining Xu, Zhuang Xiong and Daoyuan Yu, Journal of Huazhong University of Science and Technology, Vol. 24(1996) No. 09, pp. 15–17, In Chinese.
5. J.S. Hwang, T.C. Chang, Tree-axis machining of compound surfaces using flat and filleted end mills, CAD, Vol. 30 (1998) No. 08, pp. 641–647.
6. Le Xie, Rurong Zhou and Xueyu Ruan, Journal of Shanghai Jiao Tong University, Vol. 34 (2000) No. 10, pp. 1382–1384, In Chinese.
7. J.C. Choi, C. Kim, Journal of Material Processing Technology, (2000) No. 110, pp. 36–46
8. M. Shpitalni, H. Lipson, Journal of Material Processing Technology, (2000) No. 103, pp. 128–134.
9. Zhihong Liu and Xiangang Jiang, Journal of mechanical design and manufacturing, (2007) No. 03, pp. 93–95, In Chinese.

Preparation and electrical properties of ZnO/SnO_2 nanocrystalline films*

Wei Cong†

Shenyang Aerospace University,
Shenyang, Liaoning, China
†E-mail: congwei2000@163.com
www.sau.edu.cn

Chen-Zhi Peng

Shenyang, Liaoning, China
E-mail: pcz2324994@163.com

To study the effect of SnO_2 nanocrystalline coatings on the performance of dye-sensitized solar cells (DSSC) based on ZnO-nanoparticle photoanodes, SnO_2 nanocrystalline films were coated on ZnO nanoparticles by soaking ZnO photoanodes in SnO_2 solution for various durations. Analyses of the structure and electrical properties of the composite films through SEM and I-V test were then conducted. The SEM photos show that SnO_2 wraps the ZnO particles, increasing the specific surface area and dye adsorption. I-V test results indicate that ZnO film soaked into SnO_2 solution for 40 minutes would have the greatest improvement in its structure and electrical properties.

Keywords: Dye-sensitized solar cell; ZnO/SnO_2 composite films; photoelectric conversion efficiency; J-V curve.

1. Introduction

Dye-sensitized solar cells (DSSC) which has the characteristics of low cost, renewable and alternative materials diversity has become new research hot spot after silicon solar battery for scholars. Broadband gap semiconductor materials of TiO_2 and ZnO have been widely used in research of DSSC.

SnO_2 and ZnO both are wide broadband gap semiconductor materials (each forbidden bandwidth is 3.6 eV and 3.37 eV). They have good chemical and physical properties which are likely to replace traditional TiO_2 semiconductor materials. Due to the conduction band phase difference is small, this is conducive to the migration of electron. In addition, ZnO and SnO_2 have larger electron mobility which can facilitate collection of electron rapidly and shorten

* Work partially supported by grant 2-4570.5 of the Swiss National Science Foundation.

the transmission time of electron. In this paper, the composite of ZnO with SnO_2 was prepared composite film which will improve the thin film photoelectric performance of the battery.

Prepare ZnO thin film by using sol-gel method together with the spin coating on the FTO, then prepare ZnO/SnO_2 nanocrystalline composite films by dipping the ZnO films into the SnO_2 solution. Analyze the crystal face through SEM and the photoelectric property through I-V test system under different soaking time.

2. Experiment

2.1. *The preparation of ZnO nanocrystalline films*

Weigh zinc acetate 9.5 g, add 6 g PEG and 500 mL anhydrous ethanol on the mortar for grinding 20 minutes, then stewing for 30 minutes. Put them in the ultrasonic instrument for 15 minutes to get transparent gel. Then the gel is obtained by centrifugation, filtration. After that clean it with anhydrous ethanol for three times. The gel has prepared.

Spin coating method is used in 5 groups of conductive glass for coating 6 times and then put them on the incubator of 80 °C for drying 15 minutes after each coating and cooling for 15 minutes.

2.2. *The preparation of SnO_2 films*

Firstly dissolve $SnCl_2 \cdot 2H_2O$ in 100mL of anhydrous ethanol, add a small amount of acetic acid and then transfer it into the magnetic blender mixing for 5 hours to get slightly turbid solution. Secondly transfer them to reflux device under the 80°C temperature for 5 hours. Airtight stalling for 48 hours after cooling. We can get tin dioxide colloidal solution.

Put the prepared ZnO thin films into SnO_2 colloidal solution to soak for 0 minute, 20 minutes, 40 minutes, 60 minutes, 80 minutes respectively. After soaking, wash the wet membrane with anhydrous ethanol and then put them at constant temperature drying oven for 8 minutes. After that, put them in the muffle furnace for sintering under 500 °C for 40 minutes, then natural cooling.

2.3. *Battery assemble*

Put the prepared FTO glass with ZnO/SnO_2 into the 0.4 mol/L N3 to sensitize. Then using the ethyl alcohol to wash the residual dye. Choose the plating Pt electrode with 1 mm hole as the electrode. Inject the electrolyte through the hole then package all sides with adhesive tape. Cell preparation is finished.

2.4. *Characterization and photoelectric performance test*

Watch film microstructure and surface morphology with JMS-2010 transmission electric microscope. Photoelectric test through electrochemical workstation under simulation of the sun.

3. The Experimental Results and Conclusions

3.1. *SEM*

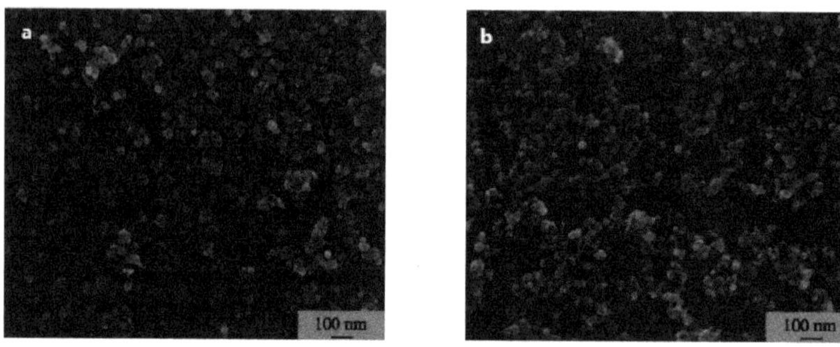

Fig. 1 SEM of ZnO(a) and ZnO/SnO2(b)

Figure 1 is the SEM of ZnO thin film and ZnO/SnO_2 thin film. We can see that the ZnO crystalline particles become more smaller after soaking, the average size is 30nm. However, the the average size of ZnO is 40nm.Smaller particle size increases the dye adsorption, which will improve the electrical performance of the battery.

3.2. *The influence of soaking time on battery photoelectric performances*

As shown in Table 1, when the ZnO thin film is soaked in SnO_2 sol for 40 minutes, it has the highest photoelectric conversion rate. Compared with ZnO thin film, short circuit current density increased from 1.642 mA/cm^2 to 1.952 mA/cm^2, open circuit voltage increased from 0.514 V to 0.516 V, fill factor increased from 49.2% to 52.0%, conversion rate increased from 0.42% to 0.52%. It has a lot to do with two factors. (1) The average particle size of crystalline has decreased after soaking in the SnO_2 for 40 minutes, which will increase the dye adsorption quantity. (2) Composite film make the chance of electronic-holes compound rate drop. Figure 2 is the band structure diagram of ZnO/SnO_2 composite films. In the composite thin film, the ZnO particles are surrounded by SnO_2 particles, which makes most dye adsorbed by SnO_2 particles. When inspired by light, electronic on SnO_2 particles will be flourished. Electronic will be injected to the SnO_2 guide line, and then transferred by ZnO to SnO_2, which will reduce the composite chance of electron-holes.

Photoelectric conversion efficiency doesn't increase continuously with the longer soaking time. This is because when ZnO is soaked in SnO_2 gel for a short time, only a small amount of SnO_2 penetrate into the pores of ZnO, which doesn't have big improvement on the film structure. When the soaking time is 40 minutes, we can obtain the highest conversion rate. When continue to increase the soaking time, conversion rate reduce instead. This is due to the film thickness is proportional to the soaking time, too long soaking time will increase film thickness. Secondly SnO_2 can also cluster with the increase of soaking time, which will reduce the photoelectric performance of the battery.

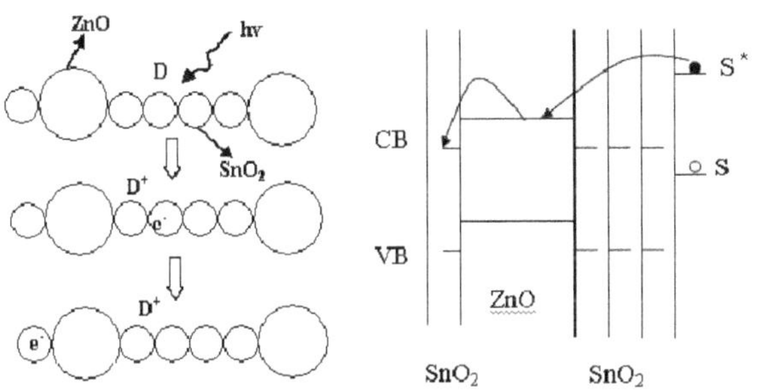

Fig. 2 Sketch map of electron transfer and energy band structure of ZnO/SnO_2 compound films

Tab. 1 Performance parameters of ZnO/SnO_2 composite films

	0 min	20 min	40 min	60 min	80 min
Jsc(mA/cm^2)	1.642	1.721	1.952	1.623	1.513
Voc(V)	0.514	0.505	0.516	0.496	0.482
FF(%)	49.2	51.3	52.0	50.8	49.5
η(%)	0.42	0.45	0.52	0.41	0.36

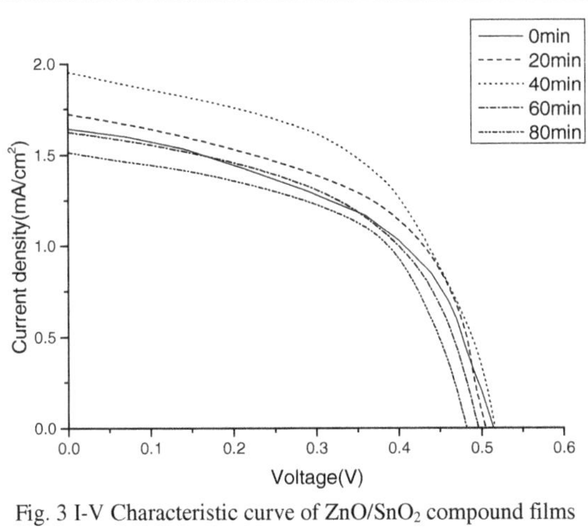

Fig. 3 I-V Characteristic curve of ZnO/SnO_2 compound films

4. Conclusion

Using sol-gel with spin coating method to prepare ZnO nanocrystalline films. Then soak them in 0.3M SnO_2 solution to prepare ZnO/SnO_2 composite nanocrystalline thin film. Analyze the film structure and I-V characteristics under different soaking time through SEM and I-V electrochemical workstation. We draw the following conclusions:

(1) SnO_2 wrap the surface of ZnO particles, which make the film's average particle size decreased, specific surface area increased and more dye will be absorbed on the films. At the same time, the addition of SnO_2 reduces the compound chance of electronic-holes in thin film, which will improve the conversion rate of thin-film battery.

(2) When ZnO thin film is soaked in SnO_2 solution for 40 minutes, we can get the best effect of photoelectric conversion. The photoelectric conversion rate has increased compared with non-compound.

References

1. McConnell RD. Assessment of the dye-sensitized solar cell.*Renewable and Sustainable Energy Reviews*, 2002, 6(3): 271–293.
2. *Grätzel* M. Recent advances in sensitized mesoscopic solar cells[J]. Accounts of Chemical Research, 2009, 42 (11): 1788–1798.
3. Alivov Yahya, Fan Z Y. Efficiency of dye sensitized solar cells based on TiO2 nanotubes filled with nanoparticles. Applied Physics Letters, 2009, 95(6): 063504–063506.
4. *Grätzel* M., Conversion of sunlight to electric power by nanocrystalline dye-sensitized solar cells. Journal of Photochemistry and Photobiology A: Chemistry, 2004, 164(1-3): 3–14.
5. Wang Z S, Kawaycgu H, Kashima Tetal. Significant influence of TiO2 photoelectrode morphology on the energy conversion efficiency of dye-sensitized solar cell. Coord. Chem. Rev., 2004, 248: 1381–1389
6. Dai S Y, Weng J S, Sui Y F et al. Dye-sensitized solar cell, from cell to module. Solar Energy Materials & Solar Cells, 2004, 84: 125–133.
7. Bandara J, Tennakone K. Interparticle charge transfer in dye-sensitized films composed of two kinds of semiconductor crystallites. Journal of Colloid and Interface Science, 2001, 236: 375–378.
8. Vinodgopal K, Bedja I, Kamat P V. Photoelectrochemical behavior of SnO2/TiO2 composite systems and its role in photocatalytic deradation of a textile azo dye. Chem. Mater., 1996, 8(8): 2180–2187.

Synthesis and properties of 1-[2-methyl-5-(4-methoxyphenyl)-3-thienyl]-2-[2-methyl-5-(4-chlorphenyl)-3-thienyl] perfluorocyclopentene*

Fang Duan and Gang Liu†
Jiangxi Key Laboratory of Organic Chemistry,
Jiangxi Science and Technology Normal University
Nanchang 330013, P. R. China
†E-mail: liugang0926@163.com

A novel unsymmetrical photochromic diarylethene 1-[2-methyl-5-(4-methoxyphenyl)-3-thienyl]-2-[2-methyl-5-(4-chlorphenyl)-3-thienyl] perfluorocyclopentene (1o), was successfully synthesized and its properties, such as its photochromic reactivity kinetics, fluorescence as well as electrochemical properties were investigated in detail. The results show that 1o exhibits remarkable photochromism, changing between colorless and blue in solution by photoirradiation. Furthermore, kinetic experiments conducted illustrate that the cyclization/cycloreversion processes of 1o were determined to be a zeroth-/first-order reactions, respectively. The cyclic voltammogram results reveal that this diarylethene exhibits evident electrochromism during electrolysis.

Keywords: Diarylethene; fluorescence; kinetics.

1. Introduction

In the past, various types of photochromic compounds have been synthesized for applications in multifunctional optoelectronic fields,[1,2] such as optical switch devices, data storage,[3–9] organic memory,[10] fluorescence sensing of specific ions,[11–14] photo-switchable probe for imaging living cells.[15] Diarylethene derivatives are the most promising candidates due to their excellent photochromic characteristics.[16] In addition, the photochromic properties of diarylethene can be predictably improved by structure variation to meet a special functional requirement.

In previous literatures, numerous photochromic properties of diarylethene derivatives have been carried out.[17,18] However, there are few publications reported photochromic unsymmetrical hybrid diarylethene derivatives with methoxyphenyl moieties. This group can make the isolation of substituent

* This work is supported from the National Natural Science Foundation of China (21262015).

electronic transfer to the conjugate system, which cloud made the density increased. It is enough to attract the attention of people.

According to the above aspects, in this work, a novel unsymmetrical and blue photochromic 1o was synthesized. Then, the photochromic properties, electrochemistry properties and fluorescence properties were discussed and investigated in detail. The photochromic reaction of 1o is shown in Scheme 1.

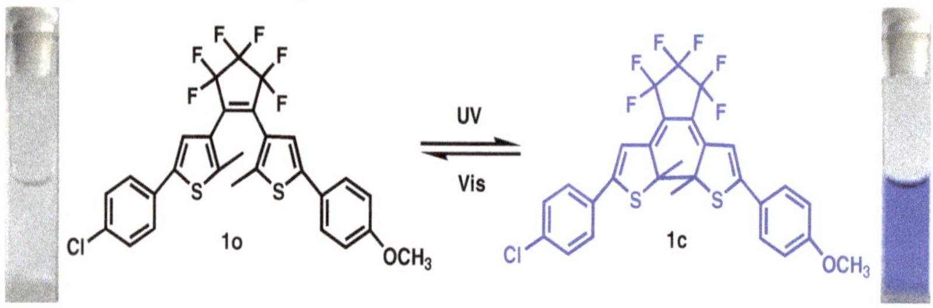

Scheme 1. Photoisomerization of 1o

2. Experiments

1o was prepared based on the method described in Scheme 2. And its synthetic steps according to the similar procedure of Pu et al.[19] The structure of 1o was characterized by ^{1}H NMR spectroscopy. NMR spectra were recorded on Bruker AV400 (400 MHz) spectrometer with $CDCl_3$ as the solvent and TMS as an internal standard. M.p.119-121°C; ^{1}HNMR(400MHz, $CDCl_3$, TMS): δ (ppm) 1.93 (s, 3H, $-CH_3$), 1.96 (s, 3H, $-CH_3$), 3.8 (s, 3H, $-CH_3$), 6.91 (s, 1H, thiophene-H), 6.93 (s, 1H, thiophene-H),7.16 (d, J = 8.0, 2H, benzene-H), 7.34 (d, J = 8.0, 1H, benzene-H), 7.36 (d, J = 8.0, 1H, benzene-H), 7.47 (d, J = 8.0, 2H, benzene-H);^{13}C NMR (100MHz, $CDCl_3$, TMS), δ (ppm):159.6, 158.9, 142.3, 141.9, 141.7, 140.9, 138.5, 133.7, 131.9, 129.2, 128.85, 127.0, 126.8, 126.1, 121.8, 114.4, 114.2, 55.4, 15.4, 1.1; IR (KBr, υ, cm^{-1}):821, 1103, 1256, 1446, 1510, 1600, 2962, 3686.

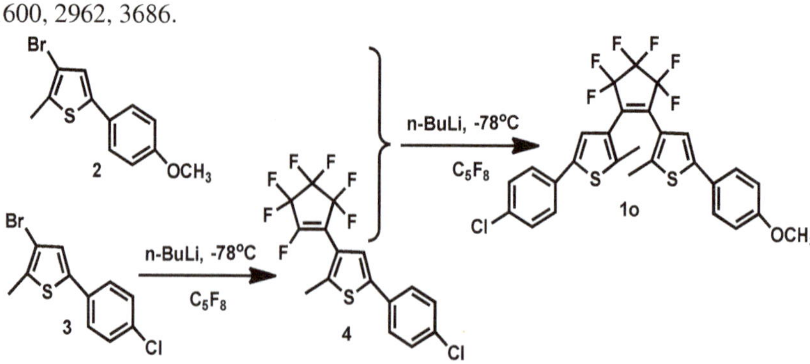

Scheme 2. Synthesis of 1o

3. Results and Discussion

3.1. *Photochromism of 1o in hexane solution*

Figure 1 showed the absorption spectral of 1o in hexane (2.0×10^{-5} mol L^{-1}) at room temperature under alternative illumination with ultraviolet and visible light. From the data, there was a sharp absorption peak at 294 nm ($\varepsilon_{max} = 2.7 \times 10^4$ mol^{-1} L cm^{-1}) corresponding to a $\pi \rightarrow \pi^*$ transition. Upon irradiation with the light of 297 nm, a new absorption band centered at 588 ($\varepsilon_{max} = 4.9 \times 10^3 mol^{-1}$ L cm^{-1}) appeared. This process could be seen with the naked eye, as the colorless solution turned blue. The change was ascribed to the formation of the closed-ring isomer 1c. Reversely, when the blue solution of 1c irradiated with visible light ($\lambda > 500$ nm), the color solution faded completely to colorless and the absorption spectrum returned to the initial state of 1o. In the absorption spectral changes, the isosbestic point for diarylethene 1 in hexane was observed at 316 nm, respectively, which clearly indicated a two-component photochromic reaction.

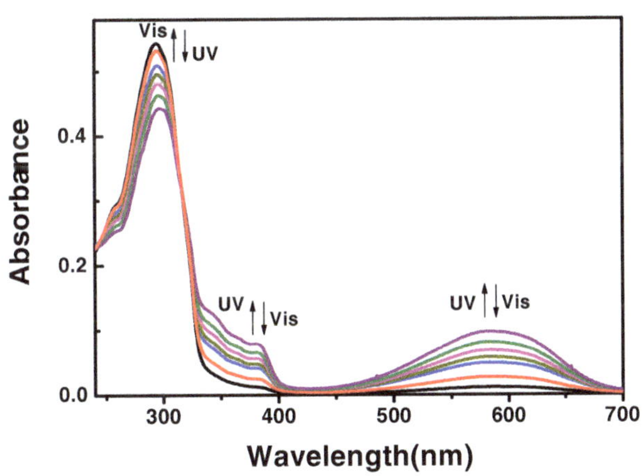

Fig. 1. Absorption spectra of 1o in hexane (2.0×10^{-5} mol L^{-1}).

3.2. *Photochromic reaction kinetics in hexane solution*

The photochromic cyclization and cycloreversion kinetics of this diarylethene in hexane were determined by UV-Vis light at room temperature. The cyclization andcycloreversion curves of 1o were shown in Figure 2, respectively. It can be seen that the relationships between the absorbance and exposal time have good linearity upon irradiation with 297 nm light. It is suggesting that the cyclization processe of 1o belongs to the zeroth order reaction. From Figure 2A we know the slope of line represents the reaction rate constant ($k_{o\text{-}c}$, 10^{-3}) of 1o in hexane solution and the value is 3.1 mol L^{-1} s^{-1}. At the same time, during the cycloreve rsion of 1c, the relationship between –log(Abs) and exposal time also behave perfect linearity. In Figure 2 (B), the cycloreversion process belongs to the first order reaction and k ($k_{c\text{-}o}$, 10^{-3}) can be easily obtained, it is 8.9 s^{-1} in hexane.

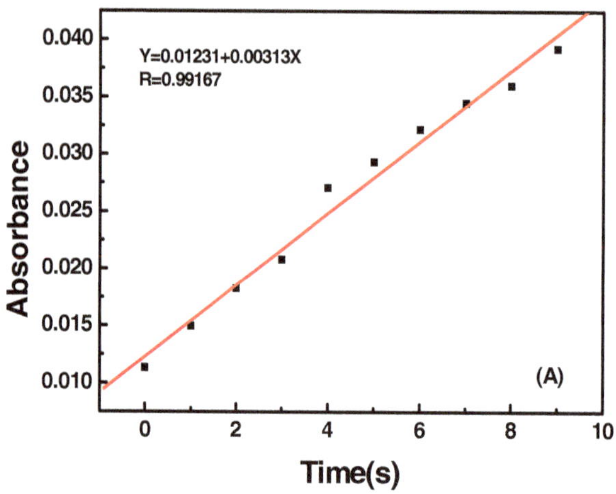

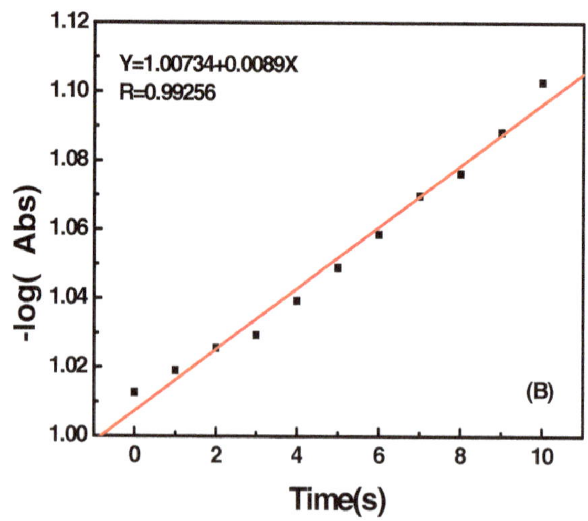

Fig. 2. The cyclization (A) and cycloreversion (B) kinetics of *1* in hexane

3.3. *Fluorescence of diarylethene 1o*

In this work, the fluorescence properties of the diarylethene1o in solution (2×10^{-5} molL^{-1}) was measured using a Hitachi F-4600 spectrophotometer, and the breadths of excitation and emission were selected 5.0 nm and 5.0 nm. As we all known, the concentration is one of the important factors which have influence on the fluorescence intensity. So the concentration dependence on the fluorescence spectrum of 1o was measured in hexane at room temperature. In Figure 3A, when the concentration arrived at 2×10^{-5}mol L^{-1}, the fluorescence intensity of the diarylethene reached the maximum. As shown in Figure 3B, the emission peak of 1o was observed at 378 nm when excited at 322 nm. Typically, the fluorescence of diarylethene could be reversibly modulated by photoirradiation during the process of photoisomerization.[20,21]

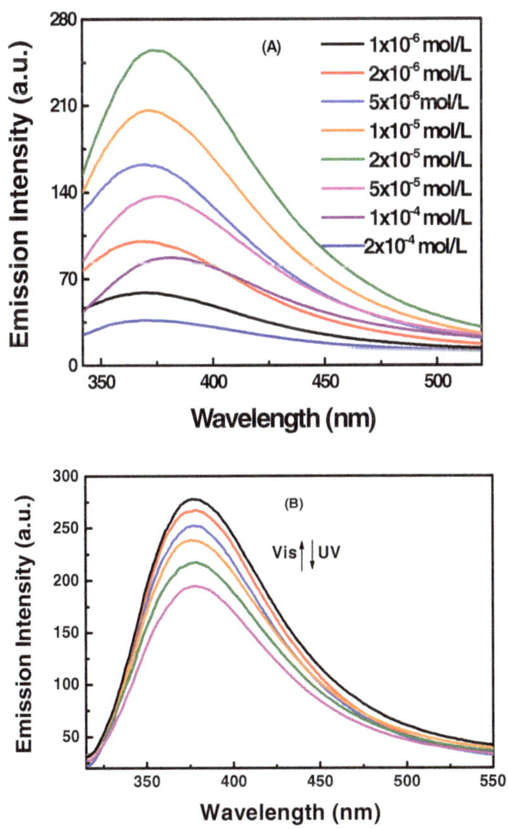

Fig. 3. Emission spectral change of *1o* in different concentrations and fluorescence spectra in hexane at room temperature (excited at 322nm)

3.4. *Electrochemistry of 1o*

As showed in Figure 4 the electrochemical property of diarylethene was determined by cyclic voltammetry (CV) at a scanning rate of 50 mV s^{-1}, the oxidation of 1o and 1c were initiated at 1.07 and 0.94 V, respectively, indicating that the oxidation process of the open-ring isomer was higher than that of the closed-ring isomer. This is attributed to the less positive potential of the closed-ring isomer caused by its longer conjugation length.[22, 23]

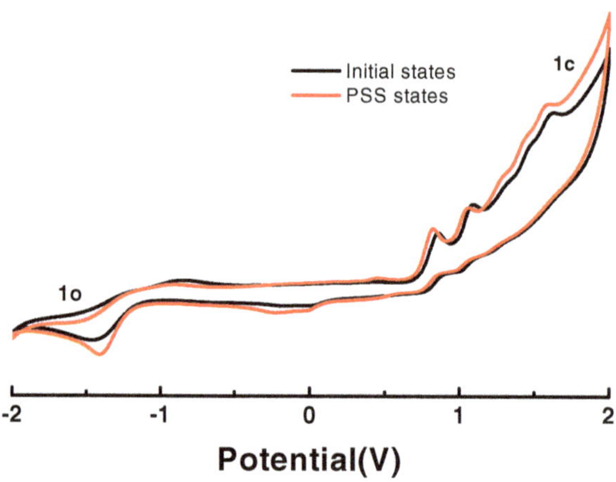

Fig. 4. Cyclic voltammetry of 1o

4. Summary

In conclusion, an unsymmetrical diarylethene was synthesized to investigate its photochromism and fluorescence properties. 1 showed good photochromism in solution and the fluorescence intensity decreased along with the photochromism upon irradiation with 297 nm light. In addition the fluorescence intensity of the open-ring isomer shows the dependence on concentration. When the concentration arrived at 2×10^{-5} mol L^{-1}, the fluorescence intensity of the diarylethene reached the maximum.

Acknowledgments

The authors are grateful for the financial support from the National Natural Science Foundation of China (21262015).

References

1. T.B. Norsten, A. Peters and R. McDonals, Reversible [7]-thiahelicence formation using a 1,2-dithienylcyclopentene photochrome, *J. Am. Chem. Soc.***123**, 7447 (2001).
2. Y. Yokoyama, Fulgides for memories and switches, *Chem. Rev.***100**, 1717 (2000).
3. M. Irie, T. Fukaminato and T. Sasaki, Organic chemistry: A digital fluorescent molecular photoswitch, *Nature* **420**, 759 (2002).
4. M. Irie, S. Kobatake and M. Horichi, Reversible Surface Morphology Changes of a Photochromic, *Science* **291**, 1769 (2001).
5. K. Uchida, M. Saito and A. Murakami, Three bits eight states photochromic recording and nondestructive readout by using IR light, *Chem. Eur. J.***11**, 534 (2005).
6. F.M. Raymo and M. Tomasulo, Electron and energy transfer modulation with photochromic switches, *Chem. Soc. Rev.***34**, 327 (2005).
7. S.Z. Pu, T.S. Yang and B.L. Yao, Photochromic diarylethene for polarization holographic optical recording, *Mater. Letter.***61**, 855 (2007).
8. A.J. Kronemeijer, H.B. Akkerman and T. Kudernac, Reversible cenductance switching in molecular devices, *Adv. Mater.***20**, 1467 (2008).
9. Z. Zhang, X. Liu and Z. Li, A smart light-controlled carrier switch in an organic light emitting device, *Adv. Function Mater.***18**, 302 (2008).
10. T. Tsujioka, T. Sasa and Y. Kakihara, Nonvolatile organic memory based on isomerization of diarylethene molecules by electrical carrier injection, *Org. Electr.***13**, 681 (2012).
11. Q. Zou, X. Li and J.J. Zhang, Unsymmetrical diarylethenes as molecular keypad locks with tunable photochromism and fluorescence viaCu^{2+} and CN^- coordinations, Chem. Commun.48, 2095 (2012).
12. S.Z. Pu, D.H. Jiang and W.J. Liu, Multi-addressable molecular switches based on photochromic diarylethenes bearing a rhodamine unit, J. Mater. Chem. 22, 3517 (2012).
13. W. Li and Y. Zhang, A triphenylamine-isophorone-based "off–on" fluorescent and colorimetric probe for Cu^{2+}, Sensors and Actuators B: Chemical 206, 640 (2015) .
14. Y. Wu, S.J. Chen and H. Tian, A novel gated photochromic reactivity controlled by complexation/dissociation with BF_3, Chem. Commun.48, 528 (2012).

15. W.J. Tan, J. Zhou, F.Y. Li, T. Yi and H. Tian, Visible light-triggered photoswitchable diarylethene-based iridium(III) complexes for imaging living cells, Chem. Asian J.6, 1263 (2011).
16. M. Irie, Diarylethenes for memories and swithches, Chem. Rev.100, 1685 (2000).
17. K. Yumoto, M. Irie and K. Matsuda, Control of the photoreactivity of diarylethene derivatives by quaternarization of the pyridylethynyl group, Org. Lett.10, 2051 (2008).
18. K. Morimitsu, K. Shibata, S. Kobatake and M. Irie, Dithienlethenes with a novel photochromic performance, J. Org. Chem.67, 4574 (2002).
19. Z.Y. Sun and S.Z. Pu, Synthesizes and properties research of 1-(1,2-dimethyl-7-azaindole-3-Yl)-2-[2-methyl-5-(4-chlorphenyl)-3thienyl] Perfluorocyclopentene, Advanced Materials Research952, 84 (2014).
20. H. Tian, B.Z. Chen, H.Y. Tu and K. Müllen, Novel bisthienylethene – based photochromic terazaporphyrin with photoregulating luminescence, Adv. Mater.14, 918 (2002).
21. S.Z. Xiao, T. Yi, Y.F. Zhou, Q. Zhao, F.Y. Li and C.H. Huang, Multi-state molecular switches based on dithienylperfluorocyclopentene and imidazo [4,5-f] [1,10] phenanthroline,Tetrahedron62, 10072 (2006).
22. S.Z. Pu, T.S. Yang, J.K. Xu, L. Shen, G.Z. Li and Q. Xiao, Syntheses and optoelectronic properties of fourphotochromic dithienylethenes, Tetrahedron 61 6623 (2005).
23. A. Perrier, F. Maurel and J. Aubard, Theoretical investigation of the substituent effect on the electronic and optical properties of photochromic dithienylethene derivatives, Journal of Photochemisty and Photobiology A: Chemistry 189, 67 (2007).

A new fluorescent sensor for Cr^{3+}, Fe^{3+} and Al^{3+} based on a photochromic diarylethene with a rhodamine unit

Ren-Jie Wang[†], Xiao-Rong Dong, Gang Liu and Shi-Qiang Cui
Jiangxi Key Laboratory of Organic Chemistry,
Jiangxi Science and Technology Normal University,
Nanchang 330013, PR China
[†]*E-mail: bio-wrj@163.com*

A new photoinduced fluorescent molecular switch based on a photochromic diarylethene with a rhodamine B fluorophore was synthesized, and its photochromic performance was examined in solution. The multi-addressable fluorescence switching characteristic was observed when induced by chemical and optical dual inputs stimulation. Upon addition of trifluoroacetic acid, it formed protonated diarylethene, which led to the color of the diarylethene turning from colorless to pink. In acetonitrile, its fluorescence emission intensity was sharply increased. Subsequent addition of triethanolamine base neutralized the protonated form and recovered the emission intensity to the original form. Furthermore, by monitoring the significant changes of fluorescence and color, the target compound was found to be selective towards Cr^{3+}, Fe^{3+} or Al^{3+}.

Keywords: Photochromism; diarylethene; rhodamine group; cation recognition; fluorescence sensor.

1. Introduction

During the past decade, a number of efforts have been made in the development of effective fluorescent sensors based on ion-induced changes in fluorescence. Such sensors are particularly attractive due to their instantaneous response, visual simplicity, and high sensitivity.[1] It is well known that metal ions involved in the structural, catalytic, and regulatory aspects of the biological system, and some metal ions serve as prognostics of human diseases.[2] For example, Cr^{3+} can cause disturbances in the glucose levels and lipid metabolism, and lead to a variety of disease, including diabetes and cardiovascular disease.[3] While excessive Al^{3+} can damage to certain human tissues and cells, resulting in disease such as Alzheimer, osteomalacia and so on.[4] In addition, chromium and aluminium are environmental pollutant and its build-up due to various industrial and agricultural activities is a matter of concern.[5] Therefore, design and synthesis of photoreactive molecules with sensitivity and selectivity to detect Cr^{3+} or Al^{3+} in aquatic or biological systems have received considerable attention in recent years.[6]

Photochromic materials have attracted much attention for their potential application in optical information storage media, biological sensor and intelligent stimuli-responsive materials.[7] Among various types of photochromic compounds, diarylethene derivatives with heterocyclic aryl rings have become the "star molecules" because of their high thermal stability, notable fatigue resistance and rapid response.[8] Therefore, there are many photo-controllable ion or anion sensors based on diarylethenes were designed and synthesized. These materials mostly incorporated diarylethenes with various ion receptor groups.[9] For example, He et al. have successfully developed a new phenanthrene-bridged photochromic diarylethene with good ions recognition ability composed of two crown ethers, which exhibited multi-responses to UV irradiation and Cu^{2+} or Hg^{2+}.[10] Tian et al. explored some new photochromic materials with ions recognition ability containing N-butyl-1,8-naphthalimide and piperazine, which showed multi-switching to UV irradiation, proton and Cu^{2+}.[11]

Rhodamine B is a good fluorescence dye and extensively used as a chemosensor due to its large absorption coefficient, high fluorescent quantum yield and easily excitation within visible wavelengths.[12] Furthermore, it can also undergo equilibrium between the colorless spirolactam (no fluorescence) form and the pink open-ring form (strong fluorescence). Currently, rhodamine derivatives have been extensively used as a kind of good visual detection of metal ions, such as detecting Hg^{2+}, Cu^{2+} and Pb^{2+} and so on, since they can be perceived by naked eyes.[13] Undoubtedly, distinct color of sensors before and after combination with ions is necessary for visual detection. The detection of Cr^{3+} or Al^{3+} is of great importance as high levels of both ions in the environment have begun to negatively affect human health. In this work, we designed and synthesized a new photochromic diarylethene, which incorporated a rhodamine unit as the ionic signal responsive group and its fluorescence can be sharply increased by Cr^{3+}, Fe^{3+} or Al^{3+}. This property facilitates us to fabricate an ions sensor for visual detection of Cr^{3+}, Fe^{3+} or Al^{3+}. The phorochromic process of the diarylethene is shown in Scheme 1.

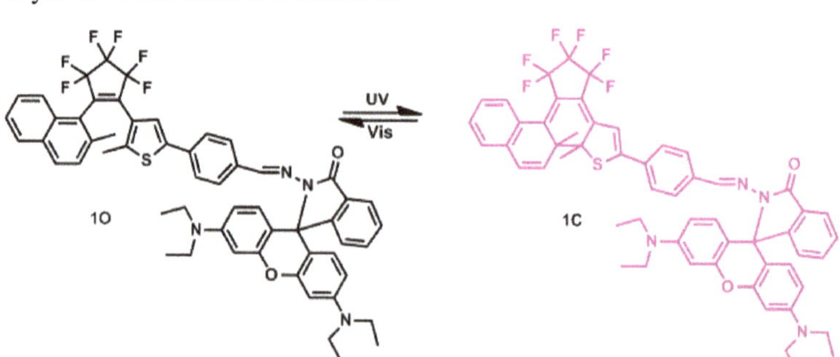

Scheme 1. Photochromism of diaryelthene 1.

2. Experimental

2.1. *Synthesis of diarylethenes 1o*

Diarylethene **1O** was prepared according to the similar procedure of reference and described in Scheme 2. The target compound **1O** was prepared by the condensation reaction of the precursor compound **2** with rhodamine B hydrazine **3**.[14] To a solution containing compound **2** (0.40 g, 0.78 mmol) and **3** (0.39, 0.85 mmol) in ethanol (50 mL) was added acetic acid (0.02 equiv). The mixture was stirred at 351 K and the reaction was detected by TLC plate. After stirring 36 h, it was cooled to room temperature and washed with dilute $NaHCO_3$ solution. After being extracted with chloroform (3 × 30 mL), dried using anhydrous magnesium sulfate, filtered, and evaporated, the crude product was purified by column chromatography on silica gel using petroleum/ethyl acetate (v/v = 4:1) as the eluent to give diarylethene **1O** (0.23 g, 32 % yield) as a brown solid. The structure of diarylethene **1O** was confirmed by elemental anaysis, ^{1}H NMR, and IR. Anal. Calcd. for $C_{56}H_{48}F_6N_4O_2S$: Calcd C, 70.42; H, 5.07; N, 5.87. Found C, 70.36; H, 5.15; N, 5.77; M.P. 127–128 °C; ^{1}H NMR (400 MHz, $CDCl_3$, ppm): δ 1.15 (t, 12H, $-CH_3$), 2.19 (s, 3H, $-CH_3$), 2.32 (s, 3H, $-CH_3$), 3.31 (q, 8H,$-CH_2$), 6.22 (d, 2H, phenyl–H, J = 8.0 Hz), 6.42 (s, 2H, phenyl–H), 6.50 (d, 2H, phenyl–H, J = 8.0 Hz), 6.78 (s, 1H, thienyl–H), 7.04 (d, 2H, phenyl–H, J = 8.0 Hz), 7.11 (d, 1H, phenyl–H, J = 8.0 Hz), 7.30 (d, 1H, phenyl–H, J = 8.0 Hz), 7.40–7.53 (m, 5H, phenyl–Hz), 7.66 (d, 1H, phenyl–H, J = 8.0 Hz), 7.79 (t, 2H, phenyl–H, J = 8.0 Hz),7.99 (d, 1H, phenyl–H, J = 8.0 Hz), 8.50 (s, 1H, –CH=N); IR (KBr, v, cm^{-1}): 529, 553, 611, 753, 780, 985, 1010, 1104, 1120, 1140, 1223, 1262, 1275, 1311, 1430, 1516, 1553, 1617, 1692, 2931, 2968.

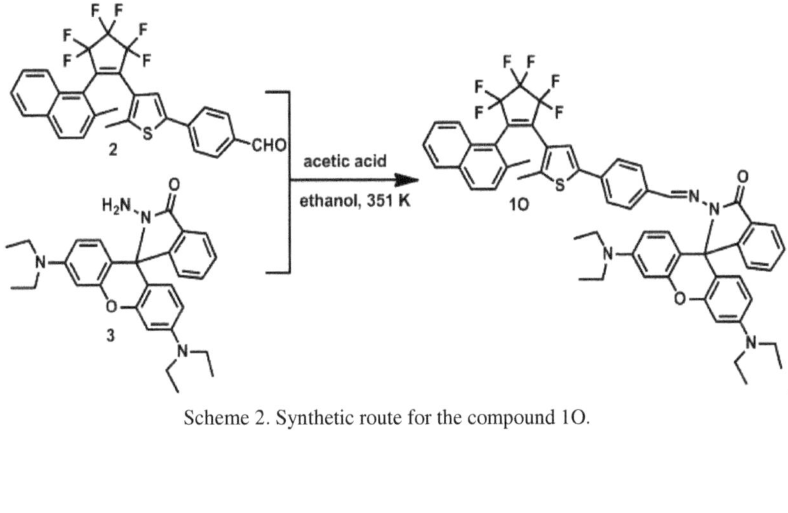

Scheme 2. Synthetic route for the compound 1O.

3. Results and Discussion

3.1. *Synthesis of Diarylethenes 1o*

The photochromic behaviors of diarylethene **1O** in acetonitrile ($C = 5.0 \times 10^{-5}$ mol L^{-1}) induced by photo-irradiation at room temperature are shown in Figure 1. As shown in Figure 1, the absorption maximum of **1O** is observed at 343 nm ($\varepsilon_{max} = 3.42 \times 10^4$ L mol^{-1} cm^{-1}) in acetonitrile, which was arisen from $\pi\rightarrow\pi^*$ transition. Upon irradiation with 365 nm light, a new visible absorption band centered at 533 nm ($\varepsilon_{max} = 2.20 \times 10^3$ mol^{-1} L cm^{-1}) emerged while the original peak at 343 nm decreased due to the formation of the closed-ring isomer **1C**, this change in absorption was manifested in the colorless solution becoming light pink. The pink colored solution became colorless upon irradiation with visible light ($\lambda > 500$ nm), indicating that the closed-ring isomer **1C** returned to the initial state **1O**. Compared to the precursor **2**,[15] diarylethene **1** exhibits a longer absorption maxima in both open-ring and closed-ring isomers than compound **2**. The bathochromic shift may be attributed to the substitute effect induced by the rhodamine unit. Furthermore, the thermal stability of the open-ring and closed-ring isomers of diarylethene **1** was tested in acetonitrile both at room temperature and 351 K. The acetonitrile solution was stored at room temperature in the dark, and exposed to air for more than a week. No change was detected in UV/vis spectra for diarylethene **1**. At 351 K, diarylethene **1** also showed good thermal stability for more than 3 h in acetonitrile.

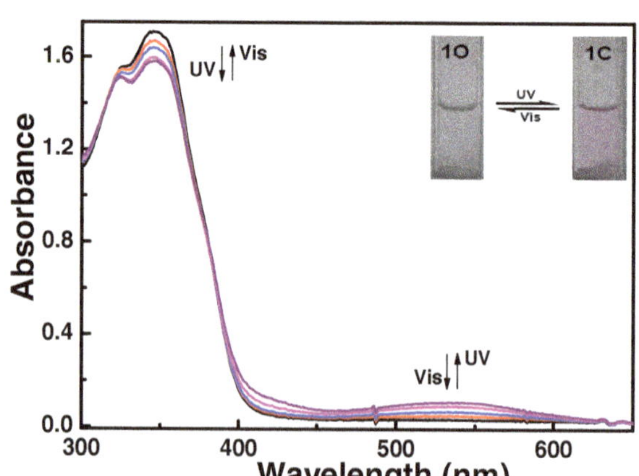

Fig. 1. The absorption spectral and color change of compound 1 in acetonitrile ($C = 5.0 \times 10^{-5}$ mol L^{-1}).

3.2. *Fluorescence properties of diarylethenes 1*

The fluorescence properties of **1O** was measured in acetonitrile (C = 2.0 × 10^{-5} mol L^{-1}) at room temperature by using a Hitachi F-4600 fluorimeter. Upon excited at 367 nm, the emission peak of **1O** was observed at 427 nm, which exhibited very weak fluorescence emission. However, the emission intensity of **1O** can be turned on through acid stimuli, which could induce the spirolactam form of the rhodamine to the open-ring form, and the open-ring form has strong fluorescent emission. Indeed, when 75.0 equiv. TFA was added to the acetonitrile containing **1O**, the fluorescent emission intensity was increased immediately with the colorless solution of **1O** became to pink due to the protonated **1O′** was generated (Figure 2). The result indicates that **1O** is very much sensitive to proton. As the fluorescence intensity reaches to the maximum, the fluorescent quantum yields of **1** were determined as 0.43 by using rhodamine B as a reference. Moreover, the fluorescent intensity of **1O′** can be quenched to the original emission spectra by adding TEA. After addition of 77.0 equiv. TEA, the emission intensity of **1O′** decreased sharply by 99%, accompanied with a notable fluorescence color change from bright yellow to dark grey. This characteristic could be potentially applied in fluorescent switches and chemical sensors.

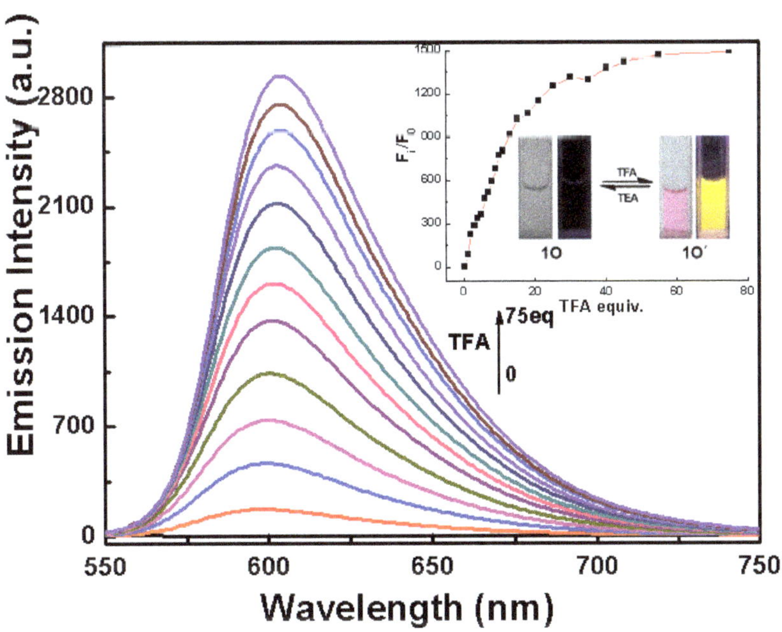

Fig. 2. Changes in emission spectral and color of diarylethene **1O** in acetonitrile (C = 2.0 × 10^{-5} mol L^{-1}) induced by the addition of TFA/TEA when excited at 565 nm at room temperature.

3.3. *Optical spectral and fluorescence changes of diarylethene 1O with response to Cr3+*

Due to metal ions can induce the spirolactam form of rhodamine to the open-ring form. In this section, the UV/Vis absorption and fluorescence changes of diarylethene **1O** with respond to Cr^{3+} were investigated by spectral analysis. The absorption spectra of **1O** in acetonitrile ($C = 2.0 \times 10^{-5}$ mol L^{-1}) were recorded at room temperature and $Cr(NO)_3{\cdot}9H_2O$ was used as the Cr^{3+} source. The absorbance change of **1O** after addition of Cr^{3+} is shown in Figure 3. In acetonitrile, the absorption maximum of **1O** is observed at 343 nm. When titrating trace amount of Cr^{3+} ion into the solution containing diarylethene **1O**, a new visible absorption peaked at 559 nm appeared due to the formation of the open-ring amide form of Rhodamine schiff base,[12,16] these change in absorption was manifested in the colorless solution becoming pink within 1 s (inset graph in Figure 3). The absorbance at the maximum wavelength (λ_{max} = 559 nm) was dramatically increased with increasing of titration amounts of Cr^{3+} ion suggesting that it has high sensitivity towards Cr^{3+}.[17] When Cr^{3+} ion was increased form 0 to 2.0 equiv., the absorbance achieved to the maximum, indicating that the naked-eye sensor for Cr^{3+} based on the diarylethene **1O** could be constructed.

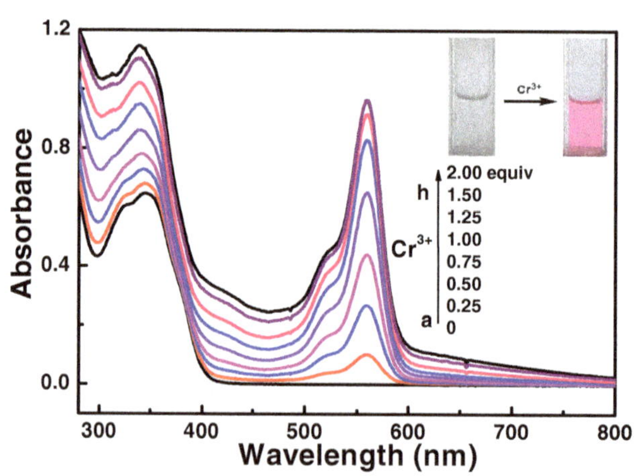

Fig. 3. The absorption spectra changes of diarylethene **1O** by stimulation of Cr^{3+} in acetonitrile (C = 2.0×10^{-5} mol L^{-1}) (a→h): 0, 0.25, 0.50, 0.75, 1.00, 1.25, 1.50, 2.00 eq. Inset: the color of **1O** before and after addition of 1.00 equiv. Cr^{3+}.

Subsequently, the enhancement effects of various amounts of Cr^{3+} on the fluorescence of diarylethene **1O** were examined at the excitation wavelength of 565 nm. Figure 4 shows the emission intensity changes with addition of Cr^{3+}

ions. A remarkable fluorescence enhancement was observed with increasing of the amounts of Cr^{3+}, when titrating Cr^{3+} into the acetonitrile solution containing diarylethene **1O** (C = 2.0 × 10^{-5} mol L^{-1}). The emission intensity at 604 nm increased by approximately 250-fold along with Cr^{3+} concentration increased from 0 to 2.0 equiv., and remained unchanged with further titration. The emission intensity is heavily dependent on the status of the diarylethene, which can be quenched by the photocyclization reaction. Upon irradiation with UV light, the emission intensity at 604 nm significantly decreased with increasing UV irradiation time due to the formation of the non-fluorescent closed-ring isomer **1C**. When arrived at the photostationary state, the emission intensity of **1C**/Cr^{3+} was quenched to 66% of its original intensity. Compared with diarylethenes with five-membered aryl rings, the fluorescent modulation efficiencies of diarylethenes **1** were significantly decreased in solution.[12] The main reason may be attributed to the low efficiency intermolecular fluorescence resonance energy transfer (FRET) from the open-ring rhodamine amide to the closed-ring diarylethene. Because the emission band of the open-ring rhodamine unit (550–700 nm) only some overlaps with the absorption bands of the closed-ring isomer **1C** (450–600 nm), which induced the FRET process occurs low efficiency between in the diarylethene moiety and the rhodamine unit stimulated by UV light. Moreover, another possible reason is that the incomplete cyclization reaction and existence of parallel conformations in the photostationary state may be responsible for this phenomenon.[18]

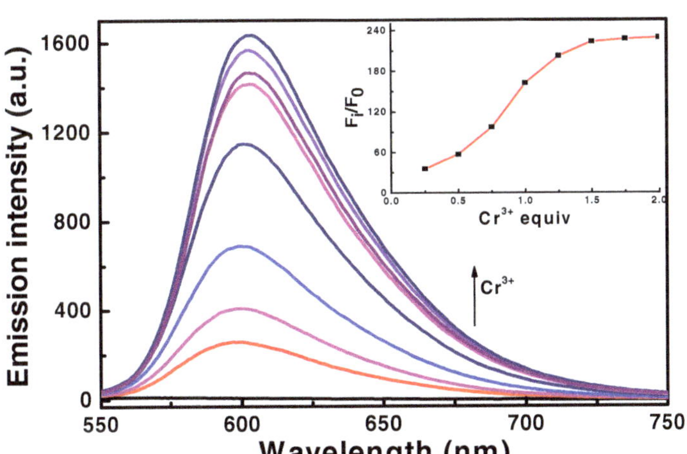

Fig. 4. Changes in emission spectral of diarylethene 1O with response to Cr^{3+} when excited at 565 nm in acetonitrile at room temperature.

In order to discuss the selectivity of diarylethene **1O** with response to Cr^{3+}, other metal ions including alkali, alkaline earth, and transition-metal ions were separately added into the acetonitrile solution containing diarylethene **1O** under the same experimental condition. Figure 8 shows the emission intensity of **1O** at 604 nm in acetonitrile in the presence of respective metal cations at room temperature when excited at 565 nm. The results revealed that **1O** responded to various metal ions with different selectivity and possessed an excellent selectivity for Cr^{3+}, Fe^{3+}, and Al^{3+}. As depicted in Figure 5, it could be easily seen that its emission intensity was essentially not influenced by addition of 2.0 equiv of Ca^{2+}, Cd^{2+}, Ba^{2+}, Co^{2+}, K^{+}, and Hg^{2+} ions, and very weakly influenced by Sn^{2+}, Cu^{2+}, Pb^{2+}, and Zn^{2+} ions. The phenomenon suggested that the fluorescence enhancement of **1O** may be induced by the ring opening mechanism of the rhodamine spirolactam moiety, which is resulted from the chelation of the rhodamine moiety of **1O** binding with Cr^{3+}, Fe^{3+} or Al^{3+}.

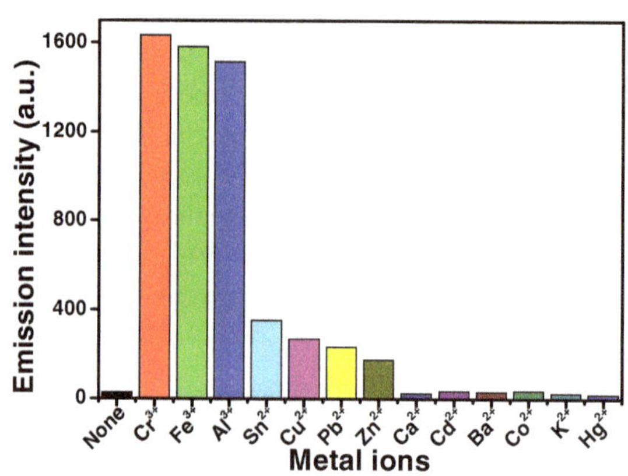

Fig. 5. Emission intensity of **1O** at 604 in acetonitrile (C = 2.0 × 10^{-5} mol L^{-1}) in the presence of respective metal cations (2.0 equiv.) at room temperature, excited at 565 nm.

4. Conclusion

In conclusion, a new fluorescent molecular switch diarylethene derivative bearing a rhodamine B fluorophore was successfully synthesized and by using it to determinate cations in acetonitrile was investigated. The results revealed that the compound response to both light and chemical inputs, which result in different fluorescent emission under sequential metal ions and protonation. Moreover, In the presence of Cr^{3+}, Fe^{3+} or Al^{3+}, the color and fluorescence of the target compound could be effectively modulated. Our results demonstrated that the diarylethene compound can be potentially used as a colorimetric and "turn-on" fluorescent probe for Cr^{3+}, Fe^{3+} or Al^{3+} detection in environmental pollutants and biological systems.

Acknowledgments

The authors are grateful for the financial support from the National Natural Science Foundation of China (21262015, 21362013 and 51373072), the Project of the Science Funds of Jiangxi Education Office (KJLD12035 and GJJ12587), and the Science Funds of Jiangxi Science & Technology Normal University (2014XJZD001).

References

1. M. Irie; T. Fukaminato, N. Tamai, T. Kawai, *Nature* **420**, 759 (2002).
2. K. Rurack, *Spectrochim Acta. A: Mol. Biomol. Spectro* **57**, 2161 (2001).
3. K. Eiichi, T. Koike, *Chemical Soc. Rev* **27**, 79 (1998).
4. H. Arakawa, R. Ahmad, M. Naoui, H.A. Tajmir-Riahi, *J. Biol. Chem.* **275**, 10150 (2000).
5. T.P. Flaten, *Brain Res. Bull* **55**, 187 (2001).
6. S.Z. Pu; T.F. Wang, G. Liu, W.J. Liu, S.Q. Cui, *Dyes Pigm* **94**, 416 (2012).
7. H. Tian, S.J. Yang, *Chem. Soc. Rev* **33**, 85 (2004).
8. G. Liu, S.Z. Pu, R.J. Wang, *Org. Lett* **15**, 980, (2013).
9. J.P. Malval, I. Gosse, J.P. Morand, R. Lapouyade, *J. Am. Chem. Soc* **124**, 904, (2002).
10. J.J. He, J.X. He, T.T. Wang, H.P. Zeng, *J. Mater. Chem. C***2**, 7531, (2014).
11. J.J. Zhang, W.J. Tan, X.L. Meng, H. Tian, *J. Mater. Chem.* **19**, 5726, (2009).
12. S.Z. Pu, D.H. Jiang, W.J. Liu, G. Liu, S.Q. Cui, *J. Mater. Chem* **22**, 3517 (2012).
13. W.Y. Lin, X.W. Cao, Y.D. Ding, L. Yuan, Q.X. Yu, *Org. Biomol. Chem* **8**, 361 (2010).
14. J.J. Zhang, W.J. Tan, X.L. Meng, H. Tian, *J. Mater. Chem.* **19**, 5726 (2009).
15. R.J. Wang, S.Z. Pu, G. Liu, S.Q. Cui, *Beilstein J. Org. Chem.* **8**, 1018(2012).
16. K. Yumoto, M. Irie, K. Matsuda, *Org. Lett* **10**, 205 (2008).
17. K.W. Huang, Y. Hong, Z.G. Zhou, M.X. Yu, F.Y. Li, X. Gao, *Org. Lett* **10**, 2557 (2008).
18. J. Mao, Q. He, W. Liu, *Anal. Bioanal. Chem.* **396**, 1197 (2010).

A review of numerical researches on fretting fatigue

Hai-Kun Shang[a] and Liang Xie[b]

Hebei Huabei Diesel Engine Co., Ltd., Shijiazhuang, Hebei, China
E-mail: [a]hckjkfb_shanghk@163.com, [b]hckjkfc@chbdp.com

Hua-Yang Zhang[†]

School of Mechanical Engineering, Beijing Institute of Technology, Beijing, China
[†]E-mail: zhhy2009bit@163.com

Fretting fatigue, which is induced by small sliding movements between contacting surfaces under cyclic loading, has become a critical failure mechanism in modern engineering structures. The modeling and analysis of fretting fatigue based on numerical methods has seen great development in recent years. This paper will review the numerical researches and advances in fretting fatigue, mainly focusing on: (i) initiation criteria for fretting fatigue cracks, (ii) modeling of the contact between different components in fretting fatigue, (iii) effects of material removal on fretting fatigue, and (iv) numerical implementation in fretting fatigue analysis using the finite element method and the extended finite element method.

Keywords: Contact modeling; material removal; numerical implementation.

1. Introduction

Fretting fatigue usually occurs between contacting surfaces which are nominally static but in fact experience small amplitude oscillatory movement [1]. Although the value of the movement amplitude has not been strictly defined, it is usually seen as less than 100 μm [2]. Due to the invisibility of the small motion, the effects of fretting on fatigue life of components are neglected from time to time, and therefore fretting may induce more severe damage to engineering components than visible damage mechanisms do.

Fretting fatigue may occur in many components, such as bolted joints [3], blade-disk dovetail connections [4] and spline couplings in gas turbine engines

---

[†] Corresponding author.

[5]. Many studies have been carried out via theoretical analysis, numerical simulation, experimental study, or combination of them in order to understand the failure mechanism. In the theoretical field, two analytical solutions are commonly used. One solution for the partial slip case was firstly proposed by Cattaneo [6] and independently completed by Mindlin [7]. The other solution including the effect of the bulk fatigue load on the Mindlin contact problem was subsequently proposed by Nowell and Hills [8]. Although analytical methods are simple, they are applicable to the ideal conditions, i.e., constant amplitude loading, infinite and idealized bodies, and elastic material properties [9]. In addition to analytical solutions, many experimental methods have also been proposed to investigate the fretting fatigue. Experimental methods are usually able to visually and accurately reflect the damage induced by fretting fatigue. However, economic cost and time consumption are very high. Under the circumstances, with the rapid development of numerical methods and computer technology, analysis of fretting fatigue based on numerical methods [10] has attracted great attention. Numerical methods are relatively high efficient and feasible compared with experimental methods. In addition, numerical methods have the advantage over analytical approaches when dealing with complex geometries and variable loading conditions.

Many aspects need to be considered in investigating fretting fatigue by using numerical methods, such as contact modelling [11] and effects of material removal [12]. Although a lot of numerical studies on fretting fatigue have been performed, to the best of the authors' knowledge, there has not been a systematic summary about fretting fatigue analysis based on numerical methods. This paper attempts to review numerical researches and advances in fretting fatigue. Firstly, a brief description of initiation criteria for fretting fatigue cracks is given. Then, contact modeling in fretting fatigue is mentioned. Effects of material removal on fretting fatigue are subsequently discussed. Finally, two numerical methods, i.e., the finite element method (FEM) and the extended finite element method (X-FEM), are discussed in detail.

2. Initiation Criteria for Fretting Fatigue Cracks

There are a lot of initiation criteria based on the critical plane. This paper mainly lists two common criteria based on the critical plane which have been used in investigating fretting fatigue.

2.1. *Smith-Watson-Topper criterion*

The critical plane SWT criterion, which is based on SWT criterion [13], was firstly suggested by Szolwinski et al. [14] to deal with the multiaxial stress and strain cycles in fretting fatigue. The criterion can be described as:

$$SWT = \sigma_{max}\left(\frac{\Delta\varepsilon}{2}\right) = \frac{\left(\sigma_f^{'}\right)^2}{E}\left(2N_f\right)^{2b} + \sigma_f^{'}\varepsilon_f^{'}\left(2N_f\right)^{b+c} \qquad (1)$$

where $\Delta\varepsilon$ is the strain range normal to the plane during the cycle, σ_{max} is the maximum stress normal to the plane at any point in the cycle, $\sigma_f^{'}$ and $\varepsilon_f^{'}$ are the fatigue strength coefficient and fatigue ductility coefficient, respectively. b and c are the fatigue strength exponent and fatigue ductility exponent, respectively. E and N_f are Young's modulus and the number of cycles to initiate a crack at a given length. The critical plane is the plane where the SWT parameter reaches the maximum value (SWT_{max}).

2.2. *Fatemi-Socie criterion*

The FS criterion was firstly proposed by Fatemi and Socie[15]. It can be written as:

$$FS = \frac{\Delta\gamma}{2}\left(1+\alpha\frac{\sigma_{max}}{\sigma_y}\right) = \frac{\tau_f^{'}}{G}\left(2N_f\right)^{b} + \gamma_f^{'}\left(2N_f\right)^{c} \qquad (2)$$

where $\Delta\gamma$ is the shear strain range during the cycle, α is the constant, σ_{max} is the maximum stress perpendicular to the chosen plane, σ_y is the yield strength, G is the shear modulus, and $\tau_f^{'}$, $\gamma_f^{'}$, b, and c are constants. The critical plane is the plane where the FS parameter reaches the maximum value (FS_{max}).

3. Contact Modelling in Fretting Fatigue

Modelling the contact numerically in fretting is similar to that in other problems to some extent. The contact surface interaction was defined by the master-slave algorithm [16] or impactor/target technique [11]. The frictional contact constraints were generally imposed by the Lagrange multiplier method [17] or the penalty method [18]. The Lagrange multiplier approach strictly enforces the stick condition between the components when the shear stress is less than the critical value. The penalty method is a more approximate approach which permits an allowable elastic slip for convergence. Although the coefficient of friction (COF) varies with the increase of cycle numbers in the practical loading conditions [19], it was generally assumed as constant in the numerical simulation for simplicity [20].

When friction is present, the coupling between the normal and the shear tractions induces the contact problem non-linear and history dependent. The incremental technique proposed by Man et al. [21] can accurately solve this problem. The above technique was applied to the contact between the rigid pad and the elastic foundation by Zografosa et al. [22]. The vertical and horizontal displacement increment for each node within the whole contact area Γ and the stick area Γ_{stick}, i.e., ΔU_Y^k and ΔU_X^k, are described as:

$$\begin{cases} \Delta U_Y^k = h(X) - Y_c^{k-1} + RemY \ (\forall node \in \Gamma) \\ \Delta U_X^k = RemX \ (\forall node \in \Gamma_{stick}) \end{cases} \tag{3}$$

where $h(X)$ depicts the outline of the pad, and Y_c^{k-1} is the vertical coordinate of each contact node in the previous increment. *RemX* and *RemY* are the remote displacement components of the pad.

The relation between the shear traction and the normal traction is written as:

$$\begin{cases} \Delta\sigma_n^k + \sigma_n^{k-1} < 0 \ (\forall node \in \Gamma) \\ \Delta\sigma_t^k + \sigma_t^{k-1} = \pm\left(\Delta\sigma_n^k + \sigma_n^{k-1}\right) f \ (\forall node \in \Gamma_{slip}) \\ \left|\Delta\sigma_t^k + \sigma_t^{k-1}\right| =< \left|\Delta\sigma_n^k + \sigma_n^{k-1}\right| f \ (\forall node \in \Gamma_{stick}) \end{cases} \tag{4}$$

where $\Delta\sigma_t^k$ and $\Delta\sigma_n^k$ are the shear traction increment and the normal traction increment, σ_t^{k-1} and σ_n^{k-1} are the total shear and normal tractions in the previous increment, f is COF, and Γ_{slip} is the slip area.

4. Effects of Material Removal on Fretting Fatigue

Fretting fatigue and fretting wear usually coexist in real situation. Wear plays an important role in cracks nucleation [23]. As fretting proceeds, embryo cracks may be worn away. Thus, material removal has an important effect on fretting fatigue. The calculation of wear depth was generally based on the Archard equation or energy. As wear proceeds, the contact surface profiles evolve. On the basis of the Archard equation [24], Ding and McColl [12, 25] computed the evolution in the incremental form:

$$\Delta h_{i,j} = k_1 \times 4 s_{i,j} \times \Delta N \times p_{i,j} \tag{5}$$

where $\Delta h_{i,j}$, $s_{i,j}$ and $p_{i,j}$ are the wear depth increment, the contact slip during a quarter cycle, and the contact pressure at node i in the jth wear increment, respectively. k_1 and ΔN are the local wear coefficient and the cycle jump, respectively.

However, the wear simulation method in Zhang's work [26] was not based on the Archard equation but energy proposed by Fouvry et al. [27]. The adaptive meshing technique was used to simulate incremental material removal. The incremental wear equation is expressed in the following form:

$$\Delta h_{i,m} = \alpha E_{i,m} \tag{6}$$

where

$$E_{i,m} = q_{i,m} \Delta s_{i,m} \tag{7}$$

where $\Delta h_{i,m}$, $E_{i,m}$, $q_{i,m}$, and $\Delta s_{i,m}$ are the incremental wear depth, the incremental dissipated energy, the shear traction and the incremental slip at contact node i in the mth increment, respectively. α is the energy wear coefficient, which is independent of load and stroke.

5. Numerical Implementation in Fretting Fatigue

Numerical implementation is almost indispensable in researches on fretting fatigue. Although BEM has been used by a few researchers to investigate fretting fatigue [3, 22], FEM and X-FEM will be mainly discussed in the following subsections.

5.1. *FEM*

FEM is a very popular numerical method, and has been widely used in many domains. FEM can be used to calculated stress fields [28]. Stress fields calculated via FEM combined with initiation criteria can be used to predict fretting fatigue life [29]. Stress intensity factors (SIFs) can also be obtained using FEM [23, 30].

Although FEM is able to capture the global scale of the structure, such as complex geometry and boundary conditions, numerous elements need to be arranged in the vicinity of the crack tip in order to accurately calculate the stress state near the crack front. It needs to remesh in order to obtain more accurate SIF due to the evolution of SIF during the crack growth [31]. Hence, the computational cost is high. Crack can only propagate along the predefined routine in FEM, i.e., the interface between elements. However, Fretting fatigue cracks are generally mixed mode cracks, and the propagation paths are unknown in advance. Therefore, FEM appears to be incapable of predicting these unknown cracks propagation.

5.2. *X-FEM*

X-FEM is a new numerical method developed in recent years. This method is based on FEM and has been widely used to investigate fretting fatigue. X-FEM overcomes shortcomings of remeshing during calculating the crack propagation in FEM. Its key merit is due to the partition-of-unity method [32] combined with special enriched functions [33] locally incorporated into the FE approximation. These enriched functions capture the asymptotic near-crack-tip behaviour and the discontinuous displacement fields [34].

The displacement approximation **u** for crack modelling in the X-FEM [34] is expressed as:

$$\mathbf{u}(\mathbf{x},t) \simeq \sum_{i \in N_{nodes}} \mathbf{u}_i(t)\Phi_i(\mathbf{x}) + H(\mathbf{x}) \sum_{j \in N_{crack}} \mathbf{a}_j(t)\Phi_j(\mathbf{x}) + \sum_{l=1}^{4} B_l \sum_{k \in N_{front}} \mathbf{b}_{lk}(t)\Phi_k(\mathbf{x}) \tag{8}$$

where $\mathbf{u}_i$ are the classical DOFs of node *i*, $\mathbf{a}_j$ are the DOFs associated with the Heaviside function $H(\mathbf{x})$, and $\boldsymbol{b}_{lk}$ are the DOFs associated with the singular enrichment functions B_l of the crack front. Φ_i, Φ_j, and Φ_k are first order shape functions. The mesh is independent of the initial crack shape. It is therefore not necessary to generate a mesh that conforms to the crack boundaries. Thus, only a single mesh, which is usually easily produced, can be used for any crack length and orientation. The computational cost is enormously reduced. Hence, X-FEM is very well suited to crack growth simulations in complex problems.

Due to the marked advantage, X-FEM has been extensively used to predict the crack propagation in fretting fatigue. Giner et al. [35] investigated the effect of the bulk load and the initial crack angle on fretting fatigue crack propagation by using X-FEM. The effect of crack lengths on SIF was subsequently investigated [36]. Pierres et al. [34] predicted 3D fretting fatigue crack propagation by using X-FEM.

6. Conclusions

Initiation criteria for fretting fatigue cracks, contact modelling, effects of material removal, and numerical implementation in fretting fatigue are reviewed in this paper. The conclusions are as follows:

(i) Defining the contact surface interaction, imposing the frictional contact constraints, and setting the coefficient of friction are concerns of contact modelling in fretting fatigue.

(ii) Material removal has an important effect on fretting fatigue.

(iii) It does not need to remesh during simulating the fretting fatigue crack propagation by using X-FEM. Therefore, X-FEM is more powerful than FEM when modelling the crack growth.

References

1. R.B. Waterhouse, Fretting fatigue, *Int. Mater. Rev.* **37**, 77 (1992).
2. J. Ding, D. Houghton, E.J. Williams and S.B. Leen, Simple parameters to predict effect of surface damage on fretting fatigue, *Int. J. Fatigue* **33**, 332 (2011).
3. A. Zografos, D. Dini and A.V. Olver, Fretting fatigue and wear in bolted connections: A multi-level formulation for the computation of local contact stresses, *Tribol. Int.* **42**, 1663 (2009).
4. M.M.I. Hammouda, P.A. Pasha and A.S. Fayed, Modelling of cracking sites/development in axial dovetail joints of aero-engine compressor discs, *Int. J. Fatigue* **29**, 30 (2007).
5. J.Ding, W.S. Sum, R.Sabesan, S.B. Leen, I.R. McColl and E.J. Williams, Fretting fatigue predictions in a complex coupling, *Int. J. Fatigue* **29**, 1229 (2007).
6. C.Cattaneo, Sul contatto di due corpi elastici: distribuzion locale degli sforzi, *Reconditi Dell Accademia Nazionale Dei Lincei* **27**, 342 (1938).
7. R. D. Mindlin, Compliance of elastic bodies in contact, *J. Appl. Mech* **16**, 259 (1949).
8. D.Nowell and D.A. Hills, Mechanics of fretting fatigue tests, *Int. J. Mech. Sci.* **29**, 355 (1987).

9. C.T. Tsai and S. Mall, Elasto-plastic finite element analysis of fretting stresses in pre-stressed strip in contact with cylindrical pad, *Finite Elem. Anal. Des.* **36**, 171 (2000).

10. M. Sabsabi, E. Giner and F.J. Fuenmayor, Experimental fatigue testing of a fretting complete contact and numerical life correlation using X-FEM, *Int. J. Fatigue* **33**, 811 (2011).

11. S. Basseville, E. Héripré and G. Cailletaud, Numerical simulation of the third body in fretting problems, Wear **270**, 876 (2011).

12. J. Ding, S.B. Leen and I.R. McColl, The effect of slip regime on fretting wear-induced stress evolution, Int. J. Fatigue **26**, 521 (2004).

13. K.N. Smith, P. Watson and T.H. Topper, A stress-strain function for the fatigue of metals, J. Mater. **5**, 767, 1970.

14. M.P. Szolwinski and T.N. Farris, Mechanics of fretting fatigue crack formation, Wear **198**, 93 (1996).

15. A. Fatemi and D.F. Socie, A critical plane approach to multiaxial fatigue damage including out-of-phase loading, Fatigue Fract. Eng. Mater. Struct. **11**, 149 (1988).

16. C.D. Lykins, S. Mall and V. Jain, An evaluation of parameters for predicting fretting fatigue crack initiation, Int. J. Fatigue **22**, 703 (2000).

17. W.S. Sum, E.J. Williams and S.B. Leen, Finite element, critical-plane, fatigue life prediction of simple and complex contact configurations, Int. J. Fatigue **27**, 403 (2005).

18. M. Nesládek, M. Španiel, J. Jurenka, J. Ružička and J. Kuželka, Fretting fatigue — Experimental and numerical approaches, Int. J. Fatigue **44**, 61 (2012).

19. O. Jin and S. Mall, Influence of contact configuration on fretting fatigue behavior of Ti-6Al-4V under independent pad displacement condition, Int. J. Fatigue **24**, 1243 (2002).

20. J.J. Madge, S.B. Leen, I.R. McColl and P.H. Shipway, Contact-evolution based prediction of fretting fatigue life: Effect of slip amplitude, Wear **262**, 1159 (2007).

21. K.W. Man, M.H. Aliabadi and D.P. Rooke, Bem friction contact analysis: Load incremental technique, Comput. Struct. **47**, 893 (1993).

22. A. Zografos and D. Dini, A combined BEM/Contact Asymptotics (BEM-CA) semi-analytical formulation for the assessment of fretting damage in bolted joints, in Mesomechanics 2009 Conference, (Oxford, England, 2009).

23. J.J. Madge, S.B. Leen, and P.H. Shipway, A combined wear and crack nucleation-propagation methodology for fretting fatigue prediction, Int. J. Fatigue **30**, 1509 (2008).

24. J.F. Archard, Contact and rubbing of flat surfaces, J. Appl. Phys. **24**, 981 (1953).

25. I.R. McColl, J. Ding and S.B. Leen, Finite element simulation and experimental validation of fretting wear, Wear **256**, 1114 (2004).

26. T. Zhang, P.E. McHugh and S.B. Leen, Computational study on the effect of contact geometry on fretting behaviour, Wear **271**, 1462 (2011).

27. S. Fouvry, T. Liskiewicz, P. Kapsa, S. Hannel and E. Sauger, An energy description of wear mechanisms and its applications to oscillating sliding contacts, Wear **255**, 287 (2003).

28. H.S. Kim, S. Mall and A. Ghoshal, Two-dimensional and three-dimensional finite element analysis of finite contact width on fretting fatigue, Mater. Trans. **52**, 147 (2011).

29. H. Lee and S. Mall, Analysis of fretting fatigue of cavitation shotless peened Ti-6Al-4V, Tribol. Lett. **38**, 125 (2010).

30. H.A. Fadag, S.Mall and V.K. Jain, A finite element analysis of fretting fatigue crack growth behavior in Ti-6Al-4V, Eng. Fract. Mech. **75**, 1384 (2008).

31. J.C. Neto, P.Wawrzynek, M. Carvalho, L. Martha and A. Ingraffea, An algorithm for three-dimensional mesh generation for arbitrary regions with cracks, Eng. Comput **17**, 75 (2001).

32. J.M. Melenk and I. Babuška, The partition of unity finite element method: Basic theory and applications, Comput. Meth. Appl. Mech. Eng. **139**, 289 (1996).

33. N. Moës, J. Dolbow and T. Belytschko, A finite element method without remeshing, Int. J. Numer. Methods Eng. **46**, 131 (1999).

34. M.C. Baietto, E. Pierres, A. Gravouil, B. Berthel, S. Fouvry and B. Trolle, Fretting fatigue crack growth simulation based on a combined experimental and XFEM strategy, Int. J. Fatigue **47**, 31 (2013).

35. E. Giner, N. Sukumar, F.D. Denia and F.J. Fuenmayor, Extended finite element method for fretting fatigue crack propagation, Int. J. Solids Struct. **45**, 5675 (2008).

36. E. Giner, C. Navarro, M. Sabsabi, M. Tur, J. Domínguez and F.J. Fuenmayor, Fretting fatigue life prediction using the extended finite element method, Int. J. Mech. Sci. **53**, 217 (2011).

CHAPTER 2

Mechanical Engineering and Mechatronics

Stress analysis of a large horizontal vessel supported on ten saddles according to three-moment theorem and Zick's method

Cheng-Hong Duan[†], You-Liang Hu

College of Mechanical and Electrical Engineering,
Beijing University of Chemical Technology
Beijing 100029, China
[†]*E-mail: duanchenghong@163.com*

For the engineering design of large horizontal vessels with uneven multiple saddles, no clear specifications are given in design codes. In this paper, an analytical method based on three-moment theorem is introduced to evaluate the shear forces and bending moments at some critical cross-sections of a large horizontal pressure vessel supported on ten saddles. Zick's method for a horizontal vessel supported on two saddles is extended to the multiple saddle arrangements. It is found that as a result of varied cross-section and uneven saddle support, the shear forces and bending moments are different at different saddle sections. Finite element calculations are also performed to prove the analytical solutions. Results show that the analytical solutions are more conservative. Experiments are also carried out to verify the reliability of the proposed method.

Keywords: Multiple saddles; horizontal vessel; three-moment theorem; zick's method; stress analysis.

Nomenclature

f	nominal design stress (N/mm^2)
f_{ac}	allowable compressive longitudinal stress (N/mm^2)
$f_{1...10}$	resultant stresses in horizontalvessel due to mode of support (N/mm^2)
l_n	length of cylindrical part$n$ of vessel
w_n	average load of cylindrical part n of vessel per millimeter length (N/mm)
W_{Fn}	concentrated force in cylindrical part n from tube and tube bundles

ω_n	area of bending moment diagram for span l_n(mm^2)
x_n	horizontal distances of the centroid of the moment area ω_n from the support n-1(mm)
y_n	horizontal distances of the centroid of the moment area ω_n from the support n(mm)

1. Introduction

For the design of horizontal pressure vessels, many codes and researchers such as ASME B&PV Code Section VIII Division 2[1] and Megyesy[2] recognize the advantages of multiple saddle supports, but no specific engineering design methods are provided. EN13445 code Section III Division 16.8 provides the design method of horizontal pressure vessels supported on multiple saddles[3]. It proposes a method for the calculation of bending moment and shearing force in the saddle region as well as equations for the strength and stability assessment of the horizontal pressure vessels with multiple saddles under uniform loadings. However, this method is only applicable to vessels with uniform cross-sections and uniform loaded saddles.

L. P. Zick[4][5] proposed the semi-empirical design method in 1951, which simplified a horizontal pressure vessel supported on two saddles as a simply supported overhanging beam under an uniformly distributed load. This method is the basis of most design guidelines. Zick's analysis is better judged on the performance it has demonstrated since it was first published, and therefore, it is also the basis of the saddle design guidelines given in pressure vessel design handbooks; see for example Megyesy[2]. A large horizontal pressure vessel supported on multiple saddles can be regarded as a statically indeterminate beam. Reactions and bending moments of each saddle can be calculated by three-moment theorem, then the stress of cylinder near and between saddles and the stress at saddles can be checked by Zick's method.

In this paper, stress analysis are carried out for a large horizontal pressure vessel with10 saddles using three-moment theorem and Zick's method. The vessel is a 91250mm long evaporator with 10 fixed tube sheets distributing along its axis. The vessel can be divided into 10 parts by tube sheets. Each part of the vessel conducts one period of evaporation. 10 saddles are not uniformly distributed. The seventh saddle is fixed and the others are slidable.

2. Basic structure and Parameters

2.1. *Basic structure*

The basic structure of the vessel with ten saddles in this paper is 91250mmlong. The diameter of part 1 to part 7 of the vessel is different from that of part 8 to part 10, therefore a conical reducer is used to connect the part 7 and part 8.

Its first seven parts are Φ8500mm in inner diameter and 14mm in wall thickness. The last three parts' diameter is Φ7000mm with wall thickness12mm. The vessel is made of stainless steel, so no corrosion allowance is considered. Tube sheets are set on both sides of each saddle, and there are pipe bundles arranged between tube sheets. Outside the vessel, stiffening rings are installed to enhance its strength and rigidity. The vessel's structure and saddle arrangements are shown in Fig.1.

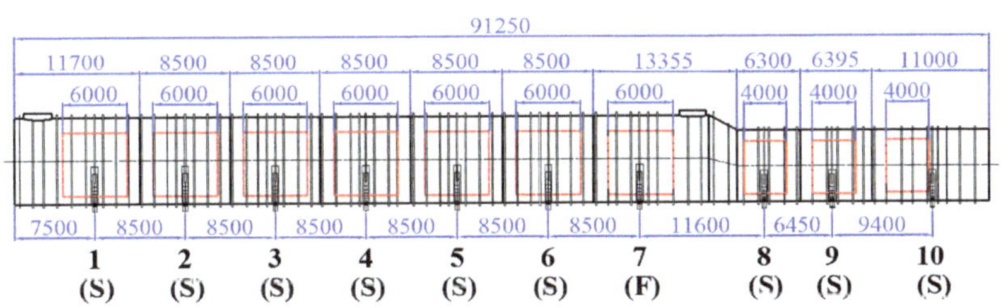

Fig. 1. Large horizontal vessel supported on multiple saddles

2.2. *Basic parameters*

Basic parameters of the vessel are listed in Table 1.

Table 1. Basic parameters.

internal pressure P_m/MPa	0.085
medium density ρ/ kg/m^3	1025
inside diameter of cylindrical part 1~7 of vessel D_{iF}/ mm	8500
inside diameter of cylindrical part 8~10 of vessel D_{iT}/ mm	7000
distance from first saddle support to adjacent end of cylindrical part A_F/ mm	7500
distance from last saddle support to adjacent end of cylindrical part A_T/ mm	5300

length of cylindrical part of vessel L/ mm	91250
inertia moment of cylindrical part 1~7 of vessel I_1, I_2,...I_7/ mm^4	1.69E+12
inertia moment of cylindrical part 8~10 of vessel I_8, I_9, I_{10}/ mm^4	8.10E+11

2.3. *Load cases and material properties*

The hydrostatic test (be filled with water) is the most dangerous case among the vessel's different load cases. So in this paper only the hydrostatic test case is considered. In the case no additional pressure exists except for hydrostatic pressure. Besides, the vessel and insulating layer's weight are included here. Heat expansion, wind and earthquake load are neglected.

Allowable stresses and Young's modulus of materials shown in Table 2 are taken at the test temperature 20°C from the Chinese code JB/T 4731-2005[6], *Steel horizontal vessels on saddle supports.*

Table 2. Material properties

Material	Temperature/°C	Allowable stress f /MPa	Young's modulus
022Cr17Ni12Mo2 (shell)	20	118	195000
Q345/ Q235 (stiffeners)	20	170/140	192000

3. Mechanical Modeling and Solution of Continuous Beam Problem

3.1. *Mechanical modeling*

Hydrostatic pressure, the vessel and insulating layer's weight are considered in the hydrostatic test case, as shown in Fig. 2(a). Load parameters of each cylindrical part are listed in Table 3.

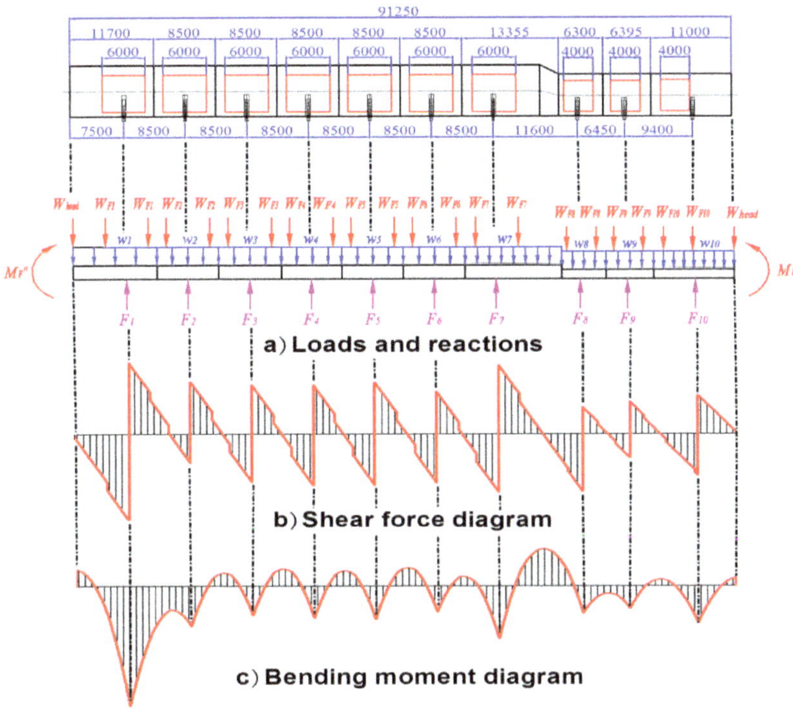

Fig. 2. Load distribution, shear force and bending moment distribution along the vessel

Table 3. Load parameters of each cylindrical part

length of first cylindrical part of vessell_1/ mm	11,700
weight of first vessel end W_{head}/ N	174,000
concentrated force in first cylindrical part from tube and tube bundles W_{F1}/ N	500,000
equivalent moment from first end to the end tangent M_F''/N·mm	2.626E+9
average load of cylindrical part 1~ part 7 of vessel per millimeter length/ N/mm	652
average load of cylindrical part 8~ part 10 of vessel per millimeter length/ N/mm	453
length of tenth cylindrical part of vessell_{10}/ mm	11,000
weight of last vessel end W_{head}/ N	116,200
concentrated force in first cylindrical part from tube and tube bundles W_{F10}/ N	175,000
equivalent moment from last end to the end tangent M_T''/N·mm	1.208E+9

3.2. *Reaction and bending moment at the saddle*

Reaction and bending moment at the saddles

$$M_{n-1}\frac{l_{n-1}}{I_{n-1}}+2M_n\left(\frac{l_{n-1}}{I_{n-1}}+\frac{l_n}{I_n}\right)+M_{n+1}\frac{l_n}{I_n}=-\frac{6\omega_{n-1}x_{n-1}}{I_{n-1}l_{n-1}}-\frac{6\omega_n y_n}{I_n l_n} \tag{1}$$

There are 10 saddles, so it's a 8 statically indeterminate problem. The longitudinal bending moment at the first saddle M_1can be calculated directly, M_1 = -1.878E+10N·mm.Similarly, M_{10}= -5.89E+09N·mm.

According to three-moment equation, i.e. Eq.(1), write the formulas for saddle 2 to 9 as:

$$M_1\frac{l_1}{I_1}+2M_2\left(\frac{l_1}{I_1}+\frac{l_2}{I_2}\right)+M_3\frac{l_2}{I_2}=-\frac{6\omega_1 x_1}{l_1\times I_1}-\frac{6\omega_2 y_2}{l_2\times I_2} \tag{2}$$

$$M_8\frac{l_8}{I_8}+2M_9\left(\frac{l_8}{I_8}+\frac{l_9}{I_9}\right)+M_{10}\frac{l_9}{I_9}=-\frac{6\omega_8 x_8}{l_8\times I_8}-\frac{6\omega_9 y_9}{l_9\times I_9}$$

Where, x_1~x_8, y_2~y_9 and ω_1~ω_9 can be calculated according to mechanics of materials. By solving these linear equations, the value of M_2~M_9 are obtained, then reactionsF_1~F_{10}at each saddle can be obtained, too. All results are listed in Table 4.

Table 4. Longitudinal bending moment and reaction in horizontal vessel at each support

Bending moment	Value / 10^9N·mm	Reaction	Value / 10^6N
M_1	18.78	F_1	10.42
M_2	6.28	F_2	5.35
M_3	4.61	F_3	6.38
M_4	4.98	F_4	6.64
M_5	5.17	F_5	6.77
M_6	4.06	F_6	5.99
M_7	8.29	F_7	8.39

M_8	4.32	F_8	5.54
M_9	3.61	F_9	3.71
M_{10}	5.89	F_{10}	5.27

3.3. *Shear force diagram and bending moment diagram*

According to the results, the shear force and bending moment diagrams are plotted in Fig. 2(b)(c). The shear force and moment at each saddle are larger than those at other locations. Both the maximum shear force and moment are at the first saddle, thus the following stress analysis will focus on the first saddle.

4. Stress Calculation and Assessment

With the calculated reaction and moment of each saddle, Zick's method is applied to evaluate the stress at some critical cross-sections of the vessel. Meanwhile, the stresses are also assessed according to PD5500: 2009[8] and corrected in accordance with Zick's method. As the maximum reaction force and moment occur at the first saddle, stress calculation and assessment will be performed only at vessel section corresponding to this saddle. Table 5 shows the parameters of the first cylindrical part.

Table 5. Parameters of first cylindrical part

included angle of saddle support θ/(°)	143
mean radius of cylindrical part 1~part7 of vessel r_F / mm	4257
analysis thickness of vessel shell 1~7 t_F / mm	14
area of effective cross-section of stiffener from horizontal vessel a / mm^2	89893
the effective width of saddle supportb_2 / mm	820.52
distance from centroid of effective area of stiffener to shell c / mm	126.29
distance from centroid of effective area of stiffener to tip of stiffener d / mm	237.71
second moment area of effective cross-section of stiffening ring I / mm^4	2.219E+09
analysis thickness of reinforcing plate t_1 / mm	50

constants C_4 and C_5 (obtained from PD5500:2009[8])	-1 and 1
constants K_1 and K_2 (obtained from PD5500:2009[8])	1
constants K_3 (obtained from PD5500:2009[8])	0.319
constants K_5 and K_6 (obtained from PD5500:2009[8])	0.6895 and 0.0087
constants K_7 and K_8 (obtained from PD5500:2009[8])	0.0411 and 0.2325

4.1. *Longitudinal stresses*

In the hydrostatic test, the maximum longitudinal stresses of the cylinder shell at the saddle are at the highest and lowest point of the cross-section. Eq. (3) and (4) are used to calculate and assess the longitudinal stresses.

$$f_3 = \frac{p_m r_F}{2t} - \frac{M_1}{K_1 \pi r_F^2 t_F} = 36.54\text{MPa} < f = 118\text{MPa} \tag{3}$$

$$f_4 = \frac{p_m r_F}{2t} + \frac{M_1}{K_2 \pi r_F^2 t_F} = -10.74\text{MPa} < f_{ac} = 39.36\text{MPa} \tag{4}$$

4.2. *Tangential shearing stress*

In the hydrostatic test, the maximum tangential shearing stress of the cylinder shell at the saddle is in the horizontal centerline of the cross-section. Revise the equation in PD5500 by Zick's method, the maximum tangential shearing stress q is obtained.

$$q = \frac{K_3 F'}{r_F t_F} = 30.21\text{MPa} < \min(0.8f, 0.06Et_F / r_F) = 38.41\text{MPa} \tag{5}$$

Where F', 5.634E+06 N, is the larger reaction of the first cylindrical part's two saddles.

4.3. *Circumferential stresses*

In the hydrostatic test case, check the circumferential stresses at the lowest point of cylinder shell cross-section and the circumferential stresses of the cylinder shell near the saddle corner, the reinforcement of plate is considered in the process. Revise the equation in PD5500 by Zick's method and refer to Eq.(6) and (7) to calculate.

$$f_5 = -\frac{K_5 F_1}{(t_F + t_1)b_2} = -14.59\text{MPa} < f = 118\text{MPa} \tag{6}$$

$$f_6 = -\frac{F_1}{4(t_F + t_1)b_2} - \frac{12K_6 F_1 r_F}{L'(t_F^2 + t_1^2)} = 137.67 < 1.25f = 148\text{MPa} \tag{7}$$

Where, L' is the sum of the extension length of the first cylindrical part and half of the distance between saddles, $L' = 11750$mm.

In the hydrostatic test case, check the circumferential stresses at the area near horizontal centerline of the cylinder shell cross-section and external rim surface of reinforcement ring. Revise the equation in PD5500 by Zick's method and refer to Eq. (8) and (9) to calculate.

$$f_7 = \frac{C_4 K_7 F_1 r_F c}{I} - \frac{K_8 F_1}{a} = -145.57\text{MPa} < 1.25f = 148\text{MPa} \tag{8}$$

$$f_8 = \frac{C_5 K_7 F_1 r_F d}{I} - \frac{K_8 F_1}{a} = 196.37\text{MPa} < 1.25f = 212.5\text{MPa} \tag{9}$$

From above stress assessment, it is concluded that the longitudinal stresses, tangential shear stress and circumferential stresses in the first cylindrical part of the vessel all meet the strength requirements. Similarly, the stress calculation and assessment can also be performed for other cylindrical parts.

5. Finite Element Analysis (FEA)

To verify the feasibility and accuracy of the above analytical method, finite element analysis of the vessel was performed and the reaction and bending moment of each saddle are extracted. The results are compared as shown in Fig.3 and Fig.4.

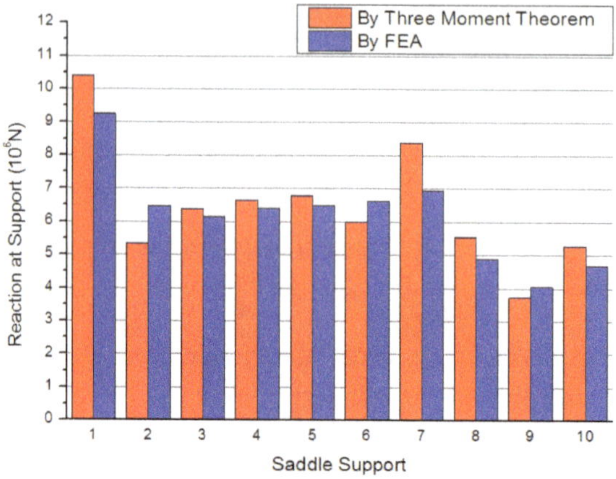

Fig. 3. Reaction distribution on saddles

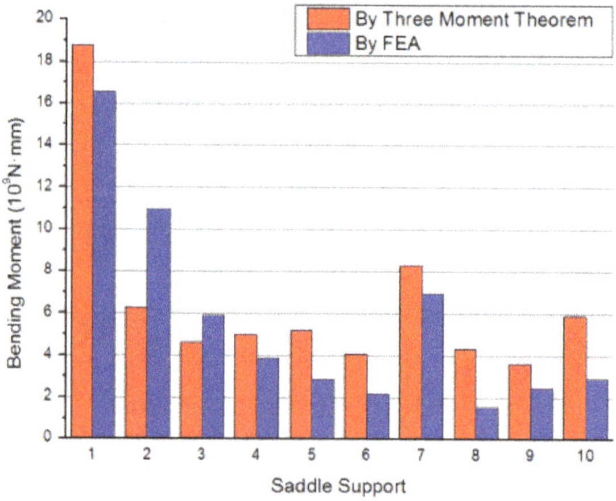

Fig. 4. Bending moment distribution on saddles

Seen from Fig.3 and Fig.4, the reactions and bending moments obtained by three-moment equation method and FEA vary in the same tendency. The values are much greater in the first and seventh cylinder part than the other parts, the maximum is in the first part. The similar results show that the application of three-moment equation method here is feasible and accurate.

It is also found that the results of three-moment equation method is slightly larger than that of the FEA. For example, the reaction obtained with the analytical method at the first saddle is 1.13 times of the FEA, so is the moment. For the seventh saddle, the reaction obtained with the analytical method is 1.21 times of FEA and the moments is 1.19 times. So if the FEA results are considered to be closer to real situation, the three-moment equation method is more conservative in engineering applications.

6. Experimental Verification

For verification, hydrostatic test is conducted after manufacture. The test set-up is shown in Fig.5.

Fig. 5. Hydrostatic test set-up of the large horizontal vessel supported on multiple saddles

In the test, the vessel is full of water, its total weight is about 6000t. During the test, the overall structure is safe without obvious deformation and damage. This test implies that the method described in this paper to design and analyze large horizontal vessel with multiple saddles is reliable.

7. Conclusions

A large horizontal pressure vessel with multiple saddles is analyzed in this paper. For special features such as unevenly spaced saddles, non-uniformly distributed loads and non-uniform cross section, no suitable method is available in the present codes for the design of the large horizontal pressure vessel.

Reactions and moments of each saddle are calculated combining Zick's method with three-moment theorem, the strength calculation and assessment of multiple saddles are carried out referring to the method for the two-saddle cylinder. Comparison with the FEA indicates the three-moment equation method for stress calculation of horizontal vessels with multiple saddles is practical, accurate and safe. In addition, experiments prove that the method presented in this paper is accurate enough to guarantee the safety of the large pressure vessel.

It is also suggested that more studies on the above method, such as saddles' settlements, should be carried out to propose corresponding standard or guidelines for the design of vessels supported on multiple saddles.

References

1. ASME. *Boiler & pressure vessel design codes an international code*, section VIII division 2 rules for construction of pressure vessel, part 4 design by rule requirements, subsection 15, design rules for supports and attachments; 2010.
2. Megyesy EF. *Pressure vessel handbook*. Tulsa, Oklahoman, USA: Pressure Vessel Publishing, Inc.; 2001.
3. *EN13445-3*. Unfired pressure vessels — part 3: design, section 16.8: horizontal vessels on saddle supports; 2009.
4. Zick LP. *Stresses in large horizontal cylindrical pressure vessels on two saddle supports*. Weld J Res Suppl 1951; 30:435–45.
5. Zick LP. *In: Stresses in large horizontal cylindrical pressure vessels on two saddle supports. Pressure vessel and piping: design and analysis — a decade of progress*, vol. 2. New York: ASME; 1985. p. 959–70.
6. *JB/T 4731:2005, Steel Horizontal Vessels on Saddle Supports* (in Chinese).
7. *Timoshenko S. Strength of materials. 2nd ed: D. Van Nostrand Company, inc.*; 1940.
8. British Standards Institution. *Specification for unfired fusion welded pressure vessels*, PD 5500. BSI; 2009.

Optimization of design parameters of a hydrogenation reactor using the finite element method

Zhan-Jun Liu[†]
Faculty of Aerospace Engineering, Shenyang Aerospace University,Shenyang,110136,China
[†]*E-mail: liuzhanjunbajin@163.com*

This paper seeks to optimize the design parameters of the horn-shaped spherical structure used in hydrogenation reactors, so as to increase reaction efficiency. Using the finite element method, an optimization analysis is conducted, and the optimized design model of the horn-shaped spherical reactor is built. The control parameters of the finite element optimization are assumed, and the optimized order reaction results are given. The maximum stress distribution of the reaction is shown. The optimal values of SMAX with the WT aimed function, along with their optimization times, are then indicated. The results are analyzed and discussed.

Keywords: Hydrogenation reactor; finite element method; optimization analysis; horn spherical structure.

1. Introduction

A certain type of hydrogenation reactor is the key equipment of the oil products processing, hydrogenation reactor is due to high pressure, in order to make more efficient reaction, the new speaker spherical structure is used. The analysis and design are based on the plastic failure criterion and fatigue failure criterion. Initial stress is caused by the mechanical load of self-limiting stress, which are divided into a general membrane stress, a local film stress and bending stress at a time. Stress analysis and design of evaluation is to consider various failure modes to determine, these failure modes include excessive amounts of plastic deformation and damage is caused by stress and creep deformation, plastic instability [1]. A stress control condition is determined based on the plastic limit load of failure criteria. In a conventional reactor design, in order to guarantee the reactor safety, the designers always try to increase the wall thickness, in order to increase its bearing capacity. With analysis and design concept, the designers design more and more optimized of the reactor structure design, by this process which can improve the stiffness of reactor design, meet the technical index and structural lightweight requirements, thus greatly to improve the economic benefit. By means of the finite element method and optimization design of organic union, it is the key equipment structure design for the future development direction.

2. The Optimization Analysis Theory of Finite Element

The basic rule of optimization problem is through the establishment of the optimization model, to use all kinds of optimization methods, under the condition of meeting the design requirements for iterative calculation, it obtains the objective function extreme value, to get the optimal design scheme. The mathematical model of optimization problems can be represented as [2]:

$$\begin{aligned} \min F(X) &= F(x_1, x_2, \dots x_n) \\ gi(X) &= gi(x_1, x_2 \dots x) \leq 0 \quad (i = 1, 2 \dots n) \\ hj(X) &= hj(x_1, x_2 \dots x_n) = 0 \quad (j = 1, 2 \dots p) \\ X &= [x_1, x_2, \dots x_n]T \end{aligned} \tag{1}$$

F (X) is as objective function, which is a function of design variables, and the design scheme can be used to evaluate the objective function optimization problem for extremum, h (X) is as constraint conditions, it is a design variable values range and scope of state variable space constraints, which is a function of design variables. X is to design a vector, formed from the design variables, which is to the optimization of design parameters, each design vector is a design, design of vector is set as design space R, to satisfy the constraints of the design of the vector collection is for the feasible region [3]. To solve the optimization problem of general numerical iterative method is such as penalty function method (SUMT), which will be converted into the constrained optimization problem, through iteration, it close to the objective function extremum. Finite element method and optimization method are the main two mathematical tool in engineering analysis, organically, to make full use of the finite element method is that numerical calculation accuracy and the optimization method of extremum efficiency, which will play a great power in engineering analysis. Due to traditional optimization methods to solve the basic premise is that the objective function and the establishment of the function equation of state variables, and for finite element model of large complex structures, if it want to come to analytical expressions of the objective function and state variables, which is difficult or even impossible, to combine the finite element method and optimization method, it must be given the objective function, the explicit function expression of the state variable function. In order to solve this problem, it can adopt the method of curve fitting, artificially formed the approximate function expressions of the objective function and state variables, objective function and state variables are curved surface fitting equation which can use the following form.

For curved surface fitting, first of all, by way of defining and randomly generated, K is set of design variables, through the finite element method and the corresponding calculation result is each objective function and state variables, in order to better for fitting, K value should be greater than the number of the unknown function coefficient. By the results of each group, it is concluded that the total weighted least squares error.

K is as the design variables on the number of groups, h (K) is by finite element method to calculate the first set of design variables, the corresponding h (K) is for approximate algorithm, it is concluded that the first K is value under the set of design variables, w (K) is for the first K set of design variables, its value can be based on the current set of design variables and the first K minimum objective function, the design of the corresponding design variables are space distance and to determine the objective function value. According to the principle of least square, minimum Z is obtained [4].

Solutions of the equations can be ai, bi, cij, thus it can be fitting in the form of the objective function and state variables, and then, using a variety of optimization methods, the extreme value of the objective function will be achieved through the optimization iterations.

3. The Optimization Design Model of Horn Spherical Structure

The main body of horn spherical structure is diameter of 346 mm, outside the reactor of wall thickness, but the mouth of the horn wall thickness has increased in the transitional position, a quarter of the finite element net shape is as shown in Fig.1. The design pressure is 24 MPa, working pressure for 22 MPa, material elastic modulus is of 210 GPa, poisson ratio is of 0.3. Design requirements are as follows: through the design of the wall thickness, it makes whole horn spherical reactor under the request of stiffness and strength satisfies, the given weight is minimum. The ball body wall thickness reference range is for 11 to 14 mm, mouth horn wall thickness reference range is for 16 to 20 mm. The material allowable stress is 260 MPa. According to the speaker spherical structure and mechanical characteristics, it is perpendicular to the side wall of the uniform pressure on the inner wall of the reactor, by 24 MPa, in horn of the mouth, the tensile stress calculation is 3.68 MPa. Selected speaker spherical ball and mouth wall thickness t1 and t2 as design variables, sigma is used to optimize the design of the structure of the equivalent stress strength, as a constraint condition, it got the optimization design mathematical model of horn spherical structure. F (X) is said the weight of the reactor, the minimum is sigma structure with finite element analysis of select data, after analysis, the main attention is spherical stress along the horn the distribution of wall thickness and size, when the check analysis is a spherical wall thickness along the horn section.

$$W t = \min f (X) \qquad (2)$$
$$X = [x_1] = [t]$$
$$\sigma \leq [\sigma]$$

4. The Optimization Analysis of Horn Spherical Structure

This paper is used ANSYS 9.0 software, through the design of the wall thickness, which make whole horn spherical reactor the given weight minimum under the request of stiffness and strength satisfies. Specific analysis process is that the initial design variables, definition unit and materials, create the model, applied load and constraints. Optimization analysis process includes that optimize the design variables, the optimization of state variables, setting the objective function, operation optimization. Part of the sequence optimization calculation results are shown in Tab.1. Display of the maximum mises stress distribution is as shown in Fig.2, the optimization of state variables SMAX changing with the number of optimization is shown in Fig.3, according to the objective function WT changing with the number of optimization, it is as shown in Fig.4.

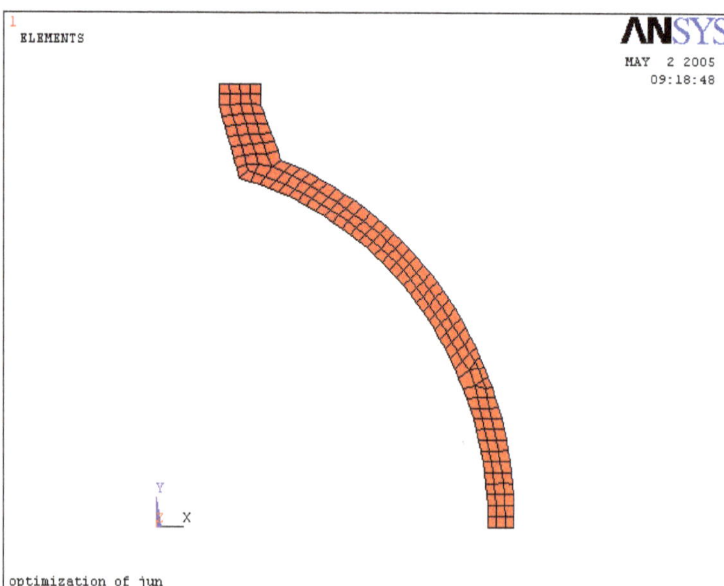

Fig. 1 1/4 model limited element net

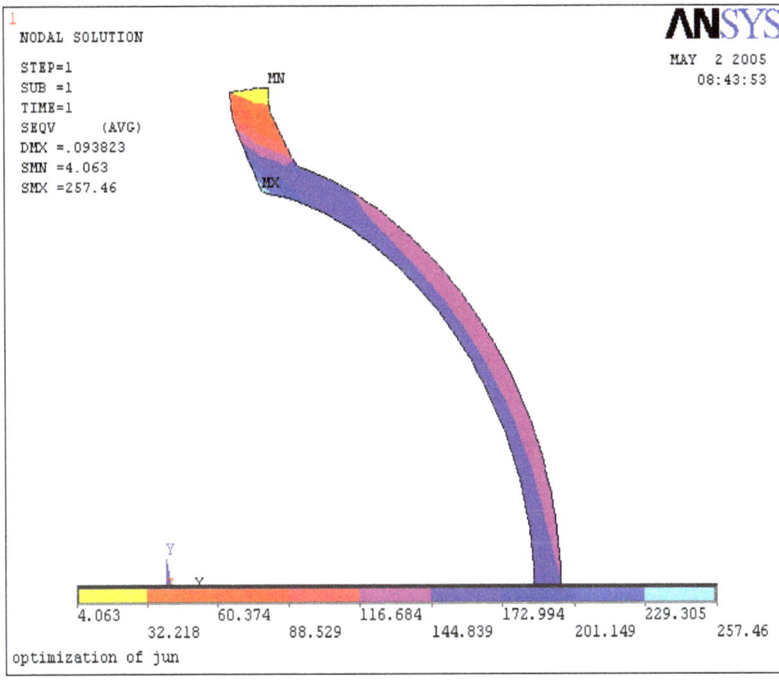

Fig. 2 Biggest Mises stress distribution

Table 1. Part excellence order result

SET1	SET2	SET3	SET4
257.46	244.22	240.26	245.12
13.00	14.00	14.00	13.54
18.00	18.48	20.00	20.00
0.19317E+08	0.20796 E+08	0.20988E+08	0.20347E+08
SET5	SET6	SET7	SET8
245.14	253.74	253.82	253.11
13.54	12.771	12.764	12.815
20.00	19.954	19.954	20.00
0.20344E+08	0.19248E+08	0.19237E+08	0.19315E+08
SET9*	SET10		
245.61	253.71		
13.503	12.77		
20.00	19.965		
0.20283E+08	0.19248E+08		

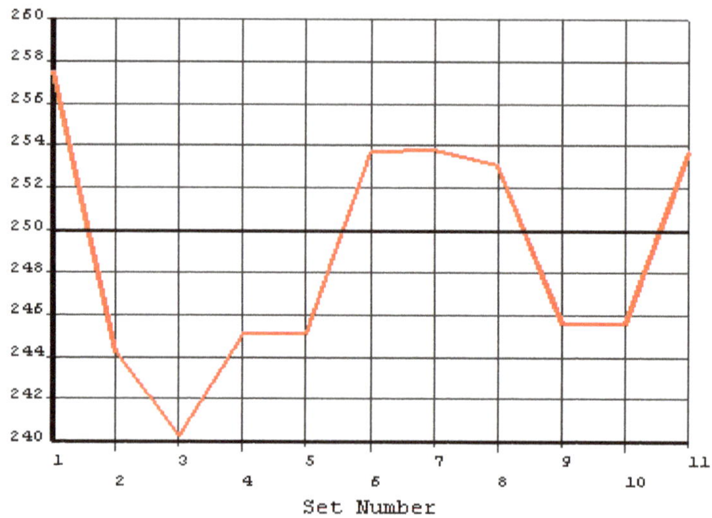

Fig. 3 Change rule of excellent alter SMAX

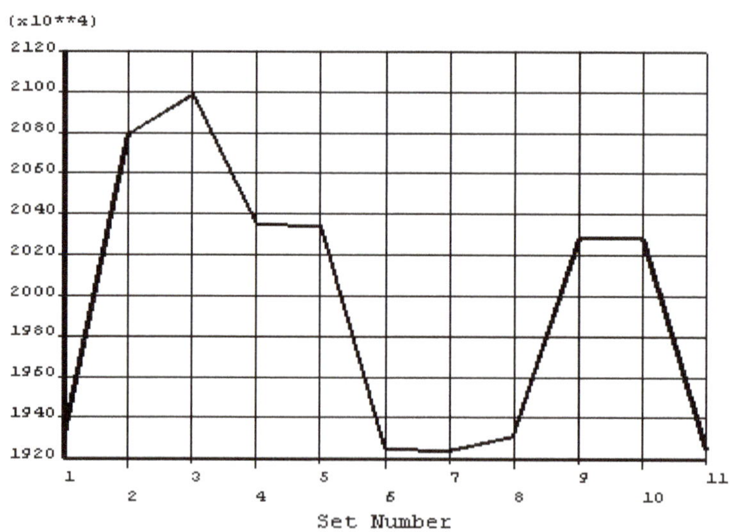

Fig. 4 Change rule of WT aimed function

5. Conclusion

From Table 1, nine is the best design sequences, including the Mises stress is 245.6 MPa, the objective function WT is fell about 6%, the optimization effect is obvious. All state variables and objective function are to approach the optimal design scheme, it proved that the finite element analysis technology is applied in the optimization design value, to overcome the traditional structure design of passive checking method, which take the initiative to seek the optimal design scheme in the feasible region, it greatly reduces the design cost and cycle, to make a horn spherical structure more reasonable.

References

1. Yu Weihui, ANSYS application in mechanical and chemical equipment, China Water Conservancy And Hydropower Press, 2006, p290–300.
2. Zhao Zhen, Duo, The friction and lubrication of metal plastic forming Technology, Beijing: Chemical Industry Press, 2004, p50–53.
3. Wen, WeiDong, Prediction Model Based On The Fuzzy Reliability, Aircraft Engine, 2004, p 322–323.
4. Liu, ZhanJun. Difficulty reliability prediction research of titanium plate stamping forming based on the fuzzy control, Beijing: Plastic Engineering Journal, 2005, p 63–66.

Research on injection mould design of fixed cover plastic parts

Zhan-Jun Liu[†]

Faculty of Aerospace Engineering, Shenyang Aerospace University, Shenyang, 110136, China
[†]E-mail: lzjlzjk@126.com

To solve the problem of forming fixed cover plastic parts, plastic materials and the molding process are analyzed with a mold scheme demonstration, structural analysis and economic analysis. The main runner, cold slug well, pull rod, and runner and gate of the gating system are also thus designed. The ejection force is calculated, and the cooling system designed. An assembly diagram of the special injection mould is given. Practical tests have shown that the shell shaped piece injection mould structure is economical, reliable, has a good machining quality and a high production efficiency.

Keywords: Plastic parts; structural analysis; economic analysis; fixed cover pieces; injection mould.

1. Introduction

The plastic cover is a kind of fixed parts, plastic parts size is not big, the minimum wall thickness is 4 mm parts, total height is 25 mm, thickness is thicker on either side of the middle gate, it can be set up in this place, so which is easy to guarantee to make the whole plastic molding parts get reliable supplement, in order to avoid the formation of the shrink mark, at the same time it can guarantee the strength of the plastic parts, also which can improve molding plastics of melt flow condition, and this place can be set up two larger ejection rod to make plastics mold release[1], it can guarantee the appearance of the parts and precision requirements. The shape of the external structure is not very complicated, but the appearance does not rule. Its appearance left with circular arc transition radius of 24 mm, throughout the whole parts of rounded corners are greater than or equal to 2 mm, it can not only avoid stress concentration, and improve the strength of the plastic parts, and also improve the melt flow and facilitate ejection. Moreover, rounded corners is for improving the strength of the mould on the plastic parts, it is also very necessary. One side of the plastic pieces, it has a circular stairs, up and down the stairs is interlinked, the size is respectively the steps of 4 mm height, it is usable embedded core when the mould design methods is to achieve the purpose of forming, the parts processing is more convenient[2]. The other side of the plastic parts is a slant of 9 ° slope transition, the circular cavity size is inside of the hole blocked. Although the slope will bring certain difficulty for molding parts size calculation, but at the same time because the rounded corners and outside circular arc transition size are larger, and it is conducive to the flow of the plastic melt of the mold cavity. The sidewalls of the parts is holes needed side core pulling in addition to diameter, side is closed for the rest of the places. The side core-pulling hole size is not big, the mold design can be inlaid in the slider, a step shaped side core-pulling parts, such institutions is also relatively simple. Plastic parts forming dimensional accuracy is not high, required by the accuracy of IT14[3]. Taken together, the structure of the characteristics conforms to the requirements for the design of the mould.

2. The Economy Analysis of Mould

The influencing factors analysis of mould price of production cost. The mold is designed into one module and two cavities in raw material, material is 45 steel formwork, due to the mold cavity with the integral, it can largely save the precious steel, to reduce the cost of the mould. Processing method is that mold cavity is appointed part in wire-cutting processing of the fixed plate, on the other plate hole or slot car basically, milling, planning, pliers and heat treatment can be realized[4]. It can greatly reduce the cost. Due to special reasons, in addition to fixed connection and positioning guide components, it is used standard parts, the rest are additional processing, mold parts processing is simpler, so the production cycle is short. With the fierce competition in the market, new products is emerged in endlessly, the update speed of old product is faster. This mold design is more scientific and reasonable, and select material cost is low, and the molds are convenient to maintenance, it can reduce maintenance costs, to extend the life cycle, so there must be a good market after the mold. The design of mould plan. The parting surface is options, due to the technical requirements of surface quality requirement, the parting surface is not set outer surface transverse. Parting surface is longitudinal Settings, by four side core working at the same time, parts are obtained by the shape[5].

3. The Parting Surface of Mould Plan

The bottom surface is a plane parts, according to the parting surface design principle, the parting surface design is on this plane, it choose the parting surface. The comparison of two schemes is that plan a is part of the side core-pulling, slider don't contact with plastic parts, plastic parts is formed only by a fixed die blocks, and dynamic model is of the set piece parts shape parts, moving mould set piece of parting surface, the higher part of the corresponding is fixed, mould set piece part is removed[6]. Scheme 2 is the slider directly involved in the molding parts, the lower part of the sliding block is consistent with the appearance of parts, parts is fixed by mould set piece of forming shape. By contrast, plan is obvious. Because of slider direct molding parts, this part must be consistent with the shape of the parts, this will bring a lot of processing, it is difficult, in addition, the size precision of parts is not easy to guarantee. And plan will not have this problem, the slider is not involved in stamping shape, slide block processing is easier, they don't have to consider with the precision problem between parts, mould blocks is decided by the dynamic model, two blocks is in a fixed mold, the product precision plays a big role[7].

4. The Design of Gating System

4.1. *The design of the main runner*

Main runner size directly affects the melt flow rate and molding time, because the main runner is with high temperature plastic melt, and injector nozzle is repeated contact, therefore, the main runner part is designed to remove the gate. In injection molding, injection machine mould pressure is very big, it is mainly effect on the gate set, so it doesn't usually put the sprue opening on the fixed template directly, but it is in a main gate set. Usually it is embedded mold after normalizing, such damage is in the process of mold, the use of easy is to replace or repair grinding[8]. The sprue bushing is the use of T8A high quality materials, gate set length is fixed, mode with the thickness of the part is consistent, the mainstream way of exports face shall not be highlighted in the parting surface. In order to make the setting material pull out smoothly, the small end of the main runner diameter should be greater than the injection machine nozzle diameter, the technical parameters is of injection machine. Main runner pits is of the entrance to the spherical, radius should be greater than the injection machine nozzle radius of ball head, and to check the technical parameters of injection machine. Usually gate set in diameter should not be too big, because the main role is by bending stress, to make the gate set easily damaged, gate set diameter also should not be too large, the purpose is to make the difference in temperature between gate set, the template is to achieve the minimum. Gate set of convex shoulders must have a circular arc transition, otherwise, it tends to damage due to stress concentration in the work. R is usually about 3 mm. Main runner half of cone angle is usually $2° \sim 4°$, big cone angle can produce turbulence or eddy current, which is easy to get involved in the air. The cone angle is too small, to make mould difficulty, which also can make the mold filling of melt, flow resistance is too large. The length of the main runner is general according to the thickness of the template. In order to reduce the pressure loss, the melt filling mold is material loss, it should as far as possible to shorten the length of the mainstream way, L is generally controlled within 60 mm. The main runner outlet should be larger than rounded corners[9].

4.2. *The cold slug well and pull rod design*

According to the need of cold slug, it should be in place of melt flow direction, and along the upper, which reaches the melt flow direction. In the runner, design should be considered when reducing the pressure in the flow passage of loss, and as far as possible to avoid the melt temperature reduced, it should also consider reducing the volume of flow channel. Its design principle is that the smallest distance will melt quickly in input cavity smoothly. Materials must be in the same temperature, and pressure conditions is from the gate to the cavity feed at the same time. As far as possible big flow, channel section is advantageous to the mold filling, and it ensures the enough pressure maintaining pressure, but from the perspective of saving material, sectional area should be as small as possible, larger cross-sectional area also increases the cool down[10].

The cross section is shape of the runner. In the design, flow channel is to reduce the pressure loss of flow passage, and hope to flow the less surface area, therefore, available, cross section area is of flow channel, and the ratio of the circumference is to represent the port efficiency. Circle and square are the highest efficiency. Due to the parting surface shape is as the plane, trapezoidal section flow channel can be used. Because the plastic melt flow is in the passage, it will be formed in the flow channel of wall solidified layer. This layer heats insulation effect, it makes the melt that department is in the center of the flow channel. Therefore, in the center of the runner, gate is located in the center of straight line.

5. The Determination of Cavity Number

Determining the cavity is injection machine clamping force. the clamping force of injection machine is set to F (N), the average pressure is for the molten plastic mold, each of the products is in the parting plane projection area, runner and gate in the points are on the surface of the projection area, the largest number of cavity is according to the statistical analysis of multiple cavity die, the average pressure melt within the cavity is mainly depends on the injection pressure, usually in the range of 25 to 40. The actual clamping force should be less than the name of the injection machine clamping force. It is determined by the injection machine mould. The nominal injection quantity of injection machine is 140g by looking up table. The volume of a single product is calculated in 37.1, the volume of the runner and gate is the biggest cavity number. Every time the actual injection quantity should be in production 0.45~0.75g. Now taking 0.6g is calculated. According to statistics, the volume of each products is required for gating system, the size of products is to estimate the 0.6V now, therefore, a practical formula is comprehensive calculation, therefore taking mould cavity number is to n = 2. Fixed cover plastic parts assembly is shown in Fig.1.

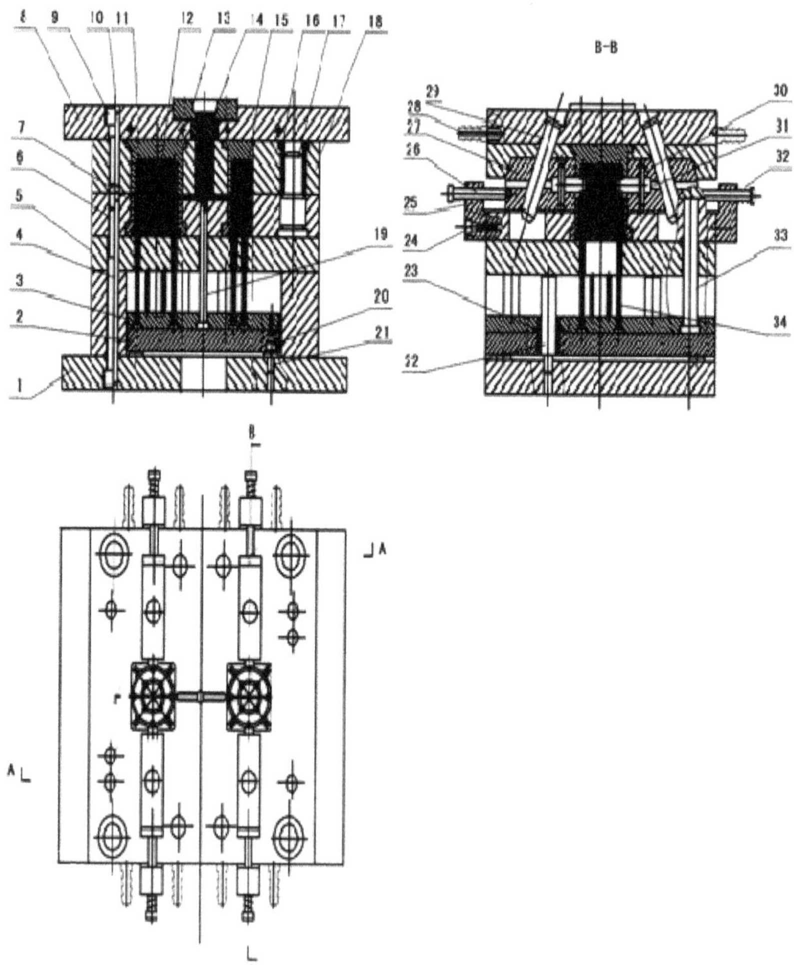

Fig. 1 Fixed cover plastic parts assembly

The section size of distributary channel mainly depends on the size of the plastic products, mould structure and processing of plastic types. In general, the increase of product size is wall thickness, because the melt is in the big cross section, small flow channel section flows in the runner when the resistance is small, therefore, large cross section flows, channel can promote the mould filling process. If shunt way is longer, the process is longer, plastic viscosity should be smaller. In addition, the products of the production of economy should also be considered, otherwise, compared with the model of product, distributary channel cross section is too big, it will affect the cooling capacity and cooling time, which is causing the waste of material and time.

6. Conclusions

It is analyzed by the technology plastics materials and molding process of fixed cover pieces, which has carried on the mold scheme demonstration, analysis of structure and economy, the forming problem of fixed cover pieces is solved, and it is designed that the main runner, cold slug well, pull rod, the runner and gate of gating system. The ejection force is calculated, and which is design of cooling system. It is given to the assembly diagram of special injection mould. Practice has proved that the shell shaped pieces injection mold structure is reasonable, reliable, good machining quality and high production efficiency.

References

1. Feng Jing, The Mixture of The Multimode System Reliability Growth Model, Computer Application Research Supplement, 2003, pp.60–61.
2. Xu Yuxiu, Complex Mechanical Fault Diagnosis Method of Fractal and Wavelet, Mechanical Industry Publishing House, Beijing, 2005, pp.24–25.
3. Weng Chaoxi, Reliability Growth, Science Press, Beijing,1993, pp.130–131.
4. Middleman S, An introduction to fluid dynamics, John Wiley and Sons Inc, New York, 1998, pp.16–17.
5. Liu ZhanJun, Difficulty reliability prediction research of titanium plate stamping forming based on the fuzzy control, Plastic Engineering Journal, Beijing, 2005, pp. 63–66.
6. Altan T, Gegel, The Fundamental Applications of Metal Forming, American Society for Metal, Metal Park, 1983, pp.20–21.
7. Dean T. A, Progress in Net Shape Forging, Progress in 4th ICTP, 1993, pp.31–32.

8. Hoffman, Edward. G. Fundamentals of Tool Design, Michigan, Society of Manufacturing Engineers, 1984, pp.59–60.
9. Guofang, Reliability Data Collection And Analysis, National Defence Industry Press,1995, p22.
10. Wen WeiDong, Prediction model based on the fuzzy reliability, aircraft engine, Beijing, 2003, pp.322–323.

Non-structural fracture blasting technique and case analysis of open pit coal mine in the arid area*

Ya-Jun Chen†, Zhi-Guo Chang and Fei Zhao

Xinjiang Institute of Engineering,
Urumqi Xinjiang830091, China
†E-mail: chenyajun_xgy@163.com

Rock breaking mechanism and optimization of blasting design of non-structural fracture are the core problems in the mining process of the open pit mine in the arid area. It is of great importance to explore effects of non-structural fracture blasting technique on the enhancement of the economic benefit of blasting, the reduction of the bulk rate, and the decrease in accidents of blasting, which is the hotspot research for the blasting at open pit coal mine in the arid area. By analyzing the non-structural fractured lighologic characteristics, plane of fracture of rock mass, attitude of fracture and blasting process in the typical open pit coal mine, we discussed the effect of non-structural jointed rock mass on blasting. Based on such results, we analyzed the key links of the blasting operation, such as detonating position, charge structure, initiation pattern etc., then we optimized the blasting design and evaluated the blasting effects. The results showed that for the mining area in the arid area with non-structural fractured rock mass development and large rock bulk after blasting, symmetric or asymmetric hole-by-hole millisecond minute difference initiation network along V-shaped oblique line was recommended according to the incline of the rock mass. As for the non-structural fractured rock mass affected by shear joints and tension joints, we suggest the adoption of zoning and subsection-based interline millisecond minute difference initiation network among multiple lines as well as deep-hole blasting with non-coupling column charge.

Keywords: Non-Structure; fracture; open pit coal mine; blasting.

1. Analysis of the Effect of Non-structural Fractured Rock on the Blasting

1.1. *Analysis of lithologic characteristics of non-structural fracture*

Non-structural fracture is jointly created by non-structural factors such as diagenesis, external forces, gravity etc, among which weathering fracture will be of general meaning. In the arid region of northwest China, the open pit coal mine with non-structural fracture development has been long exposed to a special environment, such as arid climate, less rainfall, and large temperature

* This work is supported by the National Natural Science Foundation (51464043)

difference. The weathering denudation effect is very intense. The rock mass affected by non-structure is mainly distributed at the depth of 15-20m below ground. The main features include:

Affected by oblique bedding effect, rock mass was composed of one set or several sets of lamina that were obliquely crossed with the main plane, most of which were one-way oblique bedding (Figure 1-a).

The fracture was distributed in disorder, with irregular features. The structure of the rock along the fissure surface has been changed dramatically, with the continuity and integrity being destroyed. It showed that non-structural fractured rock mass had non-directional rock development (Figure 1-b) or the rock mass was cut and broken by non-structural fracture (Figure 1-c).

Fig. 1. (a) One-way oblique bedding; (b) Non-directional non-structural fractured rock mass; (c) Rock mass cut and broken by non-structural fracture.

1.2. *Analysis of the plane of fracture of heterogeneous rock mass and attitude of fracture*

Affected by bedding cutting and non-structural fracture, the structure plane of rock mass influenced the bursting strength and bursting mode, which exerted impact on the formation of blasting crater, the quality of rock breaking and much-pile shape. The effects of plane of fracture and attitude of fracture on the rock mass blasting are listed as follows:

As for the rock mass affected by oblique bedding, blasting would be influenced by attitude plane of fracture in a fixed direction. When bedding blast, towards blasting would take place and bedding rock would be produced (Figure 2-a).

As for non-directional non-structural fractured rock mass, under the effect of explosive shock wave at short range, the dynamic stress field in the rock mass changed, leading to differences in their dynamic strength of extension, compressive strength and shear strength. Density of fracture, attitude of fracture and length of the extension of fracture were all changed. In general, along the direction of low strength of rock mass, the breaking would be more serious, leading to irregular much pile of rock bulk (Figure 2-b).

Affected by the cutting effect of non-structural fracture, the integrity of the rock has been seriously destroyed, with the strength of rock mass lower than that of rock block. The existence of fracture led to decrease in the shear strength along the fissure surface and the strength of extension perpendicular to the direction of fracture. Thus the rock was subject to dislocation or stretch along the pre-existing fracture. Consequently, the detonation gas would easily escape from the extended fissures, thus reduced the effective pressure and pressure action time of the normal breaking. Affected by tensile stress, rock block was produced in the vertical direction (Figure 2-c).

Fig. 2. (a) Occurrence of bedding rock; (b) Irregular rock bulk blasting muck pile; (c) Rock along vertical direction.

2. Case Study and Design Optimization of Non-structural Fractured Rock Mass Blasting

When the open pit coal mine, featured with non-structural development in the arid, was blasted, three occasions would occur: 1) during blasting process, disqualified bulk and medium had something to do with the pre-blasting joint fissures. The results showed that if the rock mass was broken into bulk before the blasting, the block rate after blasting would be high. It would be due to the small contact surface between the explosive and the medium, which led to relatively asymmetric energy distribution. All these resulted in big stock (Figure 3-a). 2) When the fissure surface was large enough or there were fissure surfaces that were parallel to the holes, the shock waves could be reflected on the surface. The rock breaking between the surface and the blasting cartridge would be relatively thorough. While the rock mass beyond the surfaces would be cut on the surface due to the stress reduction and newly formed fracture. Thus, the rock mass could not be thoroughly broken or rock block would not be broken and large scale bulk came into being (Figure 3-b). 3) As for weak interbed and rock mass with existing fracture, fissure surface and structural surface were both the weak part as well as the place with concentrated stress. When explosive charges blast, the damage radius along the fissure surface would be way longer than along other directions (Figure 3-c).

The above mentioned problems were involved in several respects, such as rock mass structure, rock features, fracture distribution etc. Typical cases of Xinjiang Heishan open pit coal mine and Kumishi open pit coal mine would be discussed below, respectively.

Fig. 3. (a) Large bulk; (b) Large-scale rock block; (c) Rock block ejected far away.

2.1. *Xinjiang Heishan open pit coal mine*

Lithologic character: Xinjiang Heishan open pit coal mine lies within the Toksun county, and the average elevation level of the quarry ranges from 2400-2800m. The mining lot, belonging to typical inland sub-arid desert climate zone, is featured with arid weather, being windy and less rain. The main lithology includes graywacke and medium sandstone. At the shallow bedrock which is 30m below the earth's surface, there is serious rock weathering denudation and obvious oblique bedding. Among these, at Quarry #2, the dip angle of the stratum was approximately 25°, most joints and fractures extended and distributed along the direction of dip, and obvious cutting effect was observed for non-structural fracture (Figure 1-a). At Quarry #3, the dip angle of the stratum was approximately 16-19°, the oblique bedding effect was relatively weakened, the tension joint effect was enhanced, and many craked small rocks were observed in the rock block. The fissure surface seemed rough and the width of the breach never remained, with irregular features. Fractures could exert impact on the depth of 9–12m below ground (Figure 1-b).

Blasting problems: During the blasting process, there are many problems such as too many large bulks, huge consumption of explosive charges, etc. At Quarry #2, which had obvious oblique bedding effect, only a small part of the rock block was directly produced due to the blasting effect. Most rock blocks were produced due to the blasting stress wave, which caused the dislocation and expansion of pre-existing joint fissures of the rock mass. The rock bulk produced during the blasting process would be subject to bedding ejection along the direction of dip (forward free surface). At Quarry #3, jointly affected by shear joints and tension joints, secondary vertical small fissure development was formed at upper terrace with small fracture frequency. The rock blocks after blasting were not of uniform size.

Analysis of rock breaking mechanism: Based on bursting test data, when one more free surface is added to the rock mass, the rock block rate of corresponding size will be decreased by 12–20%. The design adopted symmetric or asymmetric hole-by-hole millisecond minute difference initiation network along V-shaped oblique line. It could provide lateral free surfaces in the rock mass for the after-blasting holes. It means that two free surfaces would be added to after-blasting hole along the direction of exploding wire, and three free surfaces would be added to the other after-blasting holes. Due to the increase in the number of free surfaces due to the blasting inside rock mass, the reflected tension wave effect was enhanced, radial tensile stresses from different directions would be concentrated on the rock mass of the after-blasting hole. The block rate decreased to less than 25% and the breaking effect was enhanced.

Blasting scheme optimization: Quarry #2: firing area design includes two lateral free surfaces, the firing holes along the forward free surfaces should be more than that along lateral free surfaces. The design adopted hole-by-hole millisecond minute difference initiation network along V-shaped oblique line, which was symmetric to double lateral free surfaces (Figure 4), and deep-hole blasting with non-coupling column continuous concentrated charge. Initiation point lied in the middle of the first row on the forward free surface, and detonation started from initiation point along the stratum and went against the direction of dip in V-shaped path to propagate explosion. Initiation of explosion at a time shall not be less than 16 lines of holes. Large scale implementation shall include one less than 30 firing holes. A ground initiation network adopted the anti-parallel of nonel detonator and hole-by-hole millisecond minute difference blasting at small dan grading; a 15-dan nonel detonator had been adopted to set a delay of 350 ms in the blast holes; along the forward free surface direction, the holes in each row detonated hole-by-hole with a 50ms difference; along the V-aped direction, the holes detonated hole-by-hole with a100ms difference.

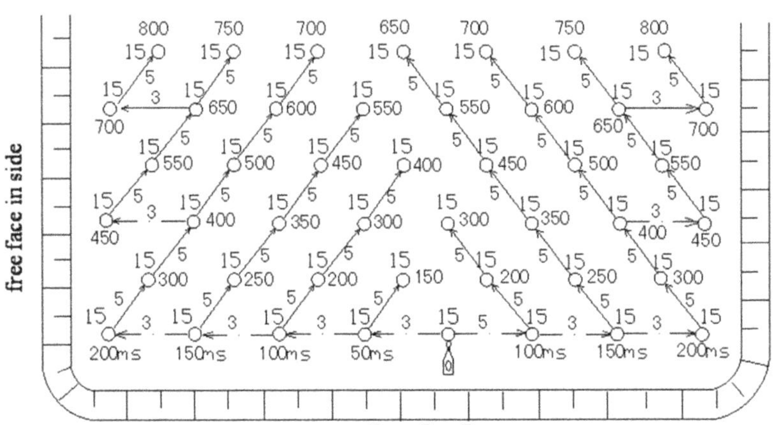

Fig. 4. Double lateral free surfaces-based symmetric hole-by-hole millisecond minute difference initiation network along V-shaped oblique line.

Quarry #3: firing area should have at least one lateral free surface, with more firing holes along the forward free surface direction compared with that along the lateral free surface. The design adopted unilateral free surface-based asymmetric hole-by-hole millisecond minute difference initiation network along V-shaped oblique line (Figure 5), and deep-hole blasting with non-coupling column deck charge)[7]. Initiation point lied at the intersection of slope crests of the two free surfaces. Detonation was propagated from outside to inside, along the oblique line. More than 6 rows of holes should be initiated at a time, with more than 7 holes in each row. Within the blasting hole, a 15-dan nonel detonator had been adopted to set a delay of 350 ms in the blast holes. Ground initiation network ensured that the holes in each row detonated hole-by-hole with a 50ms difference along the oblique line direction. Below the burst area of oblique line, the holes in the first row detonated hole-by-hole with a150ms difference along the forward free surface direction. At the top burst area, the holes detonated hole-by-hole with a 100ms difference.

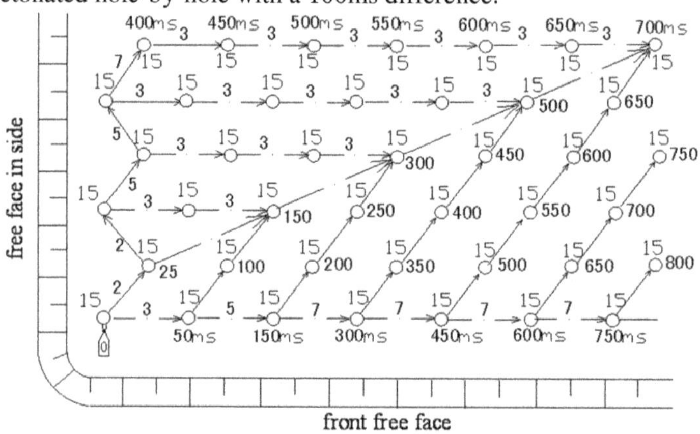

Fig. 5. Unilateral free surface-based asymmetric hole-by-hole millisecond minute difference initiation network along V-shaped oblique line.

Blasting effect evaluation: based on on-site blasting testing, at Quarry #2, since the blasting exerted impact on the rock from outside-to-inside along the anti-direction of dip, the rock block bedding ejection weakened. The ejection distance was reduced by 2-3m, and the muck pile stacked, with part being upheaval. All these were convenient for concentrated mining and loading (Figure 6-a). At Quarry #3, due to the increase in the interval time between firing holes for both forward and lateral free surfaces and the decrease in the interval time between firing holes, the oblique line effect was enhanced. There was an obvious outside-to-inside pushing wall effect. The number of large rock block decreased dramatically, with rock size reduced to less than 1.5m×1.2m× 0.7m (Figure 6-b).

Fig. 6. (a) Quarry #2 muck pile; (b) Quarry #3 muck pile.

2.2. *Kumishi open pit coal mine*

Lithologic character: Kumishi open pit coal mine is 78km southeast of the Kumishi town in the Toksun county. The main lithology includes graywacke and medium sandstone, with the dip angle of the stratum of approximately 8-11° and f=7-8. The mining lot is featured with arid weather, less rain and being windy. The rock surface suffers serious weathering and denudation. Non-structural fissure is well developed at the shallow bedrock layer 20m below the ground. Many non-structural fractures usually form conjugate X type, joint fissure effect is non-directional, and tension joint is subject to the major fracture structure.

Blasting problems: The joint effect of oblique layer weakened, the tension joints on rock mass was obvious, the muck pile bedding ejection reduced, and direct crazing of the rock mass led to large fractured ditch, with some fractured ditch crazing directly to the bottom of the step. The maximum size of the rock was 2.5×2.2×1.8m. And the secondary blasting volume increased.

Rock breaking mechanism analysis: The rock mass included many non-directional fracture groups. If small blasting cartridge is adopted to implement hole-by-hole initiation of explosion, the effect of structural plane will be more obvious due to small amount of explosive materials. The muck pile is subject to fracture and became irregular in size, thus influencing the blasting effect. However, if we increase the amount of priming charge to implement interline difference initiation of explosion, the shock wave will have long wavelength, and the minor joint will be closed soon. Thus the same blasting effect can be obtained as that of rock with low strength. The effect of joint fissure on blasting effect will be decreased.

Blasting scheme optimization: We will adopt interline milli-second minute difference safe twin type initiation network with zoning or subsection for multiple-line holes and deep-hole blasting with non-coupling column deck charge (Figure 7).

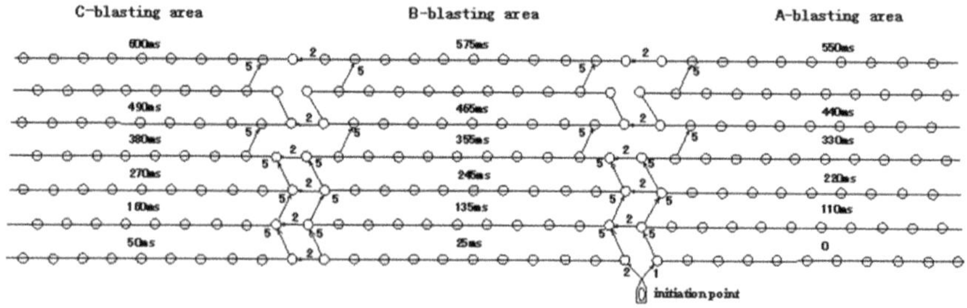

Fig. 7. Zoning and subsection interline milli-second minute difference initiation network at Kushimi open pit coal mine.

We divided the entire firing area into blasting areas, with each less than 100 holes and each line with less than 11 holes. Initiation point lied between firing area A and firing area B, and then propagated to firing area C, thus realizing oblique blasting of area-based propagation from area A to areas B and C.

In ground initiation network, lines will be connected with non-conductivity, achieving interline minute difference initiation. The holes in each line adopted detonating cord to consequently connect the nonel detonator and explosive material to trigger immediate initiation. Three sections of nonel detonators were adopted to horizontally overlap between blasting areas, so as to achieve safe compound initiation of explosion.

For each firing area, the holes of the first five lines achieved detonation with 100ms minute difference between lines. Then the holes of every two lines achieved interline minute different detonation with 100ms minute difference. Detonation with 50ms minute difference with firing range was realized for the holes in the same line.

Blasting effect evaluation: First, due to increasing amount of each charge mass of detonation, the breaking effect of detonation gas on rock has been greatly enhanced, and radial pressure stress and circular tensile stress have been improved. Second, as for the interline milli-second minute difference initiation, the rocks between the holes of the two lines lies within both the springback displacement area of the holes of the front line, and the initiative displacement area of the holes of the back line. The holes of the first line will blast first and then form an active displacement area along the free surface direction, which will provide blasting free surface for the holes of the back line. Additionally, the holes of each line in firing areas B and C are all affected by the detonation in multiple directions of pre-blasting holes from the upper firing area and the firing area itself. Explosive stress waves overlap with each other to enhance the effect, the effect of non-structural major fracture on rock mass weakens, and the rock is broken into uniform size, with the largest size being 1.5×1.2×0.8m. According to on-site data, block rate has been reduced to <18%.

3. Conclusion and Discussion

As for open pit quarry that features great influence by geological effect such as weathering, denudation and external forces, non-structural fractured rock development, and rock bulk after blasting, we recommend the following blasting design scheme:

As for the non-structural fractured rock that has obvious oblique bedding, dip angle ≥ 25°, their cracking and breaking and size and shape subject to the control and impact of larger bedding surface, we recommend the double free surfaces symmetric hole-by-hole millisecond minute difference initiation network along V-shaped oblique line and deep-hole blasting with non-coupling column continuous concentrated charge. Initiation point lies in the middle of the first row on the forward free surface, and detonation starts from initiation point along the stratum and goes against the direction of dip in V-shaped path to propagate explosion. Since the direction of blasting propagation is fixed, hole-by-hole milli-second minute difference time interval along forward free surface can be 25-50ms less than that along the oblique line, so that the propagation of explosion of the muck pile produced by blasting will be concentrated, which is convenient for mining and loading.

As for non-structural fractured rock that is affected by shear joints and tension joints, and with dip angle approximately 16-19°, we recommend to adopt the unilateral and asymmetric hole-by-hole millisecond minute difference initiation network along V-shaped oblique line and deep-hole blasting with non-coupling deck charge. To enhance propagation of explosion along oblique line from outside to inside and make sure that the blasting effect is directional, hole-by-hole milli-second minute difference time interval along the oblique line can be 25-50 ms less than that along forward free surfaces or lateral free surfaces. The purpose is to guarantee that such blasting effect can be superior to the propagation along the rock or the incline.

As for the non-structural fractured rock that has weakened oblique bedding effect and enhanced tension effect with dip angle approximately 8-11°, vertical cracking is likely to take place due to the control and influence of structural surface and tension effect, which can lead to large ditch. We recommend interline milli-second minute difference safe twin type initiation network with zoning or subsection for multiple-line holes. Deep-hole blasting with non-coupling column deck charge is also recommended. Due to increasing amount of priming explosive and the implementation of interline minute difference initiation, it leads to more major fracture and enhanced cutting effect on rock mass. The control effect is weakened. Under the stress wave overlapping effect, the size of rock block has been reduced by approximately 29%, and the blasting effect improved.

Acknowledgments

This work was financially supported by the National Natural Science Foundation (51464043).

References

1. Wang G. T, Guo Y. C. and Ma L "Study on drillhole pattern of hole-by-hole blasting technology", Opencast Mining Technology, No. 1(2012), p 21–23.
2. Shao P, Zhang Y. and He Y. N, et al "Experimental Study on Fatigue Failure of Intermittent Jointed Rock Masses Subjected to Repeated Stress Wave", Chinese Journal of Rock Mechanics and Engineering, Vol. 24, No. 22 (2005), p 4180–4184.
3. Li J. J, Duan Z. P "Experimental Study on the Effect of Jointed and Fractured Rock Masses on Blasting", Mining and Metallurgical Engineering, No. 10 (2005), p 1–3.
4. Chen Y. J. "Optimization of Blasting Design on Anistrope Rock Body", Coal Technology, Vol. 31, No. 5 (2012), p 71–72.
5. Liu G. X, Mu Y. Q. and Yang R. Y "Application of hole-by-hole blasting technology in Sandaozhuang open-pit mine", China Mining Magazine, Vol. 21, No. 8 (2012), p 495–498.
6. Chen Y. J, Wu Z. C. "Analysis of safe initiation network for large space blasting in open pit mine", China Coal, No. 4 (2011), p 102–104.

Multi-scale wavelet transform based bearing fault diagnosis*

Hui Li† and Bo Feng

Department of Electromechanical Engineering,
Shijiazhuang Institute of Railway Technology
Shijiazhuang, Hebei province, P.R. China
†E-mail: Huili68@163.com
www.sirt.edu.cn

The roller bearing characteristic frequencies usually contain very little energy, and are often overwhelmed by noise and higher levels of structural vibrations. In order to overcome the shortcomings in the traditional envelope analysis, a new approach based on the fusion of the Laplace wavelet transform and three dimensions multi-scale spectrum is proposed for detection and diagnosis of defects in roller element bearings. Laplace wavelet transform is self-adaptive to non-stationary and non-linear signal, which can detect the singularity characteristic of a signal precisely under strong background noise condition. The basic principle is introduced in detail. The bearing fault vibration signal is firstly decomposed using Laplace wavelet transform. In the end, the multi scale Laplace wavelet transform spectrum is obtained and the characteristics of the bearing fault can be recognized according to the multi-scale Laplace wavelet transform spectrum. The experimental results show that the multi-scale Laplace wavelet transform spectrum can extract the transients from strong noise signals and can effectively diagnose the bearing fault.

Keywords: Fault Diagnosis; Laplace Wavelet Transform; Bearing; Multi-scale Spectrum; Time Frequency Distribution; Signal Processing.

1. Introduction

The hertzian contact stresses between the rolling elements and the races are one of the basic mechanisms that initiate a localized defect. When a rolling element strikes a localized defect, an impulse occurs which excites the resonance of the structure. Therefore, the vibration signature of the damaged bearing consists of exponentially decaying sinusoid having the structure resonance frequency. The duration of the impulse is extremely short compared with the interval between

* This work is supported by the National Natural Science Foundation of China (Grant No. 51375319), Hebei Provincial Natural Science Foundation (Grant No. E2013421005), Key project of scientific and technical research in Hebei Education Department (Grant No. ZD2015071) and the universities in Hebei Province hundred outstanding innovative talents support plan (Grant No. BR2-110).

impulses, and so its energy is distributed at a very low level over a wide range of frequency and hence, can be easily masked by noise and low-frequency effects. The periodicity and amplitude of the impulses are governed by the bearing operating speed, location of the defect, geometry of the bearing, and the type of the bearing load.[1]

The rolling elements experience some slippage as the rolling elements enter and leave the bearing load zone. As a consequence, the occurrence of the impacts never reproduce exactly at the same position from one cycle to another, moreover, when the position of the defect is moving with respect to the load distribution of the bearing, the series of impulses is modulated in amplitude. However, the periodicity and the amplitude of the impulses experience a certain degree of randomness.[2] All these make the bearing defects very difficult to detect by conventional FFT-spectrum analysis that assumes that the analyzed signal to be strictly periodic. A method of conditioning the signal before the spectrum estimation takes places is necessary.

A commonly used method to bearing fault detection is envelope detection. To overcome the modulation problem, several signal envelope demodulation techniques have been introduced. In high-frequency resonance technique (HFRT), an envelope detector demodulates the pass-band filtered signal and the frequency spectrum is determined by FFT technique.[3] Another well-established method is based on the Hilbert transform or Teager Kaiser energy operator (TKEO).[4] The envelope analysis technique is widely accepted as a powerful tool in the detection and diagnosis of bearing fault. The resonance frequency oscillation of the impacts and the possibly suppressed impact periodicity are two modulations of the vibrations that both reduce the amplitude of the power spectrum peaks, which therefore are more likely to be suppressed below the overall noise level. A popular countermove is to remove the resonance frequency modulation with the envelope method, which consists of a band-pass filter followed by a demodulation and a fast Fourier transformation. In the traditional envelope analysis technique, the fault is identified through the peak value of envelope spectrum. However, the central frequency of the filter is determined with experience while forming an envelope signal, which will make great subjective influence on the diagnosis results.[4]

The wavelet transform provides powerful multi-resolution analysis in both time and frequency domain and thereby becomes a favored tool to extract the transitory features of non-stationary vibration signals produced by the faulty bearing.[5] The wavelet analysis results in a series of wavelet coefficients, which indicate how close the signal is to the particular wavelet. In order to extract the fault feature of signals more effectively, an appropriate wavelet base function should be selected. Morlet wavelet is mostly applied to extract the rolling element bearing fault feature because of the large similarity with the impulse generated by the faulty bearing.[6] An application of Hermitian continuous wavelet transform for the detection of cracks in geared system is presented in.[6,7]

The Laplace wavelet is a complex, single-sided damped exponential formulated as an impulse response of a single mode system to be similar to data feature commonly encountered in health monitoring tasks.[8,9] It is also applied to diagnose the bearing fault of gearbox.[10] In this paper, an alternative approach for detecting localized faults in the outer and inner races of a roller element bearing using the envelope spectrum of the Laplace wavelet transform (LWT) is investigated. The shape parameters of Laplace wavelet ensure a large similarity between the wavelet function and the generated fault impulse. The methodology developed in this paper decomposes the original times series data in intrinsic oscillation modes, using the Laplace wavelet transform. Then multi-scale Laplace wavelet transform spectrum is applied to the vibration signal of bearing fault. The techniques are demonstrated by the experiments on a gearbox with a rolling element bearing under simulated crack on the inner race or the outer race. The characteristic frequency related to the bearing fault can be effectively extracted. The experimental results show that this method based on multi-scale Laplace wavelet envelope spectrum can effectively diagnose the fault of rolling element bearing.

2. Laplace Wavelet Transform

The Laplace wavelet is a complex, analytical, and single sided damped exponential and it is given by

$$\psi(t) = \begin{cases} Ae^{-\frac{\xi}{\sqrt{1-\xi^2}}\omega_c t} e^{-j\omega_c t}, & t \geq 0. \\ 0, & t < 0 \end{cases} \tag{1}$$

where ξ is the damping factor that controls the decay rate of the exponential envelope in the time domain and hence regulates the resolution of the wavelet, and it simultaneously corresponds to the frequency bandwidth of the wavelet in the frequency domain. The frequency ω_c determines the number of significant oscillations of the wavelet in the time domain and corresponds to the wavelet center frequency in frequency domain, and A is an arbitrary scaling factor. This function is called a Laplace wavelet to emphasize its derivation is related to the Laplace transform. In particular, the Laplace wavelet has a strong similarity to the inverse Laplace transform of the transfer function for an undamped second order system. Figure1 shows the Laplace wavelet, its real part, imaginary part, and its spectrum.

It should be noted that Laplace wavelet is generated by considering features anticipated in mechanical system responses. Figure 1 also shows the projections of Laplace wavelet $\psi(t)$ onto the real and imaginary planes. These projections graphically demonstrate the relationship between the wavelet and an impulse response of a single mode system. The Laplace wavelet has real and imaginary parts which are 90° out of phase. In a manner similar to the Fourier transform, a complex wavelet is required to properly handle phase differences between the wavelet and the signal. A wavelet could be derived with only a real part to match physical data that is typically real. However, using both parts ensures a smoothness and continuity of the wavelet transform.

The continuous wavelet transform (CWT) $W_x(t,a)$ of the signal $x(t)$ is defined as a convolution integral of $x(t)$ with scaled and dilated versions of Laplace mother wavelet function, $\psi(t)$:

$$W_x(t,a) = \frac{1}{a}\int x(\tau)\psi\left(\frac{t-\tau}{a}\right)d\tau = x(t) * \psi_{ab}(t). \tag{2}$$

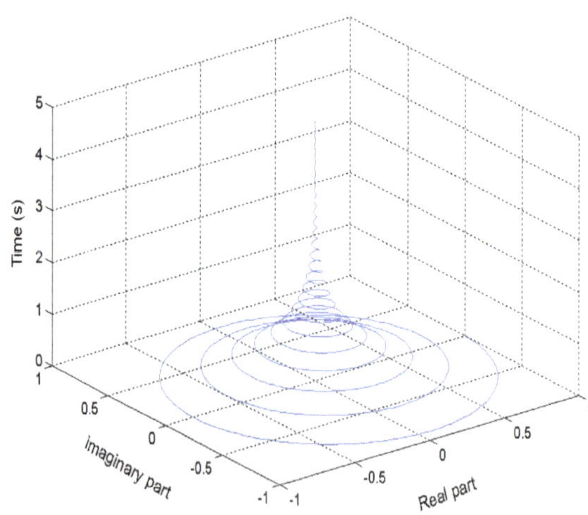

(a) The Laplace wavelet

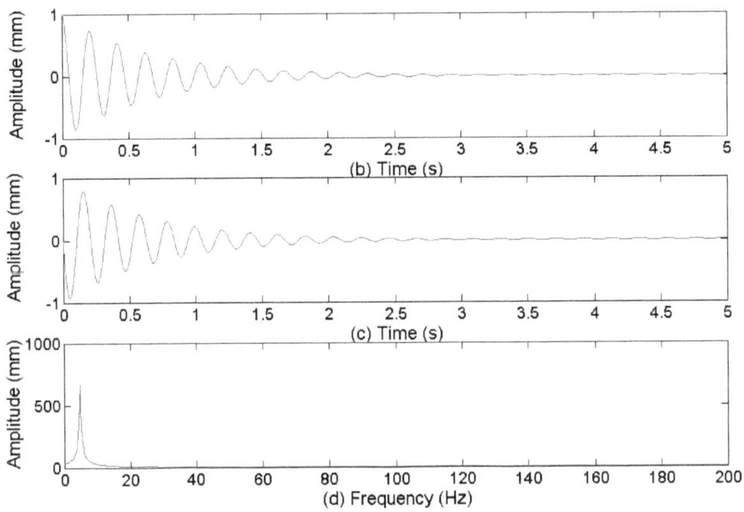

Fig. 1. (a) The Laplace wavelet, (b) the real part, (c) the imaginary part, and (d) its spectrum

When the signal $x(t)$ is periodic, Eq. (2) becomes a circular convolution, and the continuous wavelet transform (CWT) is also better calculated in the frequency domain according to

$$W_x(t,a) = \frac{1}{a} F^{-1}[X(\omega) \bullet aW(a\omega)]. \tag{3}$$

where $X(\omega)$ and $aW(a\omega)$ are the Fourier transform of the signal $x(t)$ and dilated wavelet $\psi_{ab}(t)$, respectively. The operator $F^{-1}[\bullet]$ denotes inverse Fourier transform.

Then the Laplace wavelet envelope spectrum of signal $x(t)$ is given as follows,

$$Env_x(a,b) = \sqrt{[\mathrm{Re}(WT_x(a,b))]^2 + [\mathrm{Im}(WT_x(a,b))]^2}. \tag{4}$$

The multi-scale Laplace wavelet envelope spectrum is calculated according to,

$$E_{wl}(a,f) = F[Env_x(a,b)]. \tag{5}$$

where $F[\bullet]$ denotes Fourier transform.

3. Fault Detection Based on Laplace Wavelet Transform

The original vibration signal of outer race defect is displayed in Figure 2. It is clear that there are periodic impacts in the vibration signal. There are significant fluctuations in the peak amplitude of the signal, and there are also considerable variations of frequency content. From Figure 2, we can hardly find the characteristic period of the outer race defect.

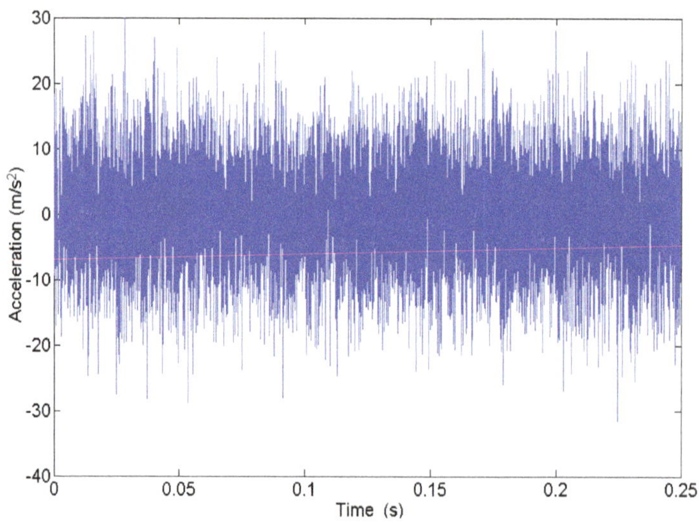

Fig. 2. Vibration signal with outer race fault

Figure 3 shows the FFT of the vibration signal with outer race fault. There is no characteristic fault frequencies component around f_{outer} and its harmonics. Therefore, classical Fourier analysis has some limitation such as unable to extract the characteristic frequencies of outer.

Figure 4 shows the traditional envelope spectrum of the vibration signal with outer race fault. There is no characteristic defect frequencies component around f_{outer} and its harmonics. Therefore, classical has some limitation such as being unable to extract the characteristic defect frequency of outer effectively under strong background noise.

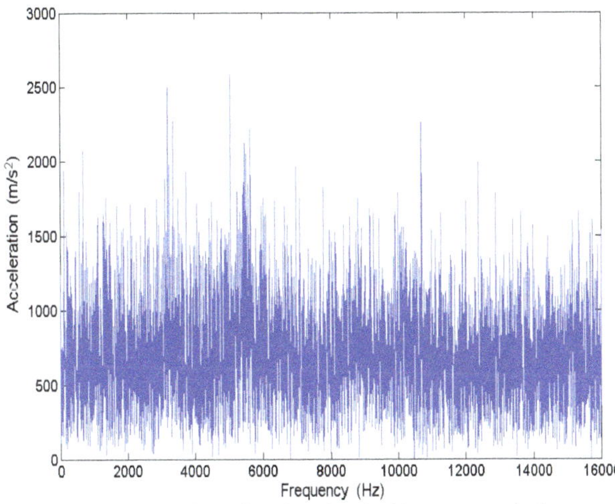

Fig. 3. FFT of the vibration signal with outer race fault

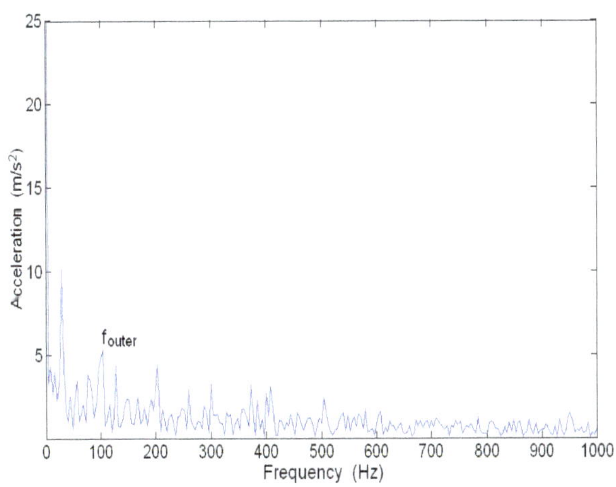

Fig. 4. Envelope spectrum of the vibration signal

The multi-scale Laplace wavelet transform spectrum is shown in Figure 5. The presences of outer race fault results in a sudden increase of vibration energy. In addition, these transient vibrations have shown a repetitive pattern with a f_{outer} (101.5Hz) interval, which corresponds to a repetitive the fault characteristic frequency of the outer race, resulting from the structural defect on the outer race. Such repetitive fault characteristic frequency reflects degradation of the outer race health condition as the defect propagated through the bearing outer race.

The multi-scale Laplace wavelet transform slice plot when scale a=3 is given in Figure 6. From Figure 6, one can see the spectrum lines at the characteristic frequency f_{outer} (101.5Hz) of the bearing with outer race fault and its high order harmonic frequencies. The simplicity of the multi-scale Laplace wavelet transform spectrum can be put down to the ability of this signal processing method to eliminate undesirable modulation and noise effects. Therefore, the multi-scale Laplace wavelet transform spectrum has shown to provide an effective tool for bearing fault detection and diagnosis.

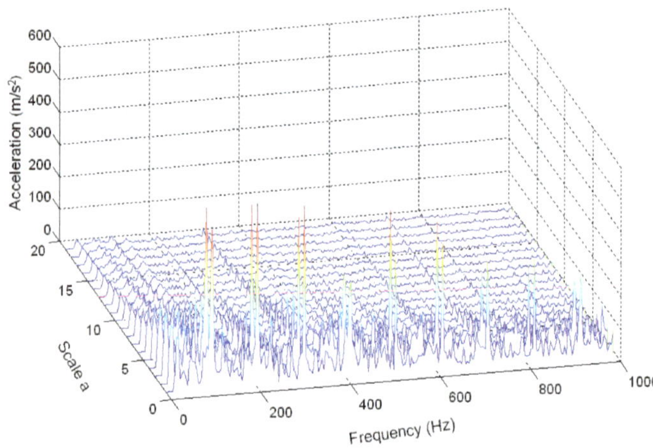

Fig. 5. Multi scale spectrum with outer race fault

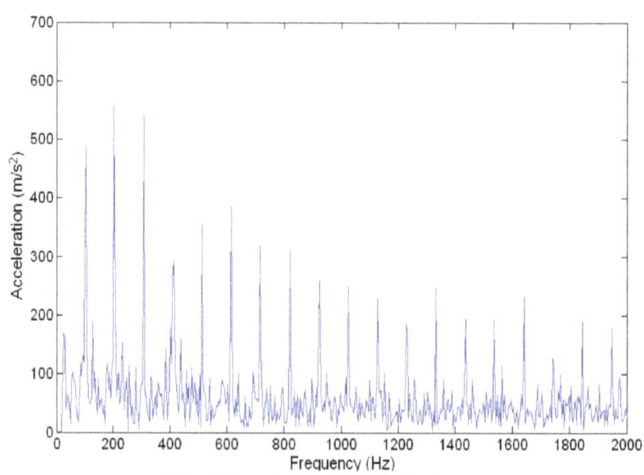

Fig. 6. Envelope spectrum (a=3)

For the defective bearing, transient vibrations caused by the rolling elements-defect interactions are clearly seen through the characteristic defect frequency and its high order harmonic frequencies. Such repetitive cycle reflects degradation of the outer race health condition as the defect propagated through the bearing outer race. Physically, impacts generated by the rolling ball-defect interactions excite intrinsic modes of the bearing system, giving rise to a train of transient vibrations at the mode-related resonant frequencies.

4. Conclusions

In this study, a method for fault diagnosis of bearing fault was presented based on a newly developed signal processing technique named as multi-scale Laplace wavelet transform spectrum. In order to extract the defect impulses buried in the vibration signal effectively, the Laplace wavelet transform is constructed. This gives a desirable ability to detect the singularity characteristic of a signal precisely. The technique excels in extracting transients, which are often the indicators of incipient defect in gearbox system. Therefore, we can recognize the vibration modes that coexist in the system, and to have a better understanding of the nature of the fault information contained in the vibration signal. This research shows that the multi-scale Laplace wavelet transform spectrum is very effective in detection and diagnosis bearing fault. According to the multi-scale Laplace wavelet transform spectrum, the characteristic of gear defect can be easily recognized. Practical vibration signal monitored from a gearbox with bearing outer fault is analyzed by the presented method. The experimental results have been shown that the multi-scale Laplace wavelet transform spectrum can effectively diagnose the bearing fault.

Acknowledgments

This work is supported by the National Natural Science Foundation of China (Grant No. 51375319), Hebei Provincial Natural Science Foundation (Grant No. E2013421005), Key project of scientific and technical research in Hebei Education Department (Grant No. ZD2015071) and the universities in Hebei Province hundred outstanding innovative talents support plan (Grant No. BR2-110). The authors are grateful for the financial support. The authors are also grateful to the editors and reviewers for their constructive comments.

References

1. R. Rubini, U. Meneghetti, Application of the envelope and wavelet transform analyses for the diagnosis of incipient faults in ball bearings, *Mechanical Systems and Signal Processing* **15**, 287–302 (2001).
2. J. Lin, L. Qu, Feature extraction based on Morlet wavelet and its application for mechanical fault diagnosis, *Journal of Sound and Vibration* **234**, 135–148 (2000).

3. P.D. McFadden, J.D. Smith, Vibration monitoring of rolling element bearings by the high-frequency resonance technique-a review, *Tribology International* **17**, 3–10 (1984).
4. Hui Li, Haiqi Zheng, Liwei Tang, Bearing Fault Detection and Diagnosis Based on Teager-Huang Transform, *International Journal of Wavelets, Multiresolution and Information Processing* **7**, 643–663 (2009).
5. N.G. Nikolaou and I.A. Antoniadis, Demodulation of vibration signals generated by defects in rolling element bearings using complex shifted Morlet wavelets, *Mechanical Systems and Signal Processing* **16**, 677–694, (2002).
6. C. Junsheng, Y. Dejie, and Y. Yu, Application of an impulse response wavelet to fault diagnosis of rolling bearings, *Mechanical Systems and Signal Processing* **21**, 920–929 (2007).
7. Hui Li, Haiqi Zheng, Liwei Tang, Bearing Fault Diagnosis Based on Amplitude and Phase Map of Hermitian Wavelet Transform, *Journal of Mechanical Science and Technology* **25**, 2731–2740 (2011).
8. Hui Li, Yuping Zhang, Haiqi Zheng, Application of Hermitian wavelet to crack fault detection in gearbox, *Mechanical Systems and Signal Processing* **25**, 1353-1363 (2011).
9. G. Strang, T. Nguyen, *Wavelet and filter banks* (Wellesley Cambridge Press, 1996).
10. Khalid F. Al-Raheem, Asok Roy and K. P. Ramachandran, et al, "Application of the Laplace-Wavelet Combined with ANN for Rolling Bearing Fault Diagnosis", *Journal of Vibration and Acoustics* **130**, 1–9, (2008).

Bridge support structure generation for 3D printing

Zhen-Hong Shen, Ning Dai† and Da-Wei Li, Chang-You Wu
College of Mechanical and Electrical Engineering,
Nanjing University of Aeronautics and Astronautics,
Nanjing, 210016,China
†*E-mail: dai_ning@nuaa.edu.cn*
www.nuaa.edu

During the printing process, vertical support structures have a high material cost and volume, while branching support structures have a low stability. We thus propose a bridge support structure generation algorithm for 3D printing. Firstly, our technique identifies the regions requiring support. Secondly, it determines a set of points that require support based on adaptive sampling. Finally it enumerates and selects new bridges through a scoring function, and connects the bridges and model with pillars. In printing experiments, we compare our approach with vertical support structure and branching support structure. The algorithm saves an average of about 15.19% of printing and about 24.41% of time.

Keywords: 3D Printing; Bridge Support; Anchoring Segment; Sweep.

1. Introduction

Three dimensional printing stacks materials to complete forming by different methods based on discrete and stacking principle. With this technique, people can print out complex parts which are difficult to be manufactured by traditional processing. But also as a result of accumulation layer-by-layer, a layer of material must be supported by its lower material. Therefore, overhangs require a disposable support structure to be printed. Otherwise, larger warp and collapse may occur. In order to address these problems, we need to properly design a support structure before processing. The design of support structure is a key technique in three dimensional printing technology, which plays an important role in guaranteeing stability and preventing error caused by deformation of the size of platforms and surface irregularities during the process, enhancing strength of a hollow part and preventing the local deformation and so on.

1.1. *Related works*

Support generation algorithms start by a determination of the surfaces in need for support. A first family of approaches consist in performing a Boolean difference between two successive slices. Chalasani et al[1] get the smallest envelope rectangular of each slice by merging the polygons, then the bounding box of the model is calculated. As input is the SLC file, information about the smallest envelope rectangular can be quickly obtained. Since the bounding box method, it requires to remove some areas that don't need support manually. Huang et al[2] perform a difference and union Boolean operation between two successive slices from top to bottom. The algorithm efficiently utilizes the ability of self-support of materials. But arbitrary polygons' Boolean operation and bi-directional offset have a certain complexity, as the number of polygons and the number of nodes increases, the time of the algorithm will increase significantly. A second family of approaches consider the down-facing facets of input mesh having an angle too steep to print correctly. Lan et al[3] first process the hanging edges and then find the facets in need for support. As the model is relatively simple, its application has some limitations. Bian et al[4] first extract the key features of areas to be supported, and local ramps are used to support the areas, but it uses a interactive way to change parameter when adding enhanced support line.

For the design of support geometry, the simplest approach consists of extruding the mesh facts requiring support downward is the method of bounding boxes, which calculates the minimum enveloping cuboid of model. In order to help support removal after processing, Strano et al[5] fill model with a relatively sparse internal structure. Kritchman et al[6] use soluble materials to print support structure. The bounding box method does not take into account the morphological characteristics of prototype parts, so it's one of the most support materials and time consuming method. In order to decrease the volume of support, a number of approaches have been proposed. Chen yan et al[7] use straight walls to support, the algorithm automatically add the supporting points and rods, but support passing parameters in the algorithm cannot be selected automatically, only can be estimated by experiments. Huang et al[8] propose a sloped wall structure, the support structure is divided into three types, which are determined by calculating the stability factor of the area polygon. Volume reduced is limited and the contact area between support and model is large. Meshmixer software uses branching support structure, which requires users to modify the original structure that is automatically generated and problems of print stability sometimes encountered. Dumas et al[9] use a bridge support structure, the material cost is less, but some horizontal bridges are directly mounted on the model, reducing the surface accuracy of the model.

We propose a bridge support structure generated algorithm. Firstly, the algorithm detects areas that need support based on the angle between facets and direction of processing (Figure 1(a)). Secondly, support points are obtained by adaptive sampling based on the area (Figure 1(b)). Finally support points are supported through a series of horizontal bridge and vertical pillars (Figure 1(c) and (d)), and experiments verify the effectiveness of the algorithm.

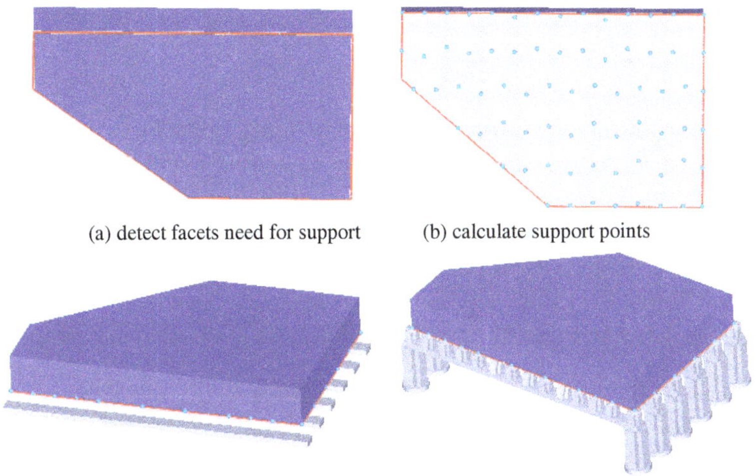

(a) detect facets need for support (b) calculate support points

(c) generate bridges (d) generate vertical pillars

Fig. 1. Overview of proposed method

2. Bridge Support Structure Generation

2.1. *Detected facets need for support*

Automatic recognition of support facets of part, that is, identify areas that need support on the part. As a STL format input, which uses a large number of triangles in the approximate surface of the model as a whole. According to the angle between surface and direction of processing, facets need support are identified. When the angle α is not greater than the critical angle β, since the function of self-support, it doesn't need support. However, you may need to add support structure when α is greater than β (red segments in Figure 2).

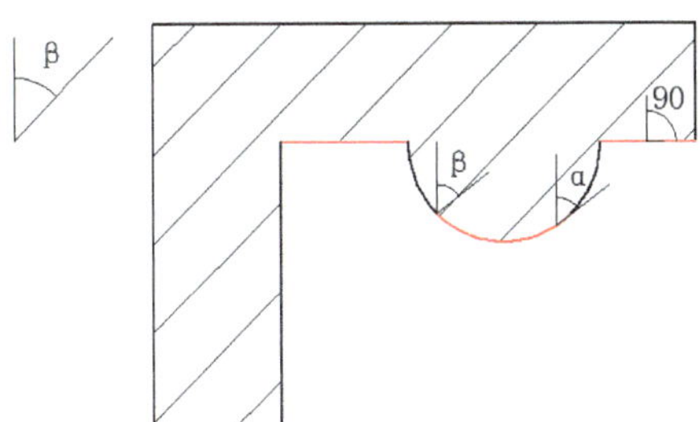

Fig. 2. Identify facets

After obtaining facets need for support $Triangles\_\sup$, we merge facets within the same region to several independent regions $Area_0, Area_i, ..., Area_n$, which meets the follow Eq. (1).

$$\begin{cases} Triangle\_\sup = \sum_{i=0}^{n} Area_i \\ Area_i \cap Area_j = \Phi, (i \neq j) \end{cases} \tag{1}$$

We use the region growing method for merging facets. Firstly, a facet in the collection is taken as a seed arbitrarily, we determine whether the seed's edge-sharing triangle is in the collection. If the answer is yes and the edge-sharing triangle hasn't been saved as a seed, then put it in the seed set and search for edge-sharing triangles with other seeds until all seeds have been searched in seed set. Repeat the above steps to merge is complete in all regions. The flowing charts are shown in Figure 3.

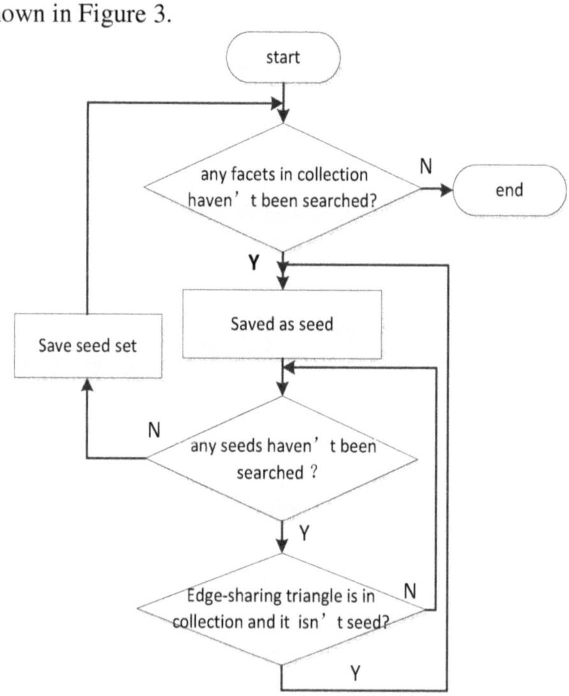

Fig. 3. Flow chart of merging facets by region growing method

2.2. *Support points*

After obtaining the support region, Adaptive sampling algorithm based on the area is used to obtain a set of support points. First calculates the enveloping rectangular of area in the XOY plane[10]. Then uses a spacing parameter to divide rectangular. In order to guarantee the stability while avoiding support points being too dense, you can do experiments to get the right result, then calculates vertexes' position in the grid. Latter we judge the number of vertexes in each grid. When the number of points in the grid is zero, insert the appropriate point, by the ray intersection, intersection point on the model is obtained and saved. When the number of points is greater than1, the center point is computed and saved.

2.3. *Bridge support structure construction*

2.3.1. *Structural analysis*

After obtaining the support points, we need to design the bridge support structure. The structure consists of a series of vertical pillars and horizontal bridges, all required points are properly supported by vertical pillars, all horizontal bridges are supported at their extremities, by vertical pillars, by other bridges, the vertical pillars connect either to the print bed, to horizontal bridges located below, or to the model itself.

Worth noting is that inclined pillars with increasing length, decline in reliability. But it is possible to print in a relatively short length. This is important to support non-collinear points on one bridge. Point o is the point requiring support, $oo2$ and $oo3$ show extreme configurations of inclined pillar, horizontal segment shows the plane bridge lies in. We restrict the maximum length of $o1o2$ 5mm. In addition, the algorithm creates bridges in several different directions in the X0Y plane, because bridges in one direction only, would give rise to large number of pillars for reliable support.

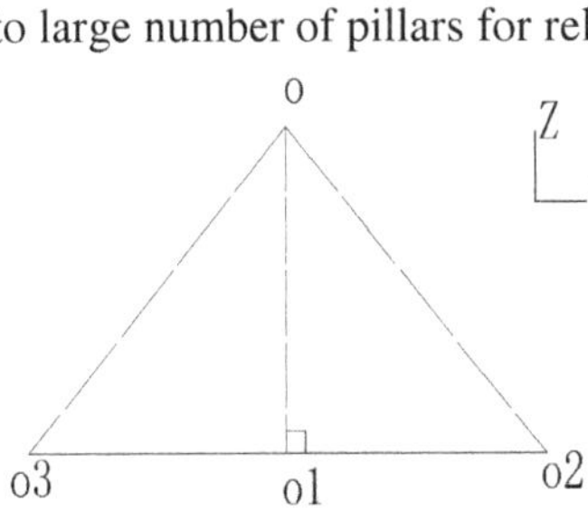

Fig. 4. extreme configurations of pillar

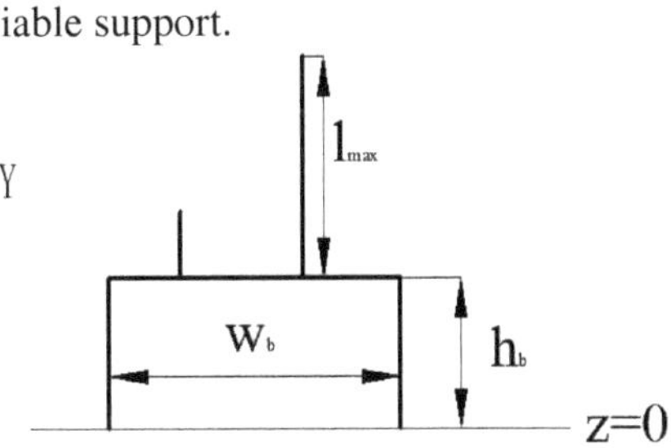

Fig. 5 horizontal bridge

Anchoring segments for points and bridges: Anchoring segment, shows the position where the bottom of the pillar which supports the point or bridge locates. Anchoring segments for points depend on the sweep direction, which are created each time a new sweep direction is selected. Anchoring segments for bridges are in the same direction with bridges. Supporting pillars from below can connect anywhere along the anchoring segments. Once connected, the bridge extends to the pillar and the anchoring segment is removed.

The algorithm enumerates and selects new bridges to build the structure. In order to objectively evaluate a horizontal bridge, we set up a scoring function:

$$S(b) = (n-2) \cdot h_b - w_b - n \cdot l_{\max}(b) \tag{2}$$

Where n is the number of pillars supported by bridge, h_b and w_b are the length and the length of bridge, $l_{\max}(b)$ is the maximum length of inclined pillar connecting above to the bridge (Figure 5).

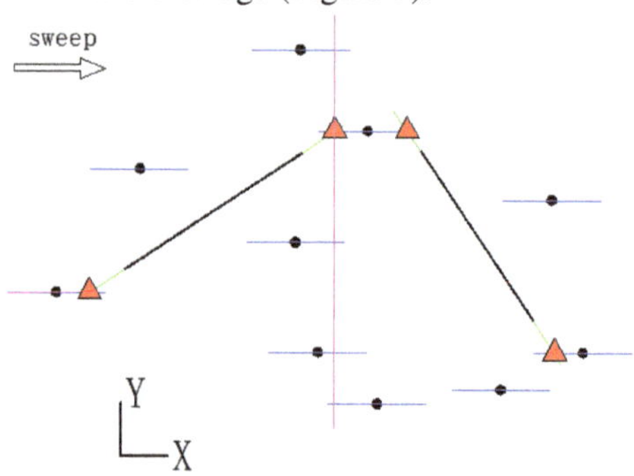

Fig. 6. Sweep strategy

2.3.2. *Construction algorithm*

The input of the construction algorithm is a set of points that need to be supported. The output of the algorithm is set of horizontal bridges and vertical pillars forming a printable bridge structure. Our approach is based on a sweep strategy (Figure 6), for the sake of clarity, we sweep a YOZ plane along the X axis. The purple line illustrates the YZ sweep plane, blue lines are the anchoring segments for points, green lines are the anchoring segments for bridges. Red triangles represent the intersections of segments, which means a bridge may be enumerated. The algorithm is roughly as follows:

1) For each sweep direction executes step (2);

2) Create anchoring segments for each point, save them into set S, make P to 0, which is a set of segments crossing the current sweeping plane; save the intersections of segments into set Q, which is a queue sorted by X coordinates from small to large. When $Q \neq \Phi$, executes step (3);

3) Make b to the first element of S, for $s \in S$ starting in b do $P \cup s \rightarrow P$; Enumerate bridge and select the bridge with the highest score. Remove the relative points;

4) Update the input: makes new bridges and rest of points as input;

5) Repeat step 1–4 until there is no bridge could meet the rating requirements;

6) Convert the structure data into entities by sweeping operation with different section elements [11–12].

3. Experimental Results and Discussion

3.1. *Experimental conditions*

The algorithm is implemented in Microsoft Visual Studio 2008 software development platform, and models are displayed through the OpenGL. All models are printed in Korea the 3DISON H700 using High Speed ABS plastic at 60mm/sec print speed, 80mm/s mm/sec travel speed, using layers of 0.2mm height.

3.2. *Critical angle calibration*

Due to the critical angle is not a fixed value, different printers and print material corresponds to the different critical angle. For this purpose, we design a set of incline planes with different angle to print, as showed in Figure 7. Experimental results are shown in Table 1, number starting from the left.

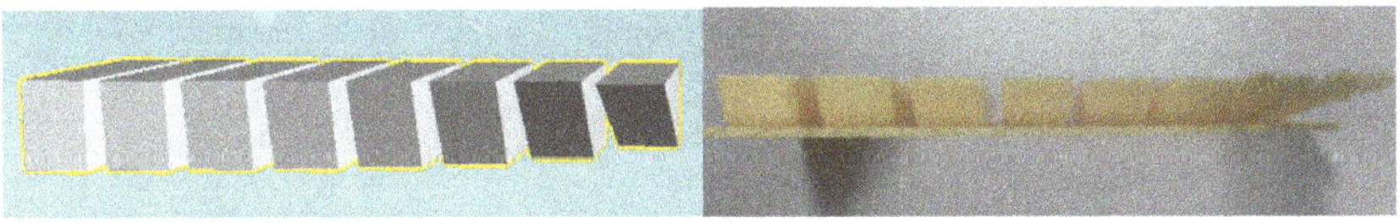

(a) model (b) printing result

Fig. 7 Models with different incline planes

Table 1. Critical angle β calibration result

Number	1	2	3	4	5	6	7	8
Angle (°)	10	20	30	40	50	60	70	80
Quality	good	good	good	good	poor	worse	worse	worse

From Table 1 we can see that when the angle is greater than 50°, the quality began to decline. With further experiments, we can find when the angle is bigger than 45°, Pilling is appear. So we make the critical angle β to 45°.

3.3. *Structure compares to others*

In order to assess the algorithm objectively, Figure 8 compares our technique to Creator-V9.1 (printer software) and Meshmixer2.6 (United States Autodesk) on Guide Board model. Table 2 and Table3 summarizes the print results. Compared to the Creator-V9.1's vertical support, proposed algorithm reduces 14.65% time and 34.75% of materials consumption, compared to the Meshmixer2.6 branching support, in terms of time and material costs reduced by 15.72%, 14.07%. In terms of print stability, our algorithm and Creator-V9.1 are more stable. As far as support removal and surfaces quality, the contact area of support and the model is small in our approach, so the quality is relatively high while reducing the difficulty of removal support.

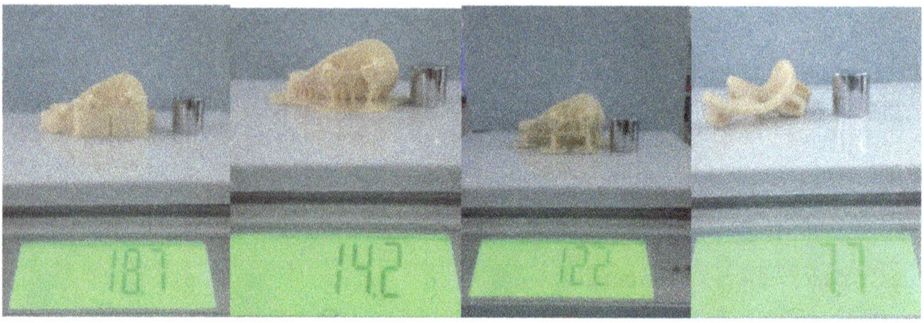

(a) Creator-V9.1 (b) Meshmixer 2.6 (c) our approach (d) Model without support

Fig. 8 Guide Board models with different support structure

Table 2. Comparison of time and total filament length

	Time (min)	Filament (m)
Creator-V9.1	157	0.8636
Meshmixer2.6	159	0.6558
Our approach	134	0.5635

Table 3. Save percentage

	Time saving (%)	Filament saving (%)
Creator-V9.1	14.65	34.75
Meshmixer2.6	15.72	14.07

4. Conclusion

As vertical support structure's large volume, branching support structure's low printing stability, we present a bridge support structure generation algorithm. Because the ability of printers to print bridges across gaps, we can construct reliable structure. Experimental results show that the method provides a reliable way to print with a reasonable material and time use.

Acknowledgements

This research is supported by:

1. Jiangsu Provincial Science and Technology Support Program (BE2014009-3).
2. Aeronautical Science Foundation of China (No. 20151652024).

3. Jiangsu Key Technology Support Program (No. BE2014009-3).
4. Jiangsu Key Laboratory of 3D Printing Equipment and Manufacturing (No. BM2013006).

References

1. Chalasani K, Jones L, Roscoe L. Support generation for fused deposition modeling[C] //Proceedings of Solid Freeform Fabrication Symposium. 1995: 229-241.
2. Xiaomao Huang, Chunsheng Ye, Jianhua Mo, et al. Slice data based support generation algorithm for fused deposition modeling[J]. Tsinghua Science and Technology, 2009, 14(S1): 223–228.
3. Po-Ting Lan, Shuo-Yan Chou, Lin-lin Chen, et al. Determining fabrication orientation for rapid prototyping with Stereolithography apparatus[J]. Computer-Aided Design, 1997, 29(1):53–62.
4. Hongyou Bian, Weijun Liu, Tianran Wang. Study on algorithm for generate support based on STL file[J]. Machinery Design & Manufacture, 2005, (7): 49–52. (In Chinese)
5. Strano G, Hao L, Everson R, et al. A new approach to the design and optimisation of support structures in additive manufacturing[J]. Advanced Manufacturing Technology, 2013, 66(9-12): 1247–1254.
6. Kritchman E, Gothait H, Miller G. System and method for printing and supporting three dimensional objects: US, 7364686[P]. 2008, 4–29.
7. Yan Chen, Shiwei Wang, Zhouwang Yang, et al. Construction of support structure for FDM 3D printers[J]. Scientia Sinica, 2015, 45(2): 259–269. (In Chinese)
8. Xiaomao Huang, Chunsheng Ye, Siyu Wu, et al. Sloping wall structure support generation for fused deposition modeling[J]. Advanced Manufacturing Technology, 2009, 42(11-12): 1074–1081.
9. Jun Hong, Wei Wang, Yuhong Zhang, et al. Research on auto-support in stereolithography[J]. Journal of Mechanical Engineering, 2004, 40(11): 134–138. (In Chinese)
10. Dumas J, Hergel J, Lefebvre S. Bridging the gap: automated steady scaffolding for 3d printing[J]. ACM Transactions on Graphics, 2014, 33(4): 98:1–98:10.
11. Bai Xu, Xiaosheng Cheng, Ning Dai, et al. The Construction of Sweep Surface Based on Precision Control[J]. Advanced Materials Research, 2014, 1006-1007: 348–351. (In Chinese)
12. Martin, R.R., Stephenson, P.C. Sweeping of three-dimensional objects. Computer-Aided Design, 22(4): 223–234, 1990.

Filling effect on stress concentration factor for finite width plate with elliptical opening*

Dong-Mei Luo†, Bi-Xin Su, Li-Can Zhuang,
Yi-Yuan Liao, Ming-Hao Lin and Ya-Fang Peng

Department of Civil Engineering, Foshan University,
Foshan, 528000, China
†E-mail: dongmei_luo@126.com

Stress concentration factors (SCFs) in finite width plates with both elliptical opening and elliptical fillings under uniaxial loads are predicted. ANSYS Parametric Design Language (APDL) is applied to build the finite element models for the plates and to perform the analysis. The influences of geometric and material parameters such as the lengths along macro-axis and minor axis of opening, the eccentricity position of the opening, and the elastic modulus of the fillings on SCFs are discussed. The results show that SCFs in the finite width plate with elliptical opening can be significantly improved by substituting the opening with hard filling, and hard filling with high elastic modulus is effective to support the maximum stress in the plate, and the results are in good agreement with those from theoretical prediction. The present study may provide designers with an efficient way to estimate the filling effect on finite width plate structures with openings.

Keywords: Stress concentration factor, Elliptical opening, Hard filling, Soft filling; Finite width plate, Maximum stress.

Introduction

Stress concentration factor (SCF) is defined as the ratio of the maximum stress in the presence of a geometric irregularity or discontinuity to the average stress away from the effect of such irregularity, and it has great practical importance because it is normally the cause of failure. Majority of the studies performed for the SCF have treated isotropic plates with circle openings. Wu and Mu [1] proposed a simple computation method based on the scale factors (SFs) to estimate the stress concentration factors (SCFs) of finite-width isotropic/ orthotropic plates/cylinders with a circular cutout and under uniaxial or biaxial tension. Yao[2] proposed a formula for estimating stress concentration factor and stress distribution near the opening, and the influence of finite width and elliptical geometry on stress concentration factor and stress has been taken into

* This work is supported by the NSFC (10772047/A020206, 11172066/A020305), Guangdong Natural Science Foundation of the P. R. China (S2011010004874).

† Work partially supported by Training Plan to Innovation and Enterprise of National University Students, Foshan University.

account reasonably. Li[3] derived the complex stress function by the affine transformation to describe accurately the stress field, and performed the stress analysis of the composite plate with circular holes by the numerical simulations, the effects of the fiber orientation on the stress concentration and the Young's module in the two main directions on the hole-edge stresses were revealed. For the composites structures with different openings, Sharma[4] proposed the general stress functions to determine the stress concentration around circular, elliptical and triangular cutouts in laminated composite infinite plate subjected to arbitrary biaxial loading at infinity. By applying the APDL[5], Tawakol A. Enab[6] established the finite element models to perform stress analysis of a finite functionally graded material (FGM) plate with an elliptic hole under uni-axial and biaxial loading conditions. Xiang[7] analyzed the orthotropic strip of finite width subjected to uniform tensile stress with an eccentric elliptical hole, and the change rule of the SCFs along with the size of eccentricity and the shape of the elliptical hole is obtained. In order to decrease the possibility of local failure occurrence, it is possible to select different reinforcements to fill the openings in a finite width plate in some applications. Yang et. Al[8] calculated the stress concentration factors (SCFs) at the tip of the elliptic hole both under hollow and filled conditions. In comparison with the hollow condition, soft materials can reduce SCF at the tip of the elliptic hole and the affected scope of the hole to the surrounding stress field significantly.

In this paper, the model for finite width plate with elliptical opening and fillings is established by the Ansys parametric design language, and the finite element analysis is performed to calculate the SCFs for the plate with the change of geometric and material parameters of the fillings. The results show that hard filling is effective to decrease stress concentration, and to improve the distribution of the maximum stress.

1. The Theoretical Analysis of SCFs

The stress concentration factor is defined as the ratio of the maximum stress σ_{max} to the mean tensile stress σ_0 produced from the axis displacement, which is shown in Eq. (1).

$$K = \frac{\sigma_{\max}}{\sigma_0} \tag{1}$$

For anisotropic infinite plate with elliptical openings, the SCF is calculated by the following formula [9]:

$$K_T^{\infty} = 1 + [2(\sqrt{E_y/E_x} - \nu_{yx}) + E_y/G_{yx}]^{1/2}(b/a) \tag{2}$$

Except for the material properties of the infinite plate, only the sizes of the opening are considered in Eq. (2), and the size of the plate is neglected. For a finite width plate with elliptical openings, it is necessary to make some modifies for the dimension difference, and some modified coefficients are described in [9]. The first one is related to the change of the finite size of the plate:

$$f_W = \frac{2+0.15(a/h)^2+(1-a/h)^3}{3(1-a/h)} \tag{3}$$

and the second one is the geometric modified coefficient

$$f_M = 1 - 0.7416(a/h)^2(\frac{K_T^{\infty}|_{a=b}}{3} - 1)^{0.7782}(\frac{b}{a}) \tag{4}$$

At last, we should consider the material modified coefficient

$$f_E = 1 - (a/h)[0.03 + 0.83(a/h)](1 - b/a) \tag{5}$$

So the modified SCF is described as

$$K_T/K_T^{\infty} = f_W \cdot f_M \cdot f_E \tag{6}$$

2. Results and Discussion

Typical geometric model applied in this paper is shown in Fig.1.

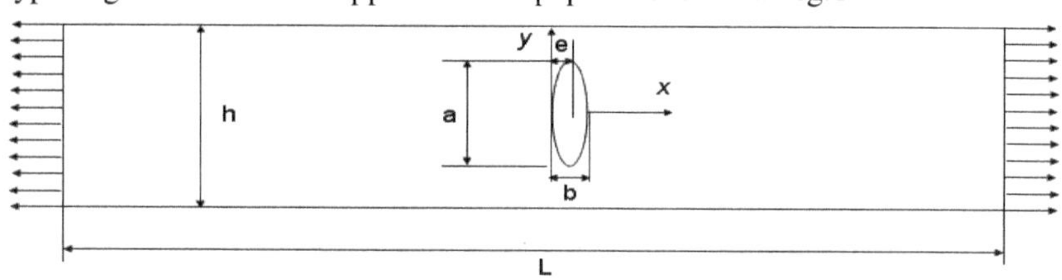

Fig.1 Typical elliptical opening for a finite width plate

Figure.1 shows a finite width plate containing an elliptical opening with different dimensions and subjects uniform loadings action at the boundary x=±*L*/2 (here, *L* is the side length of the plate along x-axis), and Cartesian coordinate system is used with origin locating at the center of the opening in the plate. The plate is modeled in two dimensions using iso-parametric quadrilateral Plane 183 elements, which have eight nodes: four vertex nodes and four midside nodes. In this paper, *L*/*h*=10, *h*=10mm, and *e* is the eccentricity of the elliptical opening.

In order to verify the reliability of the finite element analysis, the elliptic opening is taken as the filling with zero elastic modulus at first, and the typical finite mesh is shown in Figure 2, and the black colors at the tip of the elliptical opening show that the meshes around they are very concentrated. Supposing the elastic properties of the plate materials are: E = 402 GPa, ν= 0.34. The results from numerical simulation are compared with those from theoretical prediction as shown in Figure 3. It is observed that with the increase of the ratio of *a/b* (that is the ratio of length of macro-axis to length of minor-axis), the SCFs are approached to a constant for both theoretical and numerical results, and the obvious difference is only existing for the case that *a/b* is less than 2. But the result from theoretical prediction is very consistent with the numerical simulation at the length of *a*=7mm as shown in Figure 4, which shows that the numerical simulation is reliable compared to theoretical prediction for the smaller openings with the shorter length of macro-axis.

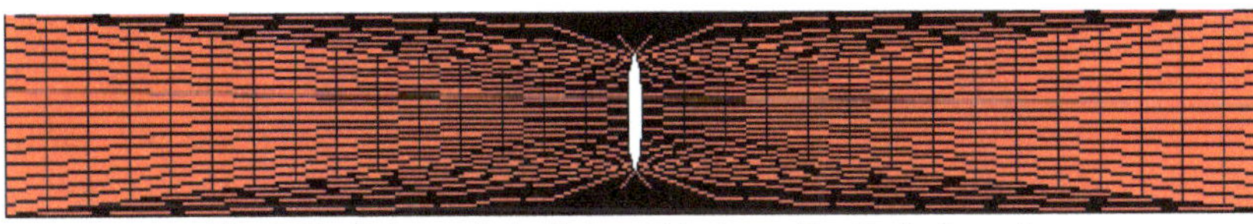

Fig. 2 The finite mesh for finite width plate with elliptical opening

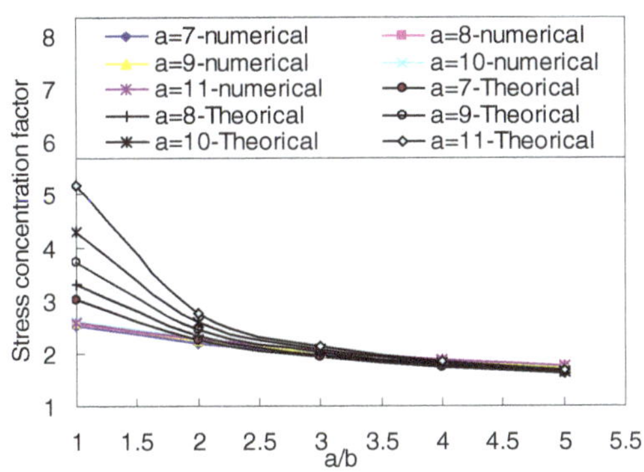

Fig. 3 Comparison of theoretical and numerical results for different lengths of macro axis

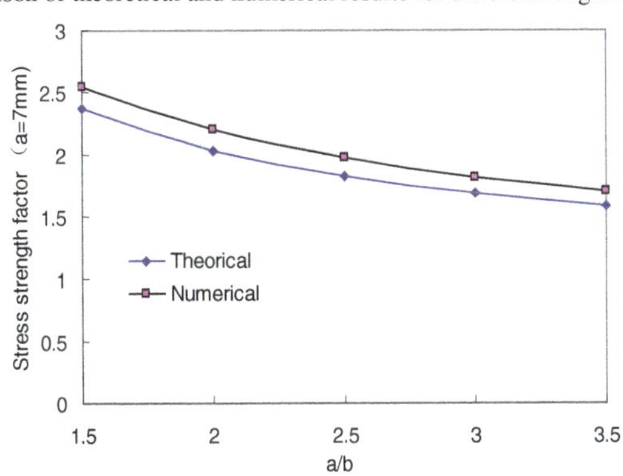

Fig. 4 Comparison of theoretical and numerical results for the length of a=7nm

Two kinds of fillings are considered in this paper, one is the filling with elastic modulus of 3.8GPa, and the Poisson ratio is 0.4, which is called soft-filling, and the other is the hard filling with high elastic modulus of 700GPa, and the Poisson ratio is 0.23. For different fillings, the changes of SCFs with the ratio of *b/a* are shown in Fig. 5, and the changes of SCFs with the eccentricity are shown in Fig. 6. We can find from Fig. 5 that SCFs decrease with the increase of *b/a*, so the circle hole is superior to elongated elliptical opening to resist on stress concentration. It shows in Fig. 6 that when the elliptical opening is far from its center, the SCFs increase. The same characteristic in Fig. 5 and Fig. 6 shows that SCFs are similar for the plate with elliptical opening and soft filling, but the SCF is nearly equivalent to 1 for hard filling, and it means that hard filling is effective to eliminate the stress concentration completely. The typical stress distribution is plotted in Figure 7, the maximum stress is 5.73MPa for elliptical opening, which is closed to the case of soft filling (5MPa), but it is only 0.45MPa for hard filling, and the maximum stress is mainly concentrated within the hard-filling. It means that the hard filling is not only effective to eliminate stress concentration, but also to decrease the maximum stress and to improve the stress distribution.

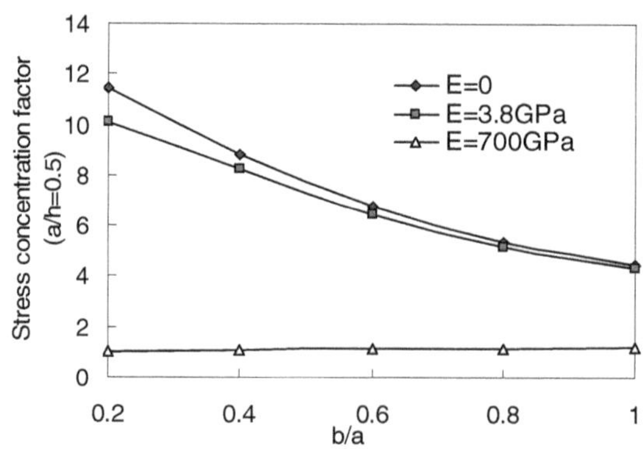

Fig. 5 Effect of filling's properties on SCFs

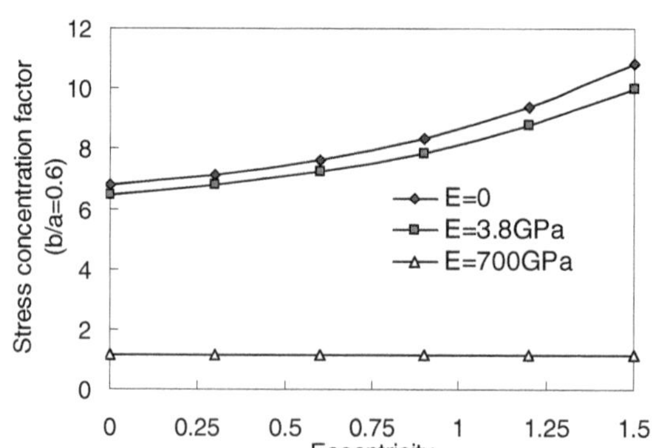

Fig. 6 Effect of filling's properties on SCFs

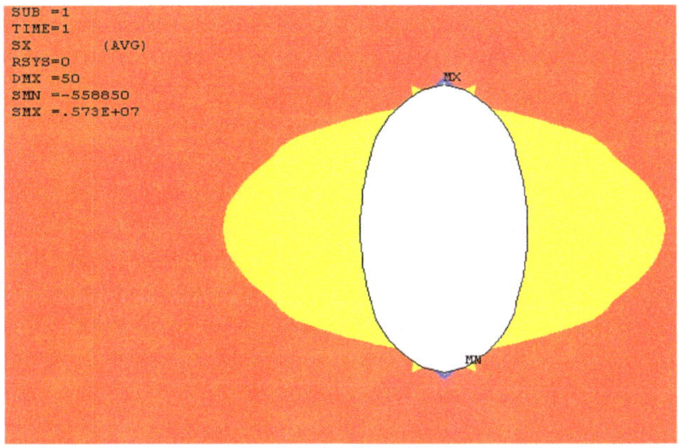

(a) Elliptical opening

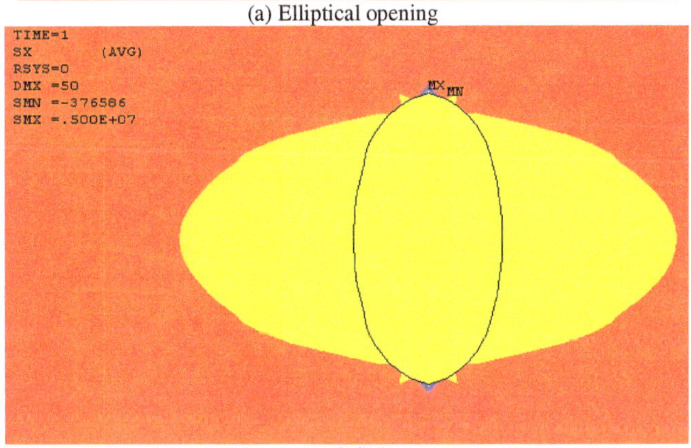

(b) Soft filling

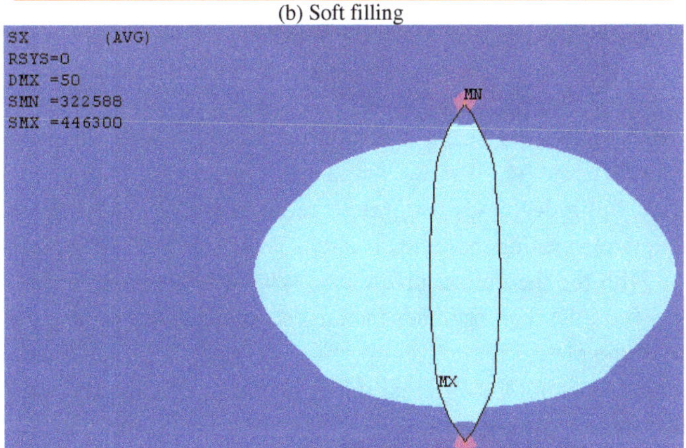

(c) Hard filling

Fig. 7 Maximum stress distribution for different fillings

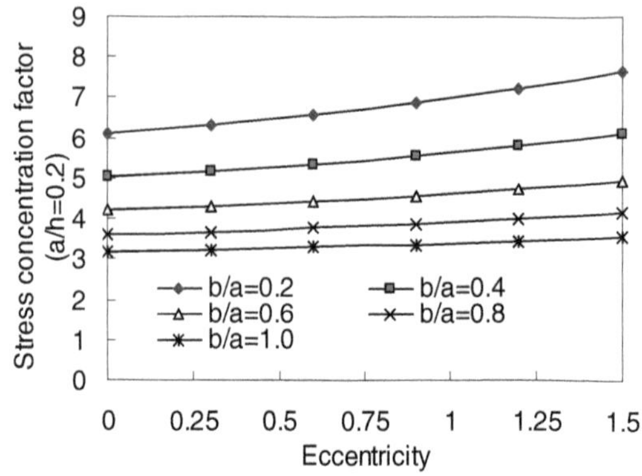

Fig. 8 SCFs Vs Eccentricity with *b/a*

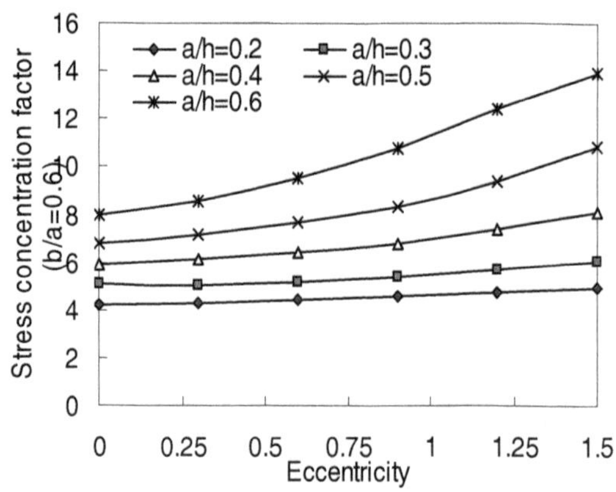

Fig. 9 SCFs Vs Eccentricity with *a/h*

Figure 8 shows the changes of SCFs for finite width plate with elliptical opening for different *b/a*. It shows that the SCF nearly has nothing to do with the eccentricity, but it decreases with the increase of *b/a*. When *b/a* is equivalent to 1, the elliptical opening becomes a circle hole, the SCF is about 3, which is consistent with the theoretical prediction. When the opening becomes a slender crack (*b/a*=0.2), the SCF becomes the largest, so the crack is dangerous to result in failure of the plate. Figure 9 shows the changes of SCFs for finite width plate with elliptical opening for different *a/h*. It shows that SCFs increase nonlinearly with the increase of the length of major-axis and the eccentricity, and the plate with a larger elliptical opening is unfavorable to resist on stress concentration.

Equations should be centered and numbered consecutively, as in Eq. (1).

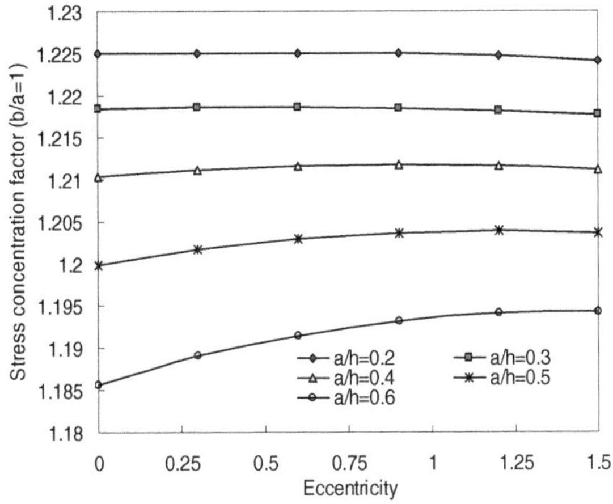

Fig. 10 SCFs Vs Eccentricity with different *a/h* for *b/a*=0.2

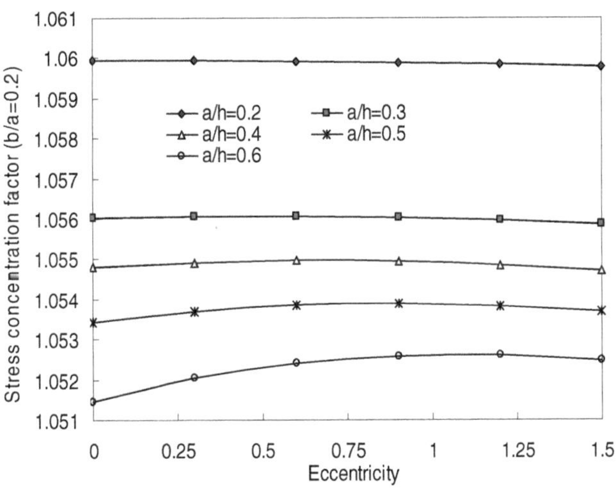

Fig.11 SCFs Vs Eccentricity with different *a/h* for *b/a*=1

The change of SCFs with *a/h* for the finite width plate with hard filling at *b/a=0.2* is shown in Fig.10, and the SCF is closed to a constant for each *a/h* except for *a/h=0.6*, and it has nothing to do with the eccentricity. As opposed to the case of elliptical opening, the SCF decreases with the increases of *a/h*, and the hard filling with a large area is beneficial to decrease stress concentration, and to bear the maximum stress for both *a/b=0.2* and *b/a=1* as shown in Fig.10 and Fig.11. Comparing to Fig.10, Fig. 11 also shows that the SCF has nothing to do with the eccentricity for different *a/h* with a circle hard filling (*b/a*=1), but the SCFs are higher than those in Fig.10. The smaller the *a/h* (the ratio of length of macro-axis of hard filling to the width of the plate), the higher the SCFs, so the SCFs reach to the largest at *a/h=0.2* for both Fig.10 and Fig.11. When *a/h*=0.6, the SCF is related to the eccentricity, and it seems to increase nonlinearly with the increase of eccentricity.

3. Conclusions

In this paper, the finite element analysis is performed to calculate the SCFs for the finite width plate with elliptic opening, and the factors to affect SCFs are analyzed. The filling with different elastic modulus is selected to substitute the opening, and the changes SCFs and the distribution of the maximum stress are discussed, some conclusions are obtained:

1)For the finite width plate with elliptical opening, the results from numerical simulation are compared well with those from theoretical prediction, and a higher stress concentration happened in the finite width plate with a slender elliptical opening (b<<a).

2)SCFs are nearly the same for the finite width plate with elliptical opening and soft filling, so the soft filling is not so effective to decrease stress concentration. Hard filling with high elastic modulus is effective to resist on stress concentration, and to improve the distribution of the maximum stress.

3)For the finite width plate with hard filling, the SCFs have nothing to do with the eccentricity, but have related to the size of the filling, and the larger hard filling is effective to decrease stress concentration.

References

1. Wu HC, Mu B. On stress concentrations for isotropic/orthotropic plates and cylinders with a circular hole, Composites Part B: Eng, 2003; 34:127–34.
2. Yao Weixing, Gu Yi, Computation of stress concentration factor and stress distributions factor for orthotropic finite width plates containing elliptical opening, Computational Structural Mechanics and Applications, 1990, 7(4), 56–64.
3. Li Cheng, Zheng Yan-ping, Tie Ying, Simulation on effect of fiber orientation on stress field of orthotropic composite plate with circular holes, Journal of Jilin University (Engineering and Technology Edition), 2008, 38(3), 544–547.

4. Sharma DS. Stress concentration around circular/elliptical/triangular cutouts in infinite composite plate. In: Proceedings of the world congress on engineering (WCE 2011), Vol. III, July 6–8, 2011, London, U.K.
5. ANSYS Engineering analysis system, Users manual, Release14.0. Canonsburg, Pennsylvania, USA: ANSYS Inc.; 2011.
6. Tawakol A. Enab, Stress concentration analysis in functionally graded plates with elliptic holes under biaxial loadings, A in Shams Engineering Journal, 2014 5, 839–850.
7. Xiang Yu, Analysis on Stress Concentration of orthotropic Strip with an Elliptical Hole, Building technology at low temperature, 2014, 5, 68–70.
8. Yang Lin Hu, Zhu Han, Liu Chun Sheng, Two-Dimensional Analysis of the Influence of Soft Material on Stress Field Around an Elliptic Hole, Journal of Tianjin University, 2010, 43(7), 601–605.
9. Peterson, R. S., Concentration factors, Wiley, 1974.
10. Dongmei Luo, Wen-Xue Wang, Yoshihiro Takao, Effects of the distribution and geometry of carbon nanotubes on the macroscopic stiffness and microscopic stresses of nano-composites, Composites Science and Technology 67, 2007, 2947–2958.

Research on the evaluation and detection method of multi-process transposition fixture locating precision*

Chen-Long Wang†, Bei-Zhi Li and Jian-Guo Yang
Mechanical Engineering Department, Dong Hua University,
ShangHai, SongJiang, China
†*E-mail: wangclzfj@163.com*
http://www.dhu.edu.cn/

Multi-process transposition fixture consists of the chuck, tray and positioning locking nail. The tray is used to install the machining workpiece and is formed as one body with the workpiece. In the corresponding process, the chuck is fixed on the machine worktable. The positioning and clamping mechanism are used to rapidly and accurately position for tray/workpiece on the chuck of machine tool in accordance to a certain process. The fixture is especially suitable for the precise installation of key parts in the micro inertial navigation system. In order to measure and control the positioning accuracy of the multi-process transposition fixture for the dual balanced flexible joint, this paper presents a method based on image processing technology, and with the aid of various tools (measuring block and machine vision calibration board), the positioning accuracy is measured.

Keywords: Multi-Process; Transposition Fixture; Positioning Accuracy; Image Processing.

1. Guidelines

The flexible joint is the key component of the dynamically tuned gyroscope, and its working performance directly influences the precision and the life of the whole gyroscope[1]. Typically, on the one hand, it is required that flexible joints should behigh-precision; and on the other hand, because of the insufficient of process, mounting and detection, the qualified rate of the monolithic dual balanced flexible joint is about 10%.But it is very important for the manufacturing precision of thinribsto determine the space position of inertial navigation system. As shown in Fig. 1, the thin rib of the integral double balanced flexible joint is formed by the interval of two tiny holes indirectly[2]. Four pairs of round hole formed the thin ribs are evenly distributed with 90° ±3′ according to the circumference. The symmetry tolerance of a pair of thin rib

---

* This work is supported byChina 863 Program (No. 2012AA041309).

hole relative to the reference axis of the flexible joint is 0.005mm. At present, in actual producing high precision flexible joints, multi-process transposition fixture which has flexible positioning, rapid multi-process transposition and other advantages are often used[3]. In order to reduce the influence of machine tools, fixture accuracy and tool wear on the machining accuracy, besides using position detection method and timely inspecting precision of flexible joint process (or location)[4], it is more important to clear the evaluation index of the positioning accuracy of multistep inversion fixture, and determine the testing method and precision about the corresponding evaluation index.

2. Evaluation of Multi-step Fixtures and Positioning Accuracy

2.1. *Multistep transposition fixture*

Multi-stepinversion fixture mainly consists of chuck, workpiece tray and flexible positioning locked nail. The chuck has four accurate distributed location teeth. The four positioning groove on tray matches up with the chuck positioning tooth while at work. In order to eliminate elastic deformation force, the configuration of the flexible positioning locked nail, and the edge of the tray positioning groove is equipped with elastic plate and deformation slot for the precise positioning, which can realize the precise and rigid positioning of the tray on the chuck[5]. According to the needs, the tray in the chuck can be positioned at any 90° precisely. To detection and access to relevant test data, the tray is positioned on the chuck four station, which named A (0°), B (90°), C (180 °), D (270°) respectively.

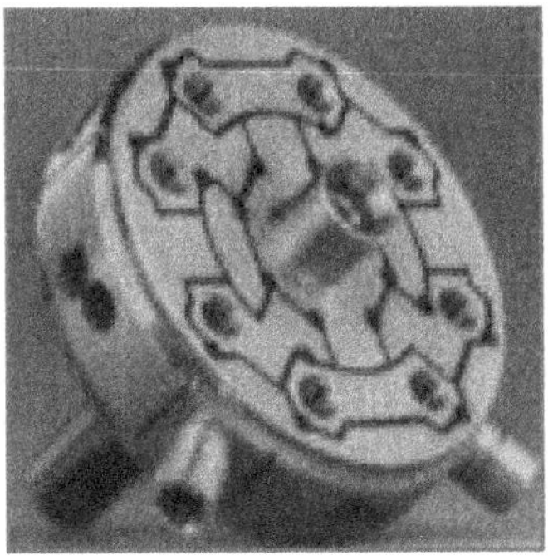

Fig. 1. Double-balanced monolithic flexible joints.

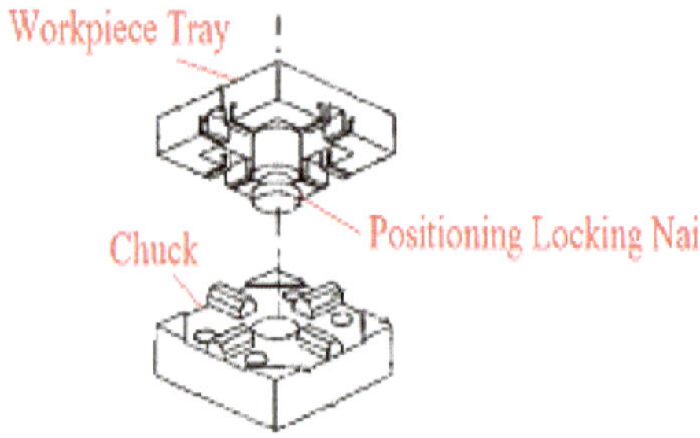

Fig. 2. Multi-step active homingbit fixture system diagram

Fig. 3. Corrective gauge

2.2. *Evaluation of fixture positioning accuracy*

According to the use of multi-process fixture and its influence on the accuracy of the flexible joint, the positioning accuracy is divided into two categories: the repeatable accuracy of the positioning and the positioning accuracy of any position. Between them, repeatable accuracy is actual center position inconsistency when the tray bearing surface repeatedly positions on a range of the chuck surface. Any positioning accuracy is the inconsistency of the actual position when any azimuth plane locating on the chuck at any range including angle and centering accuracy. In order to detect the accuracy of the index, the corrective gauge (shown as in Fig. 3) and the machine vision calibration board are adopted (shown as in Fig. 4a).One side of calibrated block is installed with multi-process active homing clamp tray, which has the same positioning teeth and clasped locking nail. Though grinding machine at the other end, the machine vision calibration board is fixed above it. Using machine vision equipment, the center position change of the square mark image is observed, and then the edge contour image of square mark is obtained through the image processing technology as shown in Figure 4b., finally, the corresponding calculation model is constructed to determine the positioning accuracy index of inversion fixture.

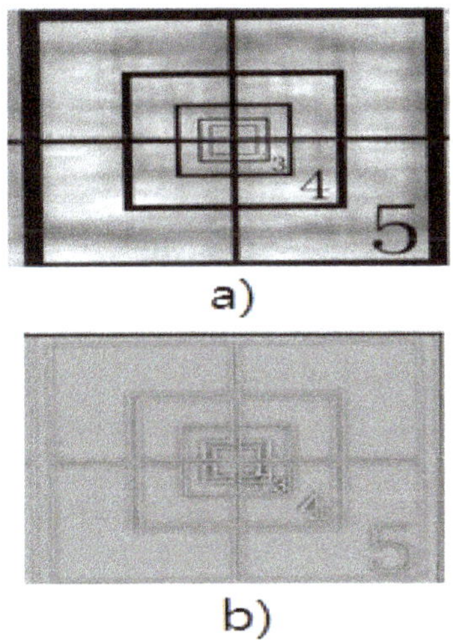

Fig. 4. Machine vision calibration plate and edge contour image

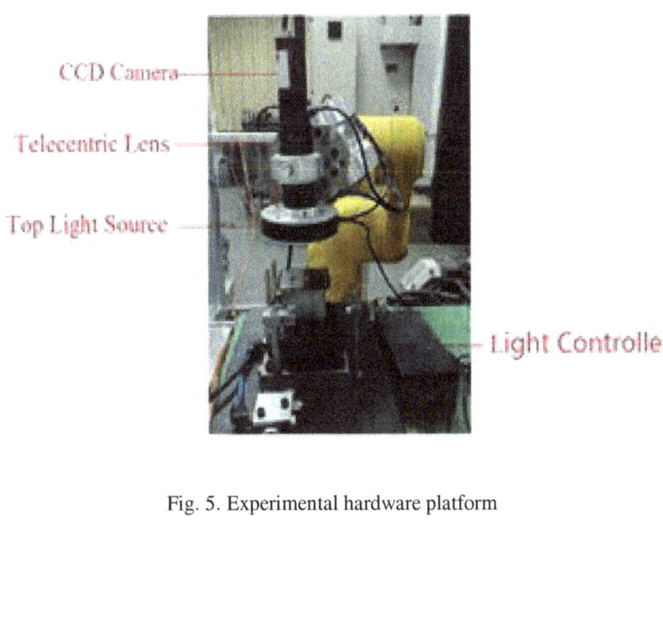

Fig. 5. Experimental hardware platform

3. Test Equipment and Instruments

The hardware part of the test includes: Navitar2X lens, F201B AVT 5 million pixel camera, LED light source (as shown in Fig. 6). The magnification factor of lens is 2 times, the theoretical pixel size[6] is 1.75μm, and the actual pixel size is 1.698248μm. The measurement accuracy of the test equipment is ±0.2μm and the measurement accuracy is 0.2μm.

4. The Detection Methods of Multi-process Inversion Fixture Positioning Accuracy

4.1. *Detection methods and test results of repeatable positioning accuracy*

- Step 1: According to the definition of repeatable positioning accuracy, the bearing surface of the tray is positioned repeatedly in a range of the chuck surface. Namely, in a corresponding position, the fixture is operated with loosing, pulling out, putting back and clamping. Then the calibration plate image is shot after clamping every time. The above operation 4 times (4 range surface) is repeated, each station is repeatedly manipulated and detected withN1 times and saving N1 pictures ($N1 \geq 5$).

- Step 2: Taking advantage of the image processing algorithm, the coordinate value of the same reference point in the n pictures are obtained.

- Step 3: After calculating the distance between any two corners of a certain position, the maximum range of the location is the repeatable accuracy.

- Step 4: the above operations are carried out on the remaining three chuck station.

The tray in the remaining three chuck station will be carried out such operations, the coordinate value of the reference square corner values can be obtained.

The relevant data are given in Table 1, which means the data of a tray in the chuck at four range plane positioning with repeated positioning.

Table 1 The data associated with the tray any repeated positioning on the clamping chuck (unit: pixel)

Measurement Times	Position A		Position B		Position C		Position D	
	X	Y	X	Y	X	Y	X	Y
1	9.14	2.5	0.98	2.91	7.47	5.26	3.01	9.19
2	9.13	2.59	0.98	2.93	7.51	5.20	2.98	9.14
3	9.10	2.59	0.98	2.94	7.53	5.22	3.03	9.11
4	9.12	2.48	0.98	2.88	7.5	5.21	2.98	9.15
5	9.12	2.49	0.97	2.9	7.47	5.16	3.01	9.16
Repeat Positioning Accuracy	0.111		0.06		0.11		0.094	
	0.188μm		0.102μm		0.178μm		0.14μm	

According to the multi-process transposition fixture definition and the data in Table 1, repeatable positioning accuracy of the test fixture is less than or equal to 0.188μm.

4.2. *Detection angle accuracy and test results*

- Step 1: In each position, the calibration board image is taken as the target, and the fixture is rotated at 90 degrees (clockwise). During imaging acquisition, the fixture is continuous rotated N_2 cycles ($4N_2$), then $4*N_2$ images are collected ($N_2 \geq 5$).

- Step 2: the image processing algorithm can be used to obtain the sub pixels of a square image[7]. One edge of the square is selected, and a linear which used as a reference line is fitted with the least squares algorithm.

- Step 3: the angle value of the reference line in the adjacent two images is calculated as the actual corner of the fixture. Compared with theoretical angle (90°), the corner error can be achieved. The error measurement results are shown in table 2.

Table 2. The measurement of the angle error of multi position and its measurement results (unit: seconds)

Cycle	Position A	Position B	Position C	Position D
1	10.17	-8.14	-22.77	18.99
2	-4.32	11.16	-20.88	20.16
3	20.52	-4.32	3.6	14.4
4	12.24	16.92	-22.9	15.56
5	10.89	15.12	-21.61	24.5
Angular Error	20.52	16.92	-22.9	24.5

According to the multi-process transposition fixture definition and angle accuracy of the data in Table 2, the measuring precision of the fixture is ± 24.5'.

4.3. *Fixture centering accuracy of detection methods and test results*

• Step 1: Fixed calibration board and data acquisition method are the same as angle accuracy detection.

• Step 2: The developed image processing algorithm is used to obtain the collection of mark squares of sub pixel. The center square in the calibration board is calculated as the reference point for the centering accuracy of the fixture. The coordinate value of the center point of the square in the four stations is shown in Table 3.

Table 3. Center point coordinates of each position square (unit: pixel)

Cycle	Position A		Position B		Position C		Position D	
	X	Y	X	Y	X	Y	X	Y
1	939.474	1564.7	661.853	1211.77	1016.67	933.24	1294.89	1288.04
2	940.079	1564.02	662.06	1212.38	1016.72	933.712	1293.93	1288.08
3	939.129	1564.08	661.607	1211.57	1015.69	933.249	1294.01	1287.94
4	940.2	1563.93	661.712	1212.7	1015.72	933.862	1293.44	1288.55
5	940.028	1564.38	661.367	1214.15	1016.26	934.125	1293.59	1288.59

To fit a cycle of four station square center, the center is the center of rotation of the clamp and the coordinate values are shown in table 4.

Table 4. Center coordinates of the circle (unit: pixel)

Cycle	Fitting Circle Center Coordinates	
	X	Y
1	978.0785	1249.021
2	978.1614	1248.839
3	977.933	1248.729
4	977.7855	1248.878
5	977.7377	1249.206

With 5 groups circle centers, the minimum circumscribed circle can be calculated. The center accuracy of the fixture is the minimum circumscribed circle diameter. According to the calculation, the center coordinate values are 977.923, 1248.993 and the radius R is 0.28μm, so the center accuracy is 0.56μm.

5. Conclusion

According to the requirements of high precision machining of flexible joint, a position measurement method based on image processing technology for the high precision multi-station inverse fixture is proposed. The method takes the square contour of fixture as reference object. Based on analyzing and calculating the position relations of different location, repeat positioning error and transposition positioning error of multi-step fixtures can be obtained. The center accuracy of the fixture is 2 ~ 3μm, and the angle error is 1 ~ 2'. Comparing with Sweden EROW fixture, it have obvious advantages to meet the requirements of high precision processing of flexible joint.

Acknowledgments

The authors acknowledge the support from China National863 High Technology Research and Development Program (No. 2012AA041309).

References

1. Chunjie Liu. Dynamically tuned gyroscope one-piece flexible abutment [D]. tongji university, 2006. (In Chinese).

2. Dean Wan, Pan Liu, Dongji Sun. Monolithic flexible joint stiffness measurement system design [J]. Metrology & Measurement Technology, 2006, 03: 15–16. (In Chinese).
3. Zhenzhen Fu. Flexible joint neck measurement and image processing methods [D]. Donghua University, 2015. (In Chinese).
4. Jun Lun Xun. Flexible positioning repositioning connection rigid clamping device: China, 200810039898.8 [P], 2008-12-10. (In Chinese).
5. Yue Long Xun. Flexible fixture repeat positioning device: China, 00217155.4 [P], 2001-03-28. (In Chinese).
6. Ruixue Xia. Image in the key techniques of measurement [D]. HeFei University of Technology, 2012. (In Chinese).
7. Meijing Zhang. Subpixel edge detection technology research [D]. Shenyang LigongUniversity, 2013. (In Chinese).

Research on static loaded test bed for metro vehicle bogie

Li-Na Ren[†], Ning Lu
School of Mechanical-electronic and Automobile Engineering,
Beijing University of Civil Engineering and Architecture
Beijing, China
[†]*E-mail: renlinalina@163.com*

As a fast and punctual mode of travel, the subway has made definite contributions to the easing of traffic pressure. Because the bogie is an important part of metro vehicles, its structure's rationality and the components' performance will directly affect the safe operation and dynamic performance of the vehicle. This paper describes the static loaded test bed for metro vehicle bogie, which includes the composition, performance, functions and features of the test bed, and it will provide some reference for the determination of metro vehicle bogie parameters.

Keywords: Bogie, Static load; Test bed; Hydraulic systems.

1. Introduction

With the rapid development of urbanization, traffic congestion has become an important factor to restrict the development of the city. As a fast and punctual travel mode, the subway has made a certain contribution to ease the traffic pressure. With the rapid development of the subway, the safety, comfort and other issues raised higher requirements, so that the subway maintenance work has higher requirements. Because the bogie is an important part of rail vehicle, bogie structure rationality and the component performance will directly affect the safe operation and the dynamic performance of the vehicle. This paper describes electro-hydraulic servo double actuators metro vehicle bogie test bed. The device is developed based on a variety of bogies at home and abroad, which fully considers the ease, compatibility, reliability and scalability of the equipment.

2. Composition of the Test Bed

The test bed consists of the following components: fixed transom door frame, bogie centering device, weight measurement units, wheelbase measurement units, hydraulic systems, electrical control systems.

2.1. *The mechanical part of the test bed*

2.1.1. *Fixed transom door frame*

Fixed transom door frame is a structural steel frame. It is made of welded steel box-type structure to ensure sufficient strength and rigidity. It meets that the plastic deformation does not occur under the condition of 600KN static pressure and ensures that the detection accuracy is not affected under the condition of the small amount of elastic deformation. It is mainly composed of the main beam and column components.

2.1.2. *Bogie centering device*

Bogie centering device is set in one side of the standard track. The role of the device is to ensure that the lateral center of the bogie coincides with the center of the static pressure device, so as to ensure the accuracy of test data.

The device is mounted on the rail base and disposed inside of both rails, the adjustment ranging of not less than 600mm. The device uses a hydraulic drive guiding mechanism, which has a positive and negative thread screw, screw nut connecting a driving wheel mechanism, dial wheel device composed of a guide rail and the light bar, to ensure that the dial mechanism to open and close at the same time along the screw axis. Bogie wheels are driven by dial pin, adjust the position of the bogie, and ultimately a two-axis center plane of the bogie coincides with loading plane.

2.2. *Measurement units*

2.2.1. *Wheelbase measurement units*

The device has 2 wheelbase measurement units, and each measurement unit is composed of a movable slider and measuring V plate. The movable slider is mounted on the rail next to the light bar through the guide mechanism, which can be moved in the horizontal direction along the light bars (X-direction), achieving the measurement process by a hydraulic motor moving; V plate is installed on the movable slider by rail, which can be moved in the vertical direction (Y direction) with respect to the mobile slider. The measuring process is that two ramps in V plate and rim fit, to ensure the central plane of the wheel of the V-block coincides with the center of gravity direction along the plane of symmetry.

2.2.2. *Weight measurement units*

Weight measurement units are core part of the test bed, and the units are the introduction of China's railway technical authorities newly developed system weighing units, which can entirely monitor the wheel weight during the bogie is loaded, weighing data bus transferring to the computer. In order to measure the weight value of four wheels when the bogie is loaded, four weighing units are mounted in each of the stations, whose shape is the same as ordinary rail. Weighing units are installed on the rail base, weighing in the range of 550mm to adapt to different bogie. The test bed is through-type test bed, fully meeting the requirements of the weighing unit for continuous operation.

2.3. *Control systems*

2.3.1. *Hydraulic systems*

Hydraulic station is the hydraulic power source of the entire system. It is installed in the vicinity of the gantry, the tubing along the column and beam receiving cylinder. The hydraulic pressure source is provided with a cooling cycle apparatus. Hydraulic station is in accordance with the relevant provisions in order to meet the requirements of the servo system for cleaning, installation, to ensure the normal operation of the various parts of the hydraulic systems.

The device is installed two ram cylinders, cylinder stroke 400mm, and single cylinder is greater than the maximum output pressure 200KN. Import displacement sensors are in the cylinder, and import force sensor is connected to the piston rod end. Pressing cylinder uses the method of electro-hydraulic servo intelligent control to achieve load control and displacement control, which can simulate vehicle coordination loaded in the case of partial load. The schematic diagram of hydraulic system is shown in Fig.1.

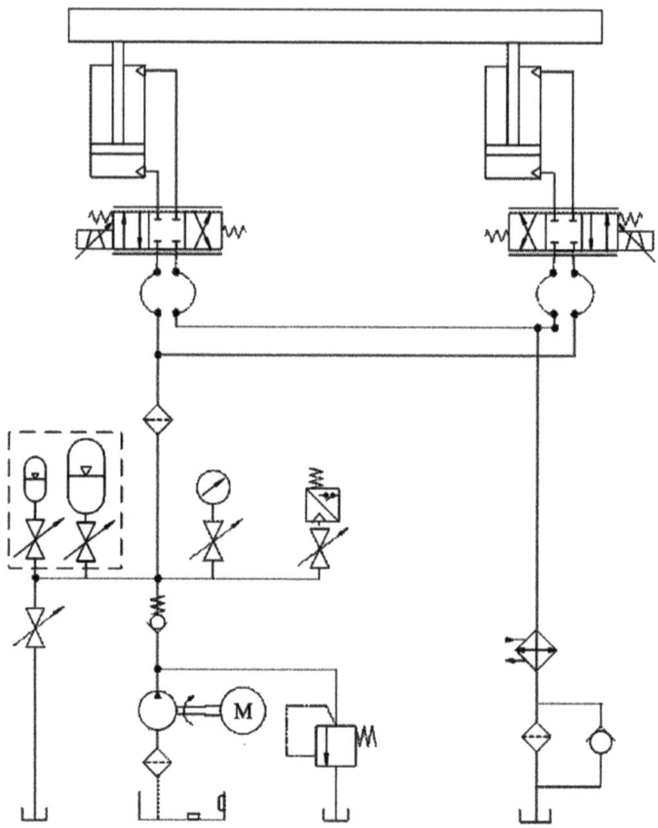

Fig. 1. Schematic diagram of hydraulic system

2.3.2. *Electrical control systems*

The control of the test bed is composed of a remote computer cabinets and site electrical control cabinet, including the hardware of the remote computer cabinet consists of: industrial control computers, monitors, UPS uninterruptible power supply, A4 printer and the corresponding buttons and lights; by means of configuration software operation cabinet to complete the test bed data storage, graphics, data reporting, query and printing. Site electrical control cabinet consists of a DC power supply, PLC modules, data acquisition modules scene, analog data output modules and other components, which mainly achieve the function of site control actuator, sensor signal detection and processing, fault information collection and more. Operation platform and control cabinet have a reliable ground, zero connection protection, terminal blocks using high-quality insulating material. Control systems have the functions of overload, overcurrent, overvoltage, overheating, under voltage, short circuit, lack of equal protection. Operation platform and control cabinets are equipped with an emergency stop button.

3. The Test Bed Functions

3.1. *Positioning and alignment*

The device has automatic positioning and alignment function for bogies to ensure that the bogie lateral centerline deviation with respect to the transverse centerline of the test equipment does not exceed 0.5mm, the relative deviation of the longitudinal centerline of the bogie longitudinal centerline of the device not more than 0.5mm. Alignment of the longitudinal, lateral deviation value is displayed in real-time device control system.

The automatic positioning alignment function of bogie on the device can be programmed to automatically adjust in place and can also be adjusted manually operated by the NC program in place. After bogie horizontal alignment, wheel in the transverse direction (along the axle parallel to the direction) in addition to the wheels — track or wheel — the wheel contact force, there shall be no other compulsory positioning.

Automatic measuring and real-time displaying size bogie wheelbase; automatic measuring and calculating the diagonal dimension; automatic measuring the distance which is bogie two lines on a plane to the rail surface; measuring bogie four angles; the specified location on the measurement from the rail bogie flat surface height measurement distance.

3.2. *Load detection*

System, which has double loading cylinder structures in the vertical direction, can simulate any load (empty load, full vehicle load) for loading test of bogie. Instant loading force can be manually entered in the computer program settings, or preset in a predetermined vehicle loader, automatically load operation, load testing process automatic real-time display in the computer. The final results are automatically stored and can be printed out.

Two loaded cylinders can achieve individually load, synchronous load, other load loading, empty load - full vehicle load cyclic loading, etc., in which a cylinder can be achieved center loaded. Cylinder loading speed can achieve step-less adjustment by a computer program.

System can detect the wheel weight and single wheel to weight difference in a simulated vehicle or any heavy load conditions, and can automatically calculate and display the total mass of the bogie. Work process is displayed instantly in the control system, and the measurement results are automatically stored and can be printed out.

4. The Test Bed Features

The test bed can be used for both metro bogie hydrostatic test and other railway bogie static load test, various simulated load testing, real-time measurement or control a variety of data, while calculation and analysis of a series of spring-padded size, axle, wheel weight difference, parallel shaft and other parameters on the basis of measured data with the assembly and testing on other relevant dimensions.

The device is the introduction of international advanced modular design concept, which uses distributed bus technology and digital multi-channel closed loop computer control policy coordination, so that the device not only has high control accuracy but also has great equipment openness and scalability. It can increase or fewer functions of the device according to user requirements. Device software development using virtual instrument technology, a modular configuration method, the device is equipped with excellent interactive features, simple interface, easy to operate.

The device control system adopts intelligent fuzzy strategy, adequately resolve the hydraulic servo system nonlinearity, hysteresis and other defects, and further improve the control precision and stability of the device.

5. Conclusions

The test bed can be used to simulate bogie no-load or full load for loading on the bogie, and can measure accurately on the bogie parameters, simple, functional, offers some reference for the determination of bogies parameters.

References

1. S C Fok, Z Wang, P D Dransfield, Digital simulation study of an adaptive controller for a pneumatic cylinder system, J. International Journal of Computer Applications in Technology. vol. 3, pp.163–171, 1995.
2. Sun XF, Fan YZ, Zhang FZ, Fault detection using an adaptive fuzzy system, J. Int Systems Science. vol. 7, pp.599–609, 2002.

3. C. Angeli, An online expert system for fault diagnosis in hydraulic systems, J. Expert System. vol. 2, pp. 115–119, 1999.
4. D.C. Karnopp, D.L. Margolis, R.C. Rosenberg, System dynamics: A unified approach, Second Ed., Wiley, New York, 1990.
5. M Lizell, Dynamic leveling a low power active suspension with adaptative control, J. Vehicle System Dynamics. Vol. 3, 1991.

Proportional pressure reducing valve design and manufacturing testing research

Kun Xie*, Ziliang An
School of Railway Transportation, Shanghai Institute of Technology
Shanghai, 201418, PRChina
**E-mail: Mrxk2000@163.com*

Qi Su
State Key Laboratory of Fluid Power and Control, Zhejiang University
Hangzhou, 310027, PRChina

Proportional pressure reducing valves have a very wide range of uses, especially in the field of engineering machinery. The valve's performance will directly affect the machine's reliability. In this paper, a series of contrast tests are conducted between home-made proportional pressure reducing valves and the Kawasaki KDRDE5K c50-31/30-102. Their performance is analyzed to find the deficiency of the designed products and corresponding measures for improvement. The improved product's usage effects are comparable to Kawasaki products.

Keywords: Proportional pressure reducing valve; Hydraulic test bench; PWM.

1. The Introduction

Proportional pressure reducing valve uses a very wide range, especially in the field of engineering machinery is a large number of applications, many domestic medium-sized excavators will use Kawasaki (model KDRDE5K c50-31/30-102), and take the form of six or eight league, used to control the walking around in the hydraulic circuit is symmetrical, movable arm, priority, etc., the stand or fall of the valve performance will directly affect the reliability of the machine. Authors refer to hydraulic related standards, this paper designed a model for drivers to use proportional pressure reducing valve, the valve is a series of compared with Kawasaki product testing, to find the deficiency of the homemade products and the improvement measures, the actual use effect can be comparable with Kawasaki products.

2. Test System Introduction

This valve test bench testing mainly includes the hydraulic test loop and the test control system of two parts. Test control system is mainly used for electric control system and the signal acquisition, mainly including the system of power supply, sensors, control of hydraulic system, etc. According to the subjects of experiment and testing purposes, using a two-way cartridge valve simulation load, by adjusting the different cut-off valve on-off switch to different test loop, by collecting different oil pressure, flow value, can measure the proportional valve of the steady-state flow characteristic, pressure gain characteristic, load flow, internal leakage and flow pressure drop characteristic such as static performance index, the circuit principle diagram as shown in the figure below. 35 mpa pump source maximum pressure, the maximum flow 180 l/min. Flow meter range 300 l/min, with digital display, the accuracy of 0.3%; Pressure sensor range 0 ~ 20 mpa, 0.25% accuracy.

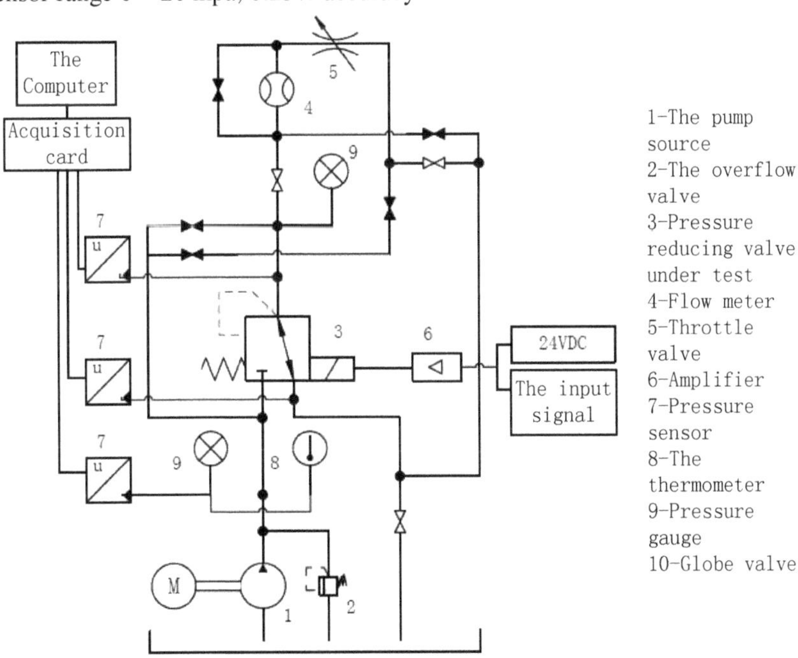

Fig. 1. Pressure reducing valve test system schematic diagram

The experimental test control system mainly includes the regulated power supply, signal generator, Zhejiang University homemade digital proportional controller and its configuration tools and software; Test circuits including pressure sensor, a signal conversion module, data acquisition card and the Labview software, etc.

3. The Static Test Results and Analysis

Successively in the process of testing the withstand voltage test, leak test, input signal — the second under constant flow pressure characteristic test, threshold test, flow — port, a pressure under constant input signal characteristic test nine test content, etc. Withstand voltage test, the leakage test, homemade valve and Kawasaki product performance.

3.1. *Under constant flow rate, the input signal – secondary pressure characteristic test*

Under constant flow rate, the input signal - secondary pressure characteristic test is the main performance of the pressure reducing valve, characterization of secondary vacuum pressure and the input signal, the linear relationship between the corresponding indexes such as linearity, dead zone and hysteresis, therefore the experiment steps are as follows: to keep the oil temperature 30 + 2°C, the input signal is 0 when adjusting subjects valve inlet pressure to the rated pressure of 3 mpa. To load all the adjustment of the input signal is from 0 to 100 % (7V), and decreased from 100% to zero, the regulating signal need to pay attention to keep strictly increasing decline, under each of the input signal to adjust load makes the flow rate is respectively 0% and 50% of the rated flow (2.5 L/min) and 100% (5 L/min), regulating the relationship between flow without guarantee strictly increasing decline under the small input signal, if

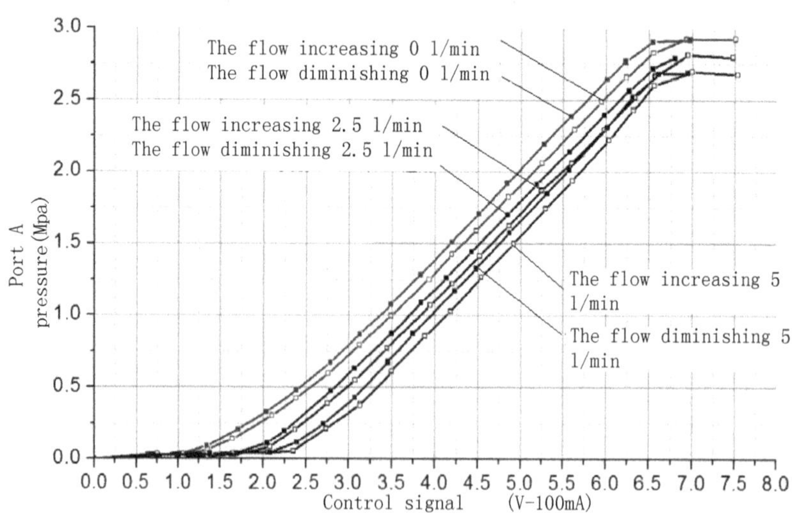

Fig. 2. Kawasaki proportional pressure reducing valve performance curve

flow cannot meet the requirements of the flow, maintain full load. After setting the flow record the signal and the discharge port A under pressure, and different flow rate from the secondary pressure curve under the input signal.

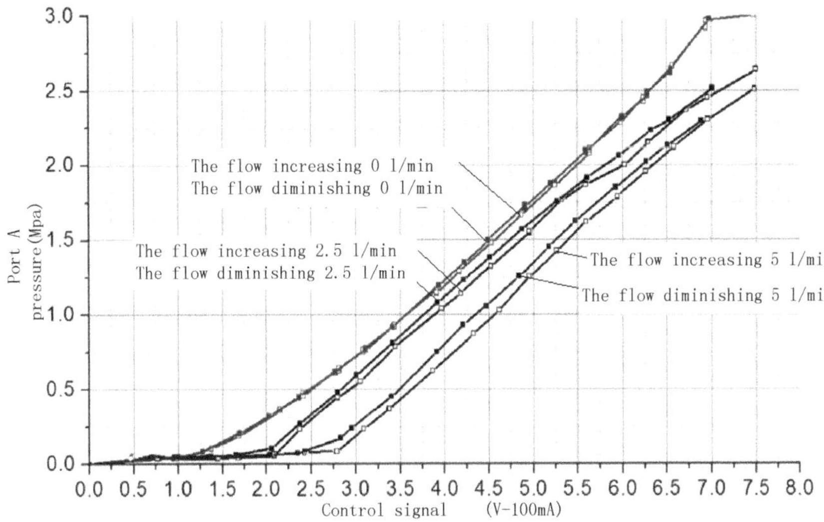

Fig. 3. Homemade proportional pressure reducing valve performance curve

From the point of test curve, flow is small, homemade pressure reducing valve close to the indicators Kawasaki, and even beyond. But with the increased flow relief valve pressure rating of the homemade, linear range, dead zones are less than the proportional valve of the Kawasaki, curve slope is less than the pressure reducing valve of Kawasaki. Analysis of the reason is electromagnet thrust is not enough, small current cannot overcome steady-state fluid power, lead to big dead zone. Hysteresis index, the deflection is homemade proportional valve friction is slightly big.

In addition, found in the testing process, homemade proportional valve in the closed load no flow, there will be a port pressure volatility, as shown in Figure 4, but as long as there is a little bit of traffic load to open, the oscillation phenomenon will disappear. In addition, in the subsequent trials found that change in Kawasaki, electromagnet, won't appear this problem, load closed port A pressure shock phenomenon disappears, as shown in Figure 5, in here and at the back of the homemade pressure reducing valve test, when need to shut down the load experiment, the following measures should be taken to prevent oscillation: will just open a little traffic load as load closed approximate treatment.

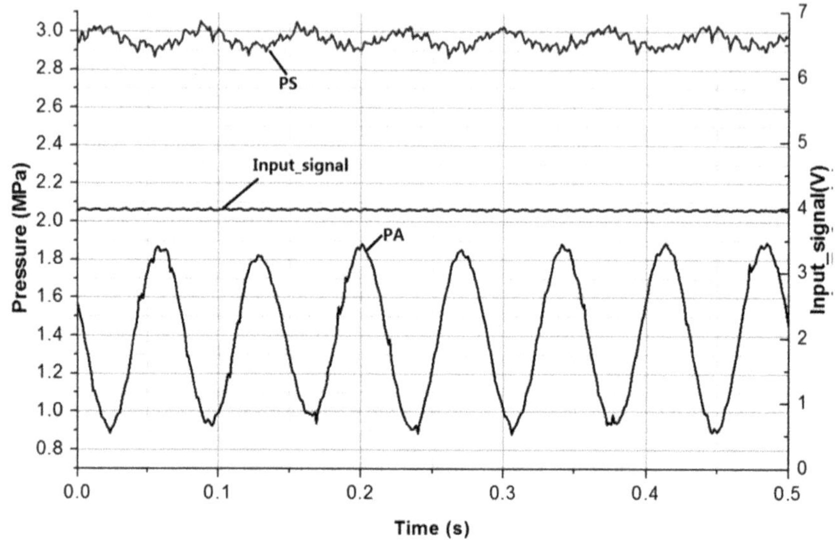

Fig. 4. Homemade valve closed load

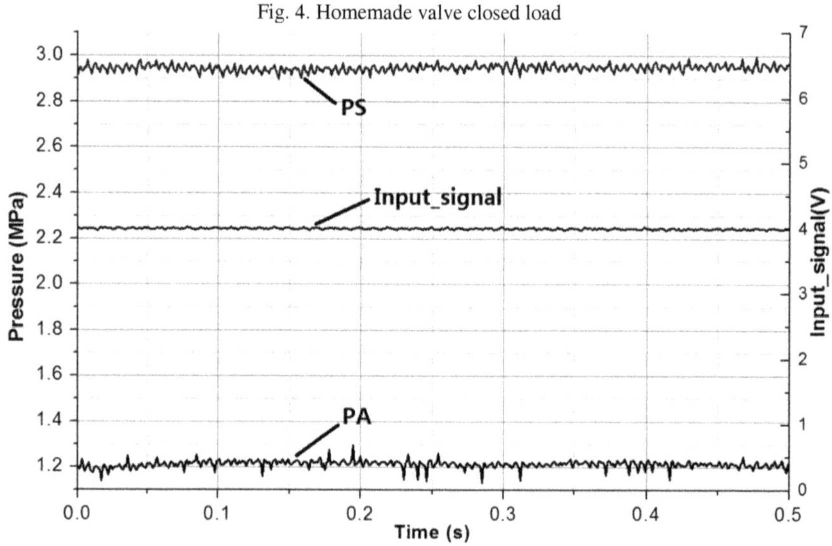

Fig. 5. Homemade valve + Kawasaki electromagnet

3.2. *The threshold test*

Threshold test is to determine subjects when reverse slope signal input valve response characteristics of proportional valve in particular near the working point of response sensitivity. Test steps are: to keep the oil temperature 30 + 2 °C, the input signal is 0 when adjusting subjects valve inlet pressure to the rated pressure of 3 mpa. Adjust the input signal is respectively the rated pressure test valve outlet pressure average of 25%, 50%, 75%, 100% (0.75, 1.5, 2.25, 3 mpa), regulating the flow of the loading valve at the same time, the 50% of the rated flow is (2.5 L/min). 0.1 Hz, superimposed on the current input signal amplitude of the triangle wave signal, the amplitude increasing since the childhood, that when the control pressure is more than 5% of the rated pressure, superposition signal amplitude threshold.

4. Dynamic Performance Test

4.1. *Step response*

Seen from the test results, Kawasaki, the stability of proportional valve pressure value is bigger than homemade valve. Single step response index, under the step signals of relatively large amplitude, both difference is not big, adjust the time around 0.4 s, overshoot Kawasaki slightly larger. Under the middle step signal, the stable value of Kawasaki a lot larger than homemade valve, Kawasaki overshoot amount is larger, response time, almost at about 0.3 s. Under the small step signal, the homemade valve appeared larger negative overshoot, some adjustment time is relatively long.

4.2. *Frequency response*

Will determine the participants valve electrical signal input and control of the pressure between the frequency response. To keep the oil temperature 30 + 2 °C, the input signal is 0 when regulating subjects valve inlet pressure to the rated pressure of 3 mpa. Regulating oil source flow is not less than 50% of the rated flow, close the loading valve, adjusting the pump export overflow valve make participants valve inlet pressure is equal to the rated pressure. Applying dc offset to make the export control of pressure is 50% of the rated pressure. Based on the dc offset stack a sine signal, under the condition of the steady state, adjust the sine signal amplitude, the control pressure amplitude to export pressure rated plus or minus 5%. Adjust frequency measuring range, the input signal and the pressure on the export control of phase lag, its range at the lowest frequency is less than 10°, the highest frequency is greater than 90°. At the same frequency range, the control pressure amplitude decrease at least 10 db, the sine signal amplitude unchanged, gradually increase the frequency, record the outlet pressure fluctuation amplitude and phase difference. Adjusting the range value of the sine input signal at the lowest frequency makes the control pressure amplitude for the exportation of plus or minus 25% of the rated pressure, repeat the above steps.

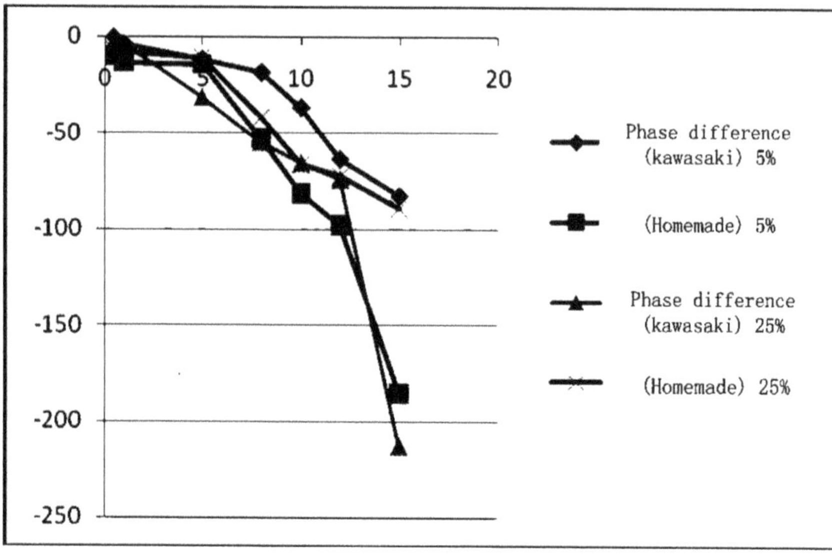

Fig. 6. Phase frequency characteristics

As can be seen from the test results, the model of proportional valve is similar to the typical model of the second order oscillation link - 3 db frequency response at about 10 Hz. The corresponding phase difference is around - 90°. Can be thought of as the proportional valve used frequency response for the below 10 Hz. Homemade proportional valve frequency response relative to the Kawasaki rate was only slightly lower, when performance - 90°, whether 5% or 25% frequency response value, are slightly below the Kawasaki proportional valve. On the amplitude frequency characteristics, input under 25%, the proportion of Kawasaki valve at around 12 Hz, shock, amplitude ratio greater than 1, it can be seen that the proportion of homemade valve damping ratio is less than Kawasaki proportional valve.

5. Improvement Measures

In order to make homemade valve and Kawasaki, pressure reducing valve has the same control features, try to ensure that both the structure of the same size, therefore increase the coil number of turns N is used to enhance the electromagnetic force. After repeated experiments and deduction, finalized the coil number of turns from before 1200 turns to 1500 turns. The improved self-control valve under the threshold test, A constant input signal flow - port A pressure characteristic test, test results and Kawasaki quite completely.

6. Conclusion

Through experimental analysis the author homemade pressure reducing valve and Kawasaki, the static and dynamic performance of the samples, according to the comparison between test data and curve is the same and differences, and find out the root cause of the lack of homemade pressure reducing valve is the electromagnetic force is not enough, then by increasing the number of ampere turns to achieve the purpose of increase the electromagnetic force, so as to improve the static and dynamic performance of the self-made pressure reducing valve, in order to achieve the independent design of domestic proportional pressure reducing valve to provide the basis. After a follow-up for more than a year of small batch live-fire test, fully meet the requirements of drivers to use homemade pressure reducing valve performance.

References

1. Lu Hu Big Lee. *Electro-hydraulic proportional control technology* (Beijing, China, 1987).
2. YuKai Yuan. *Throttling type electro-hydraulic pressure control device performance optimization research,*(Zhejiang University, Hangzhou, 1988).
3. Huang Yong. *The design of new type electro-hydraulic proportional valve and its control method research* (Changsha, Hunan university, 2007).
4. Xiangrong Zeng. *Steady fluid power compensation form single-stage overflow valve and its experimental study* (Journal of machine tools and hydraulic, China, 1984).
5. Genmao Wu, *Research on the static and dynamic per-formance of 2 way proportional throttle valve with dis - placement force feedback* (Hangzhou, China, 1985).
6. Guosheng Wang, Yun Ling, Qinghua Fan. *The design of electro-hydraulic proportional valve and the experiment study* (Journal of fluid mechanics experiment and measurement, 2004).

Study on air-fuel ratio control of electronically-controlled engines

Jian-Min Sun, Ling-Hua Gao* and Feng-Feng Yin

School of Mechanical-electronic and Automobile Engineering,
Beijing University of Civil Engineering and Architecture
Beijing, 100044, China
**E-mail: gaolinghua68@163.com*
www.bucea.edu

Accurate control of the air-fuel ratio is the most pivotal technology in modern automotive gasoline engine control. Most electronically-controlled gasoline engines run under conditions of partial load. Researching the air-fuel ratio control system of electronically-controlled engines under partial load condition is thus of great significance for reducing emissions and improving fuel economy. In this paper, after analyzing the target air-fuel ratio control, a mathematical model of the engine is first established. Then, air-fuel ratio PID control system is designed under the partial load condition, and the simulation model of the engine air-fuel ratio control is established. Finally, the control effect of the air-fuel ratio was compared with the conventional control system which didn't utilize the control strategy. Simulation results show that under partial load conditions, when using the PID algorithm, the control effect is better, control accuracy is higher, and the control performance is optimal.

Keywords: Electronically-controlled engine; Air-fuel ratio; Partial load condition; PID control.

1. Introduction

Engine control system determines the vehicle performance and emission levels directly, so the research of engine control system has been widespread concern and has made a lot of research results, and the air-fuel ratio control is one of the most basic problems of the engine control. Air-fuel ratio is the primary factor of affecting the power performance, fuel economy and emissions of gasoline engine, and the accurate control of air-fuel ratio is the key technology to the gasoline engine control of modern vehicle. The operating condition of the engine mainly include starting operation, partial load operation, full load,

* LinghuaGao is the corresponding author (e-mail: gaolinghua68@163.com).

acceleration deceleration and idle condition, and partial load operation condition is the most common conditions in car driving[1-2].

This paper mainly studies the air fuel ratio control of the engine under partial load condition. The air fuel ratio control is realized by the integrated control of the air inlet and the fuel injection quantity. In the paper, the mathematical model of the engine is established firstly, then PID control system of air-fuel ratio is designed, and the simulation model is established by mat lab/Simulink, the comparison and analysis of the air-fuel control effect of three control strategies is done; and the PID control algorithm is used to research and analyze semi-physical simulation test of the engine throttle control system.

2. Analysis of Target Air Fuel Ratio Control under Partial Load Conditions

Using electronic fuel injection technology in automobile engines, not only promote the mixed gas in the cylinder combustion, but also improve the engine's fuel utilization rate, significantly enhanced the engine power performance, effectively improve the engine's emission of gas. To a large extent, the EFI regulates the engine speed, which can meet the requirements of fast and accurate and stable [2]. After the engine speed is stable, the mixed gas in the cylinder is fully burned. Air-fuel ratio control is through the cylinder intake air quantity and the fuel injection quantity of comprehensive control, ECU according to detected the intake air quantity signal for controlling the execution unit injector fuel injection quantity, the mixed gas in the cylinder make full combustion.

Under different operating conditions, the engine's air fuel ratio is different. When the engine performance is optimal, the air fuel ratio needs to be close to or equal to the ideal air fuel ratio of 14.7, but the actual situation is not [3-6]. Engine is a complicated and changeable control system. The ratio between the inlet and the fuel injection rate of the cylinder is different in different working conditions. Curve of engine heat engine efficiency and air-fuel ratio is shown in Figure 1.

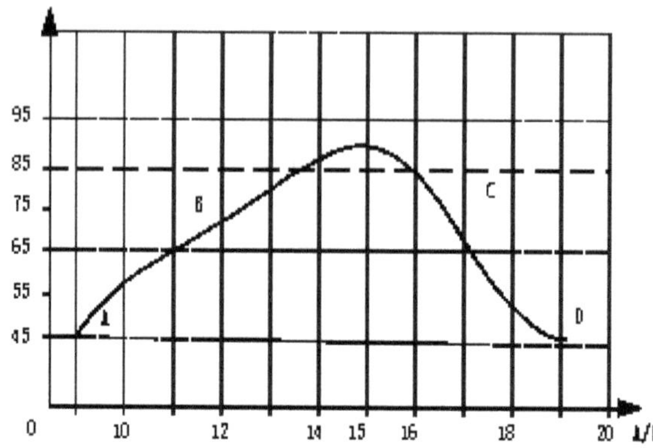

Fig. 1. Curve of engine heat engine efficiency and air-fuel ratio

From Figure 1, it is not difficult to see that the air fuel ratio in 14~16, the engine's conversion efficiency is the highest. The ratio of AB to 9~12 was CD, the conversion efficiency was lower than 75%; the air fuel ratio of BC was 12~17, the conversion efficiency was higher, the air fuel ratio was 17~19, the conversion efficiency was lower than 75% at 65% to 90%. BC said under part load conditions of heat engine efficiency and air–fuel ratio and the conditions on the output torque requirements is not high, the main consideration of the fuel economy. In this paper, the main research part load condition of air–fuel ratio, partial load condition is most engine operating condition, to meet emission standards and reduce fuel consumption is the primary goal, the air-fuel ratio selected 14.7, converter will be harmful to the exhaust gas conversion for maximum efficiency of harmless gases.

To achieve the goal of air ratio control depends on the amount of air intake and fuel injection quantity coordination and control. When the fuel injection is carried out, three parameters must be determined: the intake quantity and the fuel injection quantity, and the injection time [7-8]. The air flow sensor is used to detect the signal of air intake, and the air fuel ratio is calculated by the other sensors. The ratio of air to fuel ratio is calculated. The fuel injection rate and the fuel injection rate are calculated by using the air fuel ratio.

3. Mathematical Model's Establishment of Electronic Control Engine

Car is a very complicated and nonlinear system, to analyze its performance, we need to build a variety of models to simulate, but because the vehicle system itself is too complex, the coupling between the subsystems and subsystems has increased the difficulty of the analysis, and the simulation system has a large amount of calculation, so the establishment of a model to simplify[9-11]. Therefore, in this paper, the automotive systems as a throttle opening degree θ as input, to the speed of vehicle centroid speed u for the output of the system, focusing on the under part load conditions of air fuel ratio control strategy research. According to the derivation of the equations of automobile driving that engine input throttle θ and output rotational speed n of the relationship, to establish the mathematical model of the engine.

In normal driving, the car's driving force is Ft, the total resistance of the car is mainly including rolling resistance F_f, air resistance F_w (vehicle speed when driving), the slope of F_i (car uphill driving) and the acceleration resistance F_j (car acceleration when driving), that is

$$\sum F = F_f + F_w + F_i + F_j \quad \therefore \sum F = F_t . \tag{A.1}$$

$$F_t = \frac{T_{tq} i_g i_0 \eta_T}{r} . \tag{A.2}$$

T_{tq}-the torque, i_0–the transmission ratio of the main reducer, i_g-the transmission ratio of the variator, η_T-the mechanical efficiency of the transmission system, r-the radius of the wheel.

$$F_f = Gf \cos\theta \ . \tag{A.3}$$

f-the rolling resistance coefficient, G - the weight of the object.

$$F_w = \frac{C_D A u^2}{21.15} . \tag{A.4}$$

$C_D A$- the air resistance parameter, u- the speed

$$F_i \ = G \sin\theta. \tag{A.5}$$

$$\sin\theta \ = \frac{\mathrm{h}}{\mathrm{s}},$$

h-the slope height, s- the base of slope long, G – the weight of the object.

$$F_j = \delta m \frac{du}{dt}. \tag{A.6}$$

δ-the conversion factor ($\delta = 1.0 + \delta_1 + \delta_2 i_g^{\ 2}$, $\delta_1 \approx \delta_2 = 0.3 \sim 0.5$), m- the vehicle mass, du/dt- the driving acceleration.

According to the above analysis, the model of the vehicle traveling equation is obtained:

$$\frac{T_{tq} i_g i_0 \eta_T}{r} = Gf\cos\theta + \frac{C_D A u^2}{21.15} + G\sin\theta + \delta m \frac{du}{dt}. \tag{A.7}$$

Taking into account the road to the car, the general slope is not great, almost close to zero, cosθ≈1,sinθ≈tanθ≈i (i-transmission ratio, $i=i_g i_0$), $u=0.377nr/i_g i_0$, So the (A.1) is rewritten as (A.8):

$$\frac{F(n,\theta) i_g i_0 \eta_T}{r} = Gf + \frac{(0.377)^2 C_D A r^2}{21.15(i_g i_0)^2} + Gi + \frac{0.377\delta m r}{i_g i_0}\frac{dn}{dt}. \tag{A.8}$$

This paper takes a light passenger car as an example, the main parameters are shown in Table 1[9-12].

Table 1. The main parameters

Parameter	Symbols/unit	Parameter values
Vehicle quality	m/ kg	4025
Wheel rolling radius	r/ m	0.361
Rolling resistance coefficient	f	0.0128
Air drag parameter	$C_D A$	2.07
Transmission ratio	i_g	6.194,3.894,2.260,1.428,1.00
Main reduction ratio	i_0	4.444
	Pemax/(Kw(r/min))	66/1800
	Temax/(N·m(r/min))	220/1600

Based on Table 1, the formula (8) through Laplace transform is conversed to the transfer function of the engine throttle opening and the rotate speed:

$$H(s)=\frac{N(s)}{\theta(s)}=\frac{10.1s^3+4.9s^2+10.7s+0.9}{5.1s^4+10.1s^3+6.2s^2+3.8s+0.6}. \tag{A.9}$$

4. Air-fuel Ratio Control and Simulation of the Engine Electronic Control System under the Condition of Partial Load

The engine air fuel ratio closed loop control can effectively maintain the air fuel ratio in 14.7/1 or so, at this time it maintain high conversion efficiency, play a role in reducing emissions. In this paper, we select the ideal air fuel ratio of 14.7 for closed-loop control, the PID control algorithm, the study of electronic controlled gasoline engine in partial load condition of the target air ratio control.

The control strategy of the electronic controlled gasoline engine in partial load condition is PID control algorithm. Under this condition, the fuel efficiency of the engine is considered as the main control object, and the structure of the PID control structure is shown in Figure 2. The basic control idea is as follows: the deviation value of the target air to fuel ratio and the actual air fuel ratio is used as the output. In order to maintain the constant, the PID control needs to adjust the intake quantity.

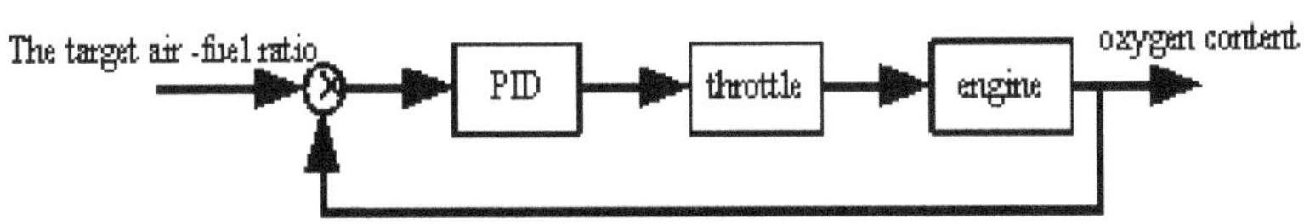

Fig. 2. PID control structure diagram of air-fuel ratio

In the partial load condition, the research and design of the PID controller are designed by PID control algorithm. Among them, the key of the whole system is to adjust the PID, KD, KP three parameters to the optimal, which need to conduct a thorough analysis of the whole system, so as to determine the optimal value of the three parameters of the PID controller [12-14]. In the actual simulation system, according to the empirical value, after repeated testing, the three parameters of PID control are: kp=17.377, ki=2.875, kd=0.031. The PID control model of the air fuel ratio under the partial load condition is established, as shown in Figure 3.

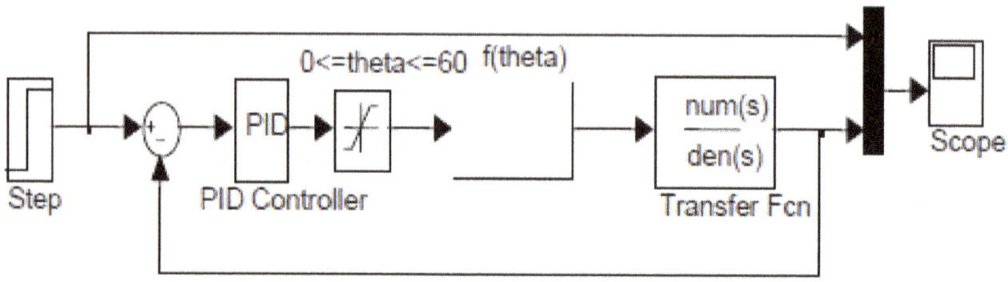

Fig. 3. PID control model of air-fuel ratio

In Figure 3, under part load conditions, establishing the air-fuel ratio PID control simulation system with Simulink, initial throttle open setting is 8.92 degree, because range of θ is between 0 and 90 degree, so the output of the PID controller be limited amplitude protection. In this condition the upper limit set for the throttle opening degree of about 2/3, in the simulation system, theta is 60 degrees were applied to the conventional control (control) and PID control simulation system of step signal, as the input signal of the model, to observe the response curve as shown in Figure 4.

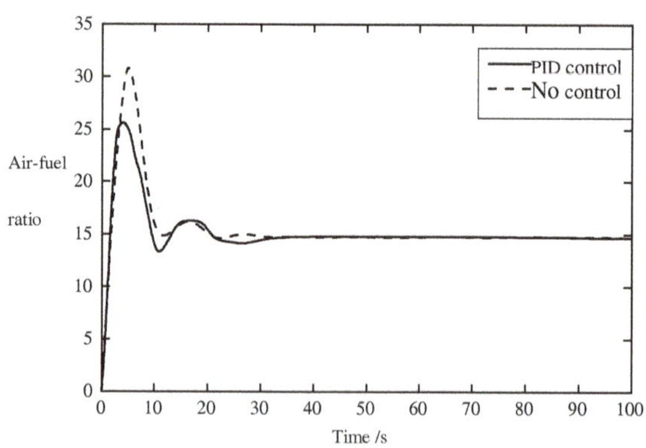

Fig. 4. Response curve of the air-fuel ratio of PID control system

From Figure 4 it can be seen that the dynamic response of the engine under the condition of partial load is not very ideal. The maximum overshoot of the conventional system has exceeded 30, the maximum overshoot is over 100%, and the adjustment time is longer. PID control can adjust the integral parameters to reduce the overshoot, so that the adjustment time is shortened, compared to the two, the dynamic performance of PID control, whether it is a super tune or adjust the time, have been improved.

5. Conclusion

This paper mainly studied air-fuel ratio control of the engine under the condition of partial load. The mathematical model of engine as a controlled object, simulation model of air-fuel ratio control is established and simulation analysis is done. Based on the throttle control experiment system of the electronic control engine, the study about throttle PID control of electronic control engine is marched.

(1) The air-fuel ratio accurate control is the key link to improve fuel economy and performance and reduce emissions of engine, and the partial load condition is one most conditions that electronically controlled gasoline engines running, so the research of air-fuel ratio control system of electronically controlled engine under partial load condition is of great significance to reduce emissions and to improve fuel economy.

(2) According to the car moving equation, the engine system is simplified, the engine mathematical model that throttles percentage as input and speed as output is established, and the transfer function of engine throttle percentage and speed is reached.

(3) The engine under partial load conditions, the dynamic response effect of air-fuel ratio PID control is not very ideal. But compared with the conventional system that did not adopt PID control, PID control reduced the overshoot by adjusting the parameters of the integral; shorten the adjusting time and accordingly improved dynamic performance.

Acknowledgments

The authors wish to express sincere gratitude to the reviewers for their invaluable suggestions and comments.

References

1. Hong Chen. Automotive Control: the State of the Art and Perspective. ACTA AUTOMATICA SINICA.2013, 39(4): 322–346.
2. Hou Z X, Wu Y H. The research on air fuel ratio predictive model of gasoline engine during transient condition. IEEE International Conference on Mechatronics and Automation, Harbin, 2007.
3. Hashimoto E, Ishiguro T, Yasui Y, Akazaki S. High reliability electronic throttle system design. In: Proceedings of the 2003 SAE World Congress. Detroit, USA: SAE, 2003.2003-O1–0708.
4. Wang, T., Lin, X, Feng X, et al. Research in arg, injection Revision control of gasoline engine based on the electronic Throttle during transient condition [C] 2, International Conference on Computer, Mechatronics, The Control and Electronic Engineering. Changchun, China, 2010: 285–287.
5. Liming Zhou. Research and Design on Speed Control of Gasoline Engine. 2011.
6. Yujia Zhai, Ding-Li Yu. Neural network model-based automotive engine air/fuel ratio control and robustness evaluation. In: Engineering Applications of Artificial Intelligence 22 (2009) 171–180.
7. Z.X. Hou, Y.H. Wu. Predictive control for air fuel ratio based on adaptive expand particle swarm optimization[C] Proceedings of the 7th World Congress on Intelligent Control and Automation June 25–27, 2008, Chongqing, China.
8. Hou Z X, Wu Y H. Multi-step predictive model of air fuel ration for gasoline engine based on neural network[C] International Technology and Innovation Conference, 2007.
9. Fan Yuan. The Design and Emulation of PID Control System for the Automobile Engine. Tech Information Development & Economy. 2007 (17).12.203–205.
10. Hua Zou. Analysis and research on control strategy of electronic-controlled gasoline engine. Wuhan University of Technology. 2005.

11. Takao Kenji, Yamamoto Toru, Hinamoto Takao. A Design of Model Driven Cascade PID Controllers Using a Neural Net-work Network. IEICE Transactions on Fundamentals of Electronics, Communications and Computer Sciences, 2004(9): 2322–2330.
12. Cesare Alippi. A Neural Network Based Control Solution to Air Fuel Ratio for Automotive Fuel Injection System[J]. IEEE Transactions on System Man and Cybernetics Part C, 2003, 33 (2): 106–113.

CHAPTER 3

Electronics Engineering and Microelectronics

A novel array structure for the meter-wave radar

Xiao-Qin Hu[†], Liang-Fa Hua
Radar Technique Department of Huangpi NCO School of Wuhan Early Warning Academy, Wuhan, 430345, China
[†]E-mail: huxiaoqin-0529@sohu.com, lfhua@sina.com

Fei Xu
Airforce Ordinance Equipments Military Representative Bureau of PLA, Beijing 100166, China

The elevation angle estimation for meter-wave radar is easily affected by the indirect signal of the low elevation signal. However, the indirect signal is coherent with the direct signal. A novel array structure was designed to improve the estimation performance for the meter-wave radar. The presented array consists of two different non-uniform arrays without ambiguity, and the algorithm was amended. The narrow beamwidth is obtained to differentiate the closed signals using the whole array. The coherent direct angle and indirect angle can be estimated exploiting the partial sensors. Simulation experiments' results show that compared with the uniform linear array, the new array structure greatly improves the estimation performance for meter wave radar.

Keywords: Meter-wave radar; Array structure; Non-uniform array; Ambiguity.

1. Introduction

Meter-wave radar whose wavelength is longer has many advantages including anti-stealth, anti-radiation missile (ARM) compared with the microwave radars. Meter-wave radar draw our attention again since 90s 20th. However, meter-wave radar has wide beamwidth, low angle precision, poor range resolution, especially poor height-finding precision because of the influence of the multipath reflection. In the meantime, these sources cannot be distinguished because of the small elevation difference when several signals lie in the same range unit. For meter-wave radar, it is difficult to estimate accurately the elevations of all sources without the influence of the coherent multipath reflection signal. This is a hard nut to crack for the application of three-dimensional meter-wave radar.

Many researches have been done to improve the accuracy of elevation or height for meter-wave radar. A composite multipath model [1] in the meter-wave radar height-finding applications considering the product factor, which is based on the function of the product factor and signal arrival angles, was proposed. Elevation angle and multipath fading coefficient joint estimation algorithm in meter-wave radar was studied [2]. This method retains antenna aperture and does not require a special antenna array configuration. Angles measurement of meter-wave radars by mainlobe multipath jamming suppression was presented [3]. The weight of the new approach was obtained through certain constraints on the sum-difference beams. An auto-calibration angle estimation algorithm for meter-wave radar in the presence of mutual coupling was studied [4]. Compared with many existing array calibration methods, the proposed algorithm retains antenna aperture and avoids iterative processing in the presence of coherent multipath signals and unknown mutual coupling matrix. The improved algorithm for meter-wave radars to target at low altitude was proposed [5]. The linear transform was applied to the received data of every subarray. The assumptive arrays of meter-wave radar are all uniform linear arrays in the above researches. Therefore, the improved estimation performance is limited.

A novel array structure is proposed for meter-wave radar to improve the estimation performance based on the characteristic of height-finding for meter-wave radar. The presented array is constituted of different non-uniform arrays without ambiguity. The narrow beamwidth is obtained to differentiate the closed signals using the whole array and the coherent direct angle and indirect angle can be estimated exploiting the partial sensors. To obtain better estimation performance, firstly these uncorrelated sources are estimated using the whole array, and then estimate the low angle signals based on filtering the information of uncorrelated sources. Simulation results show the effectiveness of the proposed array structure and the estimation performance is improved largely compared with the uniform linear array.

This paper is organized as follows. In Sec.2, the meter-wave radar model is discussed. Section 3 presents the novel array structure for meter-wave radar and design two new arrays. The performance evaluation conducted by computer simulations is shown is Sec.4. Finally, Section 5 concludes the paper.

2. Array Model of Meter-wave Radar

The meter-wave radar received signals have the reflection signals from the ground or sea surface, besides the direct signals. Based on the reflection surface roughness, the reflection signals can be divided into the specular reflection and diffuse reflection. The representative specular reflection model is used in this paper. To model the received signals more accurately, the curvature of the signal path due to the curvature of the earth itself must be taken into account. The multipath geometry for a curved earth and some parameters are given in Fig.2.

Target range R_d can be obtained from the time delay. In order to solve other parameters, firstly obtain the grazing angle ϕ according to the law of cosine.

$$\phi = \cos^{-1}\left(\frac{(h_t + R_e)^2 + (h_r + R_e)^2 - R_d^{\ 2}}{2(h_t + R_e)(h_r + R_e)}\right) \tag{1}$$

Where $\cos^{-1}$ denotes the arccosine, R_e and is the effective radius of the imaginary earth.

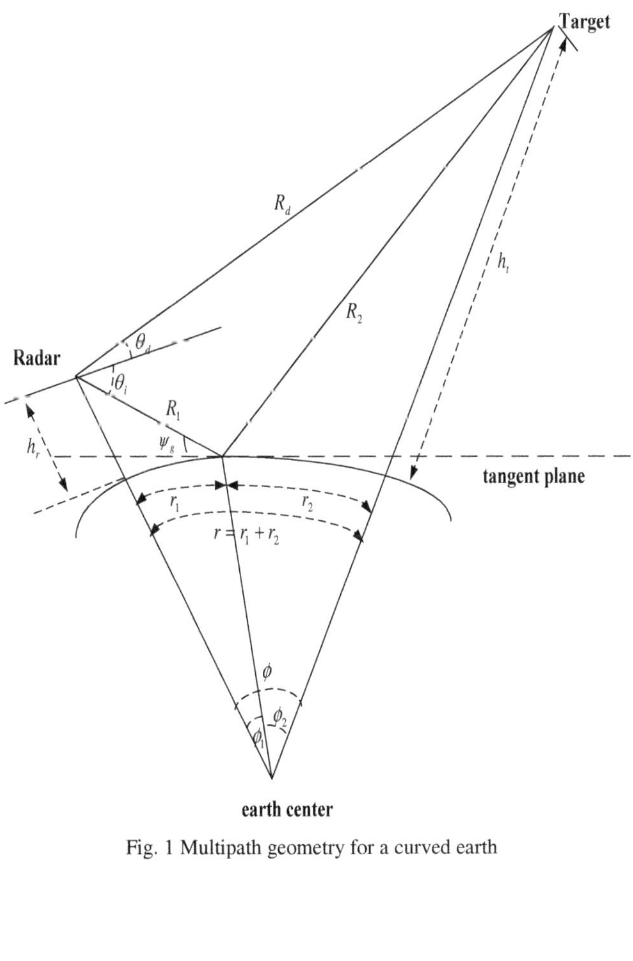

Fig. 1 Multipath geometry for a curved earth

The horizon distance is computed by $r = R_e\varphi$. Because the heights of signals and height of antenna are much smaller than the radius of earth, the distance from the radar to the point of reflection r_1 can be found by solving the following cubic equation:

$$2r_1^3 - 3rr_1^2 + \left(r^2 - 2R_e\left(h_r + h_t\right)\right)r_1 + 2R_e h_r r = 0 \tag{2}$$

Next, we solve for ϕ_1, ϕ_2 using r_1

$$\phi_1 = r_1 / R_e \tag{3}$$

$$\phi_2 = \left(r - r_1\right)/R_e \tag{4}$$

Using the law of cosine obtains

$$R_1 = \sqrt{R_e^2 + \left(R_e + h_r\right)^2 - 2R_e\left(R_e + h_r\right)\cos\phi_1} \tag{5}$$

$$R_2 = \sqrt{R_e^2 + \left(R_e + h_t\right)^2 - 2R_e\left(R_e + h_t\right)\cos\phi_2} \tag{6}$$

$$R_d = \sqrt{\left(h_t + R_e\right)^2 + \left(h_r + R_e\right)^2 - 2\left(h_t + R_e\right)\left(h_r + R_e\right)\cos\left(\phi_1 + \phi_2\right)} \tag{7}$$

The direct angle and indirect angle are expressed by

$$\theta_d = \sin^{-1}\left(\frac{h_t - h_r}{R_d} - \frac{R_d}{2R_e}\right) \tag{8}$$

$$\theta_i = -\sin^{-1}\left(\frac{h_r}{R_1} + \frac{R_1}{2R_e}\right) \tag{9}$$

The relationship of the elevation angle and height of signal is obtained using the above equations.

3. A Novel Array Structure

1.1. *Arrangement method*

The novel array is consisted of two non-uniform linear arrays without ambiguity. One non-uniform array has narrow beamwidth to differentiate the closed signals. The other non-uniform array can estimate coherent sources so as to estimate the direct angle and indirect angle. The two chosen non-uniform arrays have more same elements as far as possible and the number of all elements will decrease.

The non-uniform linear array without ambiguity that differentiates the closed signals is named as $Y1$, and the element position is $y_1 = 0.5\lambda[n_1, n_2, \cdots, n_{M1}]$, where λ is the wavelength and $M1$ is the number of sensors in array $Y1$. The non-uniform linear array without ambiguity that estimates coherent sources is denoted by $Y2$ and the element position is $y_2 = 0.5\lambda[m_1, m_2, \cdots, m_{M2}]$, where $M2$ is the number of sensors in array $Y2$. Subtracting the repeated sensors of array $Y1$ and array $Y2$, the element position of novel array Y is $y = 0.5\lambda[t_1, t_2, \cdots, t_M]$, where M is the number of the new array.

1.2. *Ambiguity analysis*

It is very significant to analyze the ambiguity for non-uniform array. According to the document, the array $Y1$ is unambiguous, then

$$\rho t1 = \{\rho_2 \cap \rho_3 \cap \cdots \cap \rho_{M1}\} = \{0\} \tag{10}$$

where, $\rho_i = \{0, \pm 2k_i / n_i\}$. $(i = 2, 3, \cdots, M)$ k_i is the positive integer corresponding to n_i and $k_i < n_i$.

Because $Y2$ is unambiguous and then

$$\rho t2 = \{g_2 \cap g_3 \cap \cdots \cap g_{M2}\} = \{0\} \tag{11}$$

where, $g_j = \{0, \pm 2l_j / m_j\}(j = 2, 3, \cdots, M2)$, l_j is the positive integer corresponding to m_j and $l_j < m_j$.

Because

$$\{\rho t1 \cap \rho t2\} \subseteq \rho t1 = 0 \tag{12}$$

We have $0 \subseteq \rho t2$, so $\rho = \{\rho t1 \cap \rho t2\} = 0$. It indicates that the novel array Y is unambiguous.

1.3. *The amendment of algorithm*

Conventional MUSIC algorithm [6] is used to estimate elevation angles of signals. The algorithm is amended to improve the estimation precision of low angle signal.

Assume the estimation elevation angles of independent sources are $\hat{\boldsymbol{\theta}} = \left[\hat{\theta}_1, \hat{\theta}_2, \cdots, \hat{\theta}_N\right]$ using the whole array, where N is the number of independent sources. The estimation precision is high because of the narrow beamwidth with the whole array. The angles of low elevation signals are estimated after filtering the information of estimated independent sources. The covariance matrix of whole data and the covariance matrix of the independent signals are denoted by $\boldsymbol{R}$ and $\boldsymbol{R}_1$ respectively. Then the covariance matrix of low elevation signals can be given by

$$\boldsymbol{R}_2 = \boldsymbol{R} - \boldsymbol{R}_1 \tag{13}$$

$\boldsymbol{R}_2$ only contains the information of low elevation signals. The direct angle and indirect angle can be estimated by eigen decomposition of $\boldsymbol{R}_2$ without the influence of independent sources.

1.4. *The design example*

Assume the non-uniform arrays $Y1$ and $Y2$ are obtained by the minimal redundancy array [7] and spacing combination array [8] respectively. The number of elements of each array is ten, the senor position $y_1 = 0.5\lambda[0,1,3,6,13,20,27,31,35,36]$ and $y_2 = 0.5\lambda[0,1,3,4,6,7,9,10,12,13]$. The novel array is named as array one and the sensor position is $y = 0.5\lambda[0,1,3,4,6,7,9,10,12,13,20,27,31,35,36]$. The number of elements is 15 filtering the repeating elements.

Array two is designed based on the above theory. Array $Y1$ in array two is same as that in array one. The sensor position of array $Y2$ is $y_2 = 0.5\lambda[0,3,5,8,10,13,15,18,20,23]$. The number of array two is 16 and sensor position is $y = 0.5\lambda[0,1,3,5,6,8,10,13,15,18,20,23,27,31,35,36]$.

Fig.2 gives the sensor position of array one and array two, the valid at x axis and y axis denotes the ratio of element position and half of wavelength.

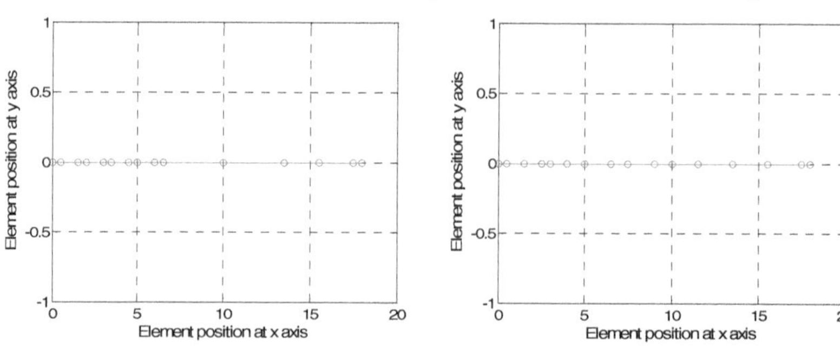

(a) Array one (b) Array two

Fig. 2 The sensor position of array one and array two

4. Simulation Results

The designed array one and array two are placed vertically. The radar frequency is 300MHz, and radar height is 8m. The target range is 130km. A model for curved earth is considered, and the effective earth radius is 4/3 actual earth radius. The number of snapshots is 64. There are three independent sources arriving at heights 4000m, 12500m and 14200m. The direct angles are 1.320°, 1.320°, 5.070°, 5.830° and the indirect angle is -1.330° corresponding to the low elevation angle 1.320°.

Compared to the uniform linear array with the same number of elements, the estimation performance of the presented arrays is given in fig.3 including probability of success and RMSE (root mean square root) versus SNR (signal noise ratio). One hundred independent experiments are done.

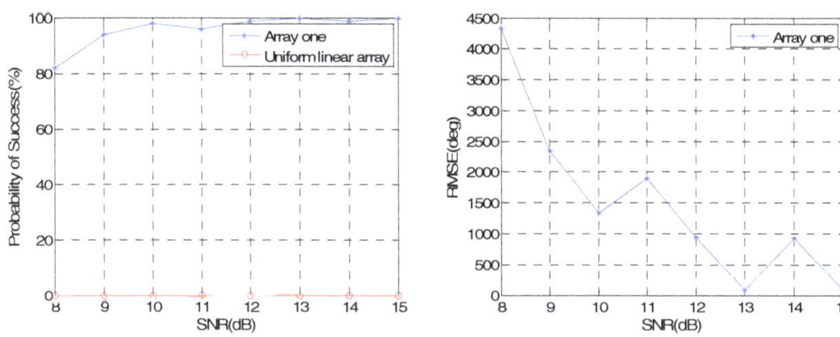

Probability of success RMSE

(a) Array one

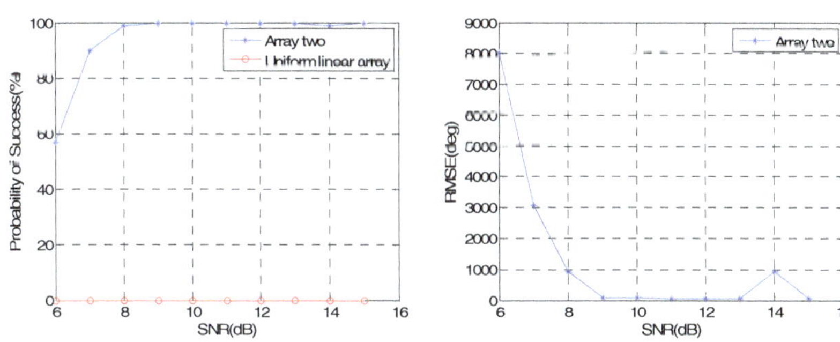

Probability of success RMSE

(b) Array two

Fig. 3 Estimation performance versus SNR

Two closed uncorrelated sources, direct signal and indirect signal of low elevation signal all can be estimated using the proposed array one and array two when the SNR is larger than 10dB. The RMSE is smaller than 1000m. The estimation performance of array two is superior compared with array one. The uniform array is invalid in these simulation parameters.

5. Conclusions

In this research, a novel array is presented to improve the estimation performance for meter-wave radar considering the influence of coherent indirect signal with low elevation angle. The new array is consisted of two non-uniform arrays without ambiguity. Based the designed array, the conventional algorithm is amended to improve the angle resolution of low elevation angle. Two novel setting arrays are given in this paper. In practice, according to the actual requirement more arrays can be obtained using this idea.

References

1. Xiaoqin Hu., Jianwen Chen., & Yongliang Wang. (2008). Research on meter-wave radar height-finding multipath model. *Chinese Journal of Radio Science*, **23**, 651–657.
2. Jun Liu., Zheng Liu., Yunfo Liu. (2011). Elevation angle and multipath fading coefficient joint estimation algorithm in VHF radar. *Journal of Electronics& Information Technology*, **33**, 33–37.
3. Yingjun Zhao., Rongfeng Li., Yongliang Wang. (2013). Angles measurement of meter-wave radars by mainlobe multipath jamming suppression. *Journal of Huzhong University of Science and Technology Nature Science,* **41**, 51–55.
4. Jun Liu., Zheng Liu., Rong Xie. (2011). Auto-calibration angle estimation algorithm for VHF radar in the presence of mutual coupling. *Chinese Journal of Radio Scienc*e, **26**, 1046–1051.
5. Xiangdong Wu., Shouhong Zhang., Mei Dong. (2009). Improved algorithm for meter-wave radars to target at low altitude. *Journal of System Simulation,* **13**, 4056–4058.
6. Schmidt R O. (1986). Multiple emitter location and signal parameter estimation. *IEEE Trans. on AP,* **34**, 276–280.
7. Chen Hui, Wang Yongliang. (1999). Performance improvement in estimation direction-of-arrival by array geometry arrangement. *Acta Electronica Sinica,* **27**, 97–99.
8. Moffet A T. (1968). Minimum redundancy linear arrays. *IEEE Trans. on Antennas Propagat,* **16**, 172–175.

Research on the typical technology of electromagnetic forming

Zhan-Jun Liu[†]

Faculty of Aerospace Engineering, Shenyang Aerospace University, Shenyang, 110136, China

[†]E-mail: liuzhanjun5555@163.com

The electromagnetic forming process is not affected by the moving parts' mechanical inertia, allowing electromagnetic forming to achieve rapid processing. This makes it easy to mechanize and automate. Electromagnetic forming is often used for flat work pieces and tube formation. The electromagnetic forming puts medium pressure on the workpiece with a magnetic field, which involves no mechanical contact. This technology is used in the process of deforming the curve of tube billets. The main influencing factors of the process are also analyzed.

Keywords: Electromagnetic forming; Tube blank; Tube billet; Mechanization.

1. Introduction

In the electromagnetic forming process, most of the deformation is in pulse pressure disappears. Under the effect of pulse pressure, the metal is obtained great acceleration, and got a lot of kinetic energy, metal forming is mainly depended on the kinetic energy to achieve. The electromagnetic forming is with pure electromagnetic properties, its forming process is not affected by mechanical inertia of the moving parts, so the electromagnetic forming can achieve rapid processing, which can work hundreds of times per minute. And the intensity of magnetic pulse can be accurately controlled, which is easy to realize mechanization and automation. The magnetic field can penetrate the nonconductive material, so to have a non-metallic coating or in forming the part with container[1]. The electromagnetic forming is medium pressure on the workpiece with magnetic field, which is no mechanical contact. So the workpiece does not produce friction, which is no lubrication. The electromagnetic forming is a single-mode forming, which simplified the mould manufacturing, to increased the processing flexibility. The electromagnetic forming parts are with high precision, residual stress low, and the spring back is small. The process of materials are best used in below 0.15um, the high resistance materials need to be adopt the high frequency equipment, the driven piece of structure is used on the tooling. Electromagnetic forming is often used in flat workpieces and the forming of tube blank, the typical technology mainly includes is as follows.

2. The Deformation Analysis of Electromagnetic

Along with the deformation of tube, its deformation increases gradually, the central part of the deformation is the largest at the end of the central part. Circular tube is bulging, the tangential deformation uniformity is directly related to the pipe length. As shown in Fig.1, the circular tube billet of bulging tangential strain is distribution along the axial direction. Processing the stitch is length of minor, the strain distribution is not uniform. Processing stitch length is bigger, the strain distribution is more homogeneous. As shown in Fig.2, the free bulging of the strain circular tube is distribution of tangential strain, and radial strain approximately is equal in value, and the axial strain is compared with smaller. As shown in Fig.3, the high-speed photography is for shrinkage tube forming of tube cross section deformation process. In figure representation is the corresponding changes of magnetic flux density, according to the forming coil diameter, it is the tube billet initial radius, which is for pipe section change process. Shrink tubing is shrinked to about half of the initial radius, which can be considered axisymmetric deformation[2], the deformation speed is of 950 m/s. Deformation within the most is produced in the first wave of current, on the other hand, thin-walled tube is forming under low charging energy, which do not produce the change of wall thickness, deformation process produced tiny wrinkles. In order to prevent the tube reducing wrinkles, the axial is prestretching stress in circular tube, the results show that the additive not wrinkle of the blank, when the axial tensile stress limits the deformation increasing greatly. Without axial tensile stress, the deformation wrinkling is 8% or so, when the axial tensile stress is plused of the material yield stress, the maximum deformation of the blank can be up to 20%.

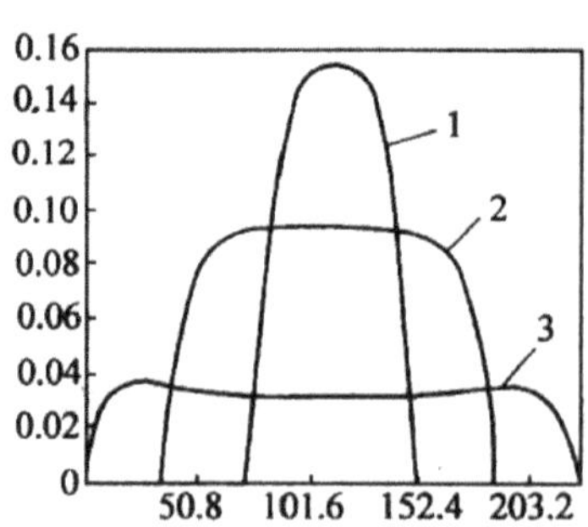

Fig. 1 Tube bulging tangential strain distribution

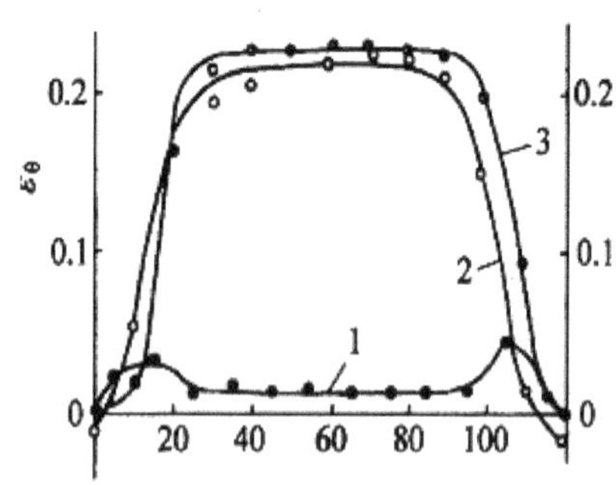

Fig. 2 Pipe free bulging strain distribution

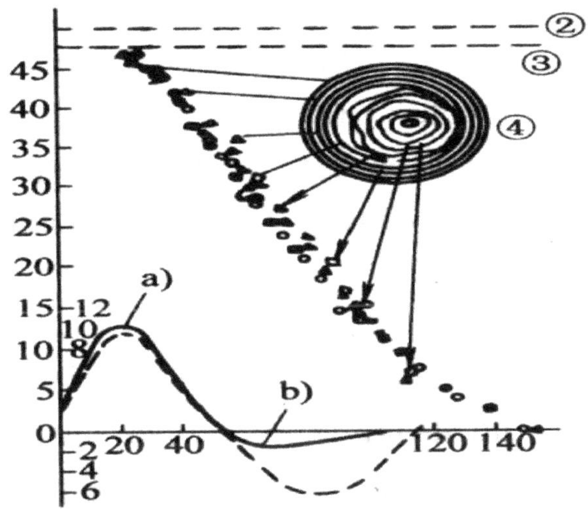

Fig. 3 Shrinkage tube cross-section shape change

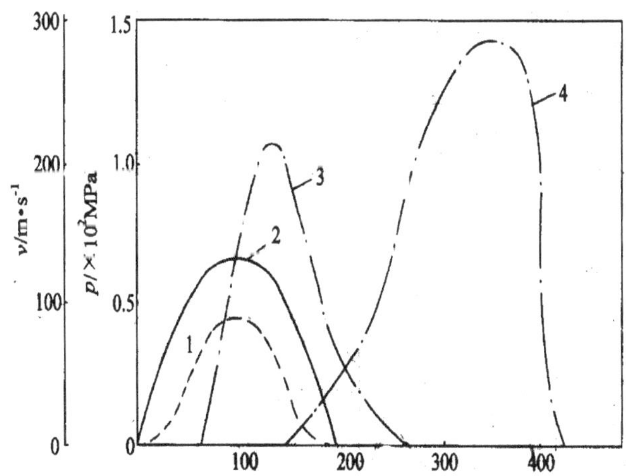

Fig. 4 Differences in the process of moving

2.1. *The flat blank deformation process*

As shown in Fig.4, flat blank deformation is in the process of the movement of the differences[3]. By the chart shows, the rate of movement of the differences between the initial time and motion is quite different, because the planar coil is the magnetic field force of distribution, the central part is the minimum of magnetic field force, magnetic field force of radius of the central is maximum, so the blank radius of central point is to move first, and the inner part of the blank is driven in the middle of radius of material, when deformation occurs, so the movement of the initial moments is later, the late deformation is the center part of the largest deformation. Flat blank deformation is as shown in Fig.5, logarithmic strain - forming is height distribution. Fig.6 and Fig.7 are different coil winding method for forming blank deformation process. Because there are two peak pressure, so Fig.7 is more uniform deformation[4]. Fig.8 is the four coil component forming of the final contour shape, the figure is shown, it closer to the cone, the forming height is the largest, forming is the highest efficiency, the forming efficiency is the most poor, failed to forming of conical.

2.2. *The tube billet is bulging*

Working coil is in the tube, when the pulse current is through the coil, a magnetic pressure pushs the tube outside billet mold. This method can be used for bulging and connection technology. Reducing is tube billet. Within the workpiece, the coil is inside the magnetic pressure to the workpiece of the compression modulus, which is often used for forming and connection process. The flat blank is forming. A spiral flat coil is placed on a flat blank, when the pulse current is through the coil, the electromagnetic puts parts into the mould. This method can be used for flat blank forming, finishing and cutting. Electromagnetic forming is very suitable for tube and pipe, tube and plate connection technology may realize the connection of tube and the core shaft[5].

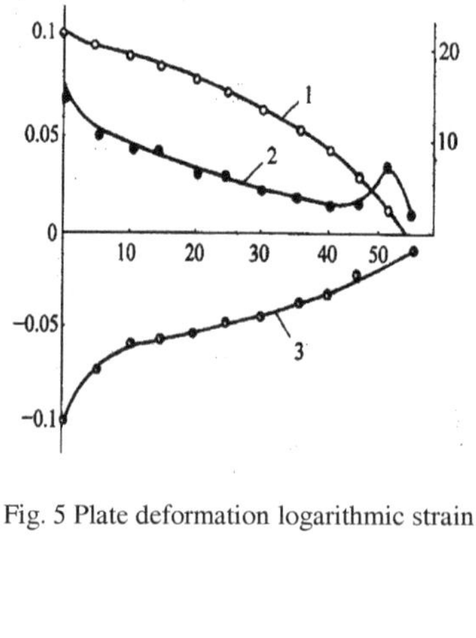

Fig. 5 Plate deformation logarithmic strain

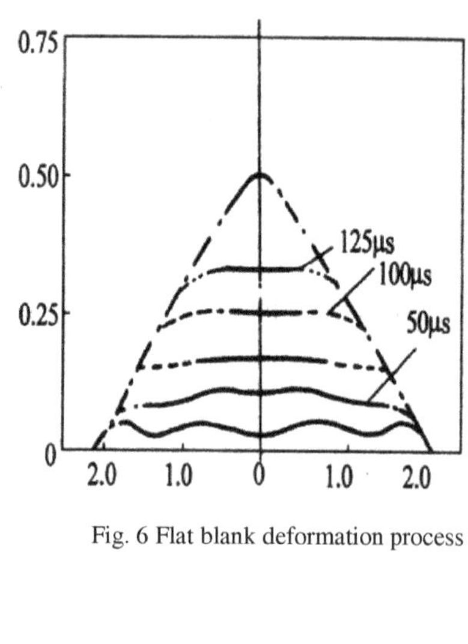

Fig. 6 Flat blank deformation process

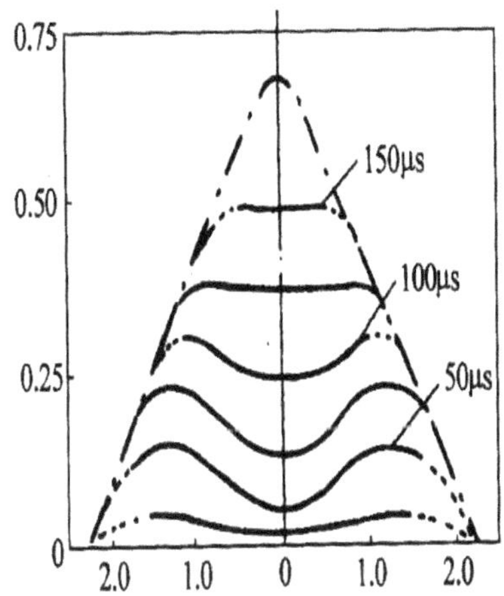

Fig. 7 Deformation process of outer diameter

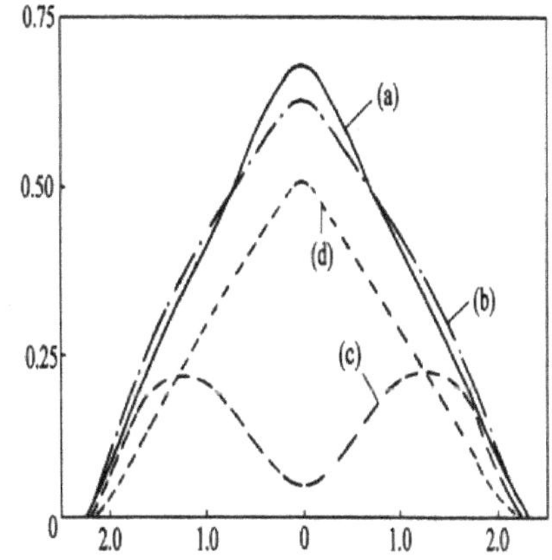

Fig. 8 Final outline of component forming

3. The Following Is the Main Influencing Factors of Analysis of Slapping

As shown in Fig. 9, the battery energy is relationship with pipe outline changes at any point. The pipe deformation increases with the increase of energy[6].

In the pipe clamping frame, deformation difference is very small, but the difference outside the pipe end is bigger, which uses steel protective casing, pipe is almost parallel extensions, the protect casing with copper, aluminum, or bronze, tube is flaring into cone, with nylon protect casing, pipe is expansion into bell mouth shape, and this trend becomes more obvious with the increase of energy, protect casing resistance is smaller, the degree of deformation of the workpiece flaring is smaller, and high resistance increased the radial magnetic flux density of workpiece, and which makes the bell shape even more significant.

As shown in Fig.10, when using copper and steel protective casing, pipe deformation is forming coil of taper, the relationship between the figure can be seen, when taper coil large, because of the small coil diameter decreases, and the gap increases between the coil and the workpiece, the magnetic pressure is less, so parts taper is larger. Also it can be seen from the picture as the previous analysis, protective material influences the shape of pipe end.

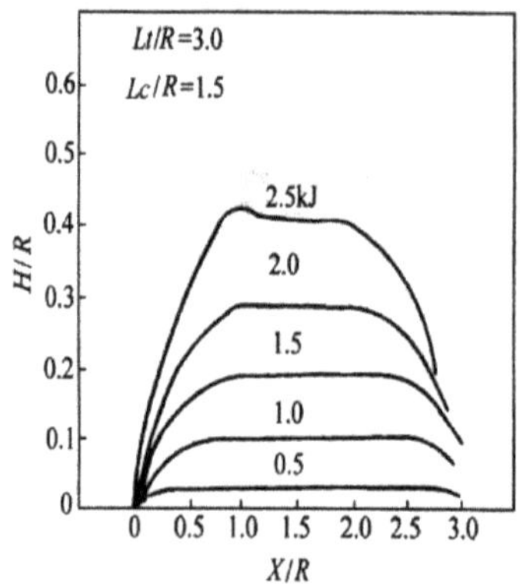

Fig. 9 Charging energy affect flaring contour

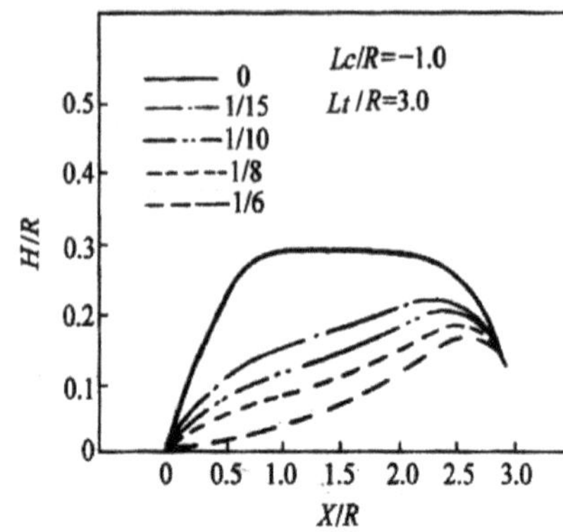

Fig. 10 Forming coil taper affect flaring contour

4. Conclusions

The electromagnetic forming is not affected by mechanical inertia of the moving parts, so the electromagnetic forming can achieve rapid processing, which is easy to realize mechanization and automation. Electromagnetic forming is often used in flat workpieces and the forming of tube blank, The electromagnetic forming is medium pressure on the workpiece with magnetic field, which is no mechanical contact. The typical technology is mainly achieved. It is gived to the curve of tube billet deformation process. The following analysis is the main influencing factors of slapping.

References

1. Jiang Hongwei; Chun-Feng Li, Zhao Zhiheng, Hai-Ping Yu; The latest progress of the electromagnetic forming technology [J]; Materials Science and Technology, 2004.
2. Zhao Zhiheng; Chun-Feng Li; Hai-Ping Yu; Pipe length on the influence of electromagnetic forming coil discharge current [J]; Materials Science and Technology, 2008.
3. Yang Jun, Wen-Feng Chen; Discusses the forming efficiency in electromagnetic forming [J]; Electric Processing and Mould, 2003.
4. Sun Huiping; Wei-Gang Guo; Xiao-Feng Wu; Chen Lei; Electromagnetic forming technology in the study of the application of the sheet forming progress [J]; Machine with hydraulic, 2007.
5. Zhao Zhiheng, Li Jianhui; Electromagnetic Forming Force [J]; Journal of Plastic Engineering, 2001.

6. Wei Ouyang, Shang-Yu Huang; The research and application of the electromagnetic forming technology [J]; Journal of plastic engineering, 2005.

Design of combined-shaped metamaterial invisible cloak

Ye Liu[a], Zhu-Ying Li[b], Xiang Yu[c],
Wen-Kang Cao[d] and Ya Zhao[e]
Scientific Research Department,
Naval University of Engineering, Wuhan, China
E-mail: [a]ly20040608@126.com, [b]lzyhjgc@163.com,
[c]yuxiang898@sina.com, [d]cao-mol.cwk@163.com,
[e]winterxue1986@126.com

Based on coordinate transformation, a combined-shaped metamaterial invisible cloak with the deduced electromagnetic tensors is designed. The results of simulation software based on finite element methods (FEM) are obtained on the condition that the wave origin is a transverse electric wave (TE wave) with a frequency of 200 MHz. The simulation results show that the invisible cloak can change the direction of the incoming wave, and that the wave can almost return to its original characteristics when it propagates away from the cloak. The field intensities of the free space outside the cloak stay almost invariant, with little scattering. The combined-shaped cloak is thus considered to be effective in causing objects to be invisible to electromagnetic waves.

Keywords: Metamaterial; Invisibility cloak; Combined-shape; Coordinate transformation.

1. Introduction

Metamaterial electromagnetic invisible cloaks have been considered a novel invisible device at the beginning of this century. The invisible cloaks are designed by coordinate transformation and conformal transformation, which are proposed by Professor J. B. Pendry and Professor Ulf Leonhart respectively. And the above two professors published their scientific research works at the same issue of American Science magazine[1, 2] at 2006. At the same year, J. B. Pendry's research team have finished the microwave experimental validation of the first two-dimension cylindrical electromagnetic metamaterial invisible cloak[3] in the world. This scientific work and invention has been evaluated by American Science magazine as one of the year's top ten scientific and technology development in 2006. From then on, metamaterials have been studied intensively by scientists and engineers all over the world. During the almost ten years, there has been rapid development to the scientific research related to metamaterials, including the design of configuration[4-6], the exploitation of functions[7], the mechanics of realization[8, 9] and so on. Based on the previous efforts and scientific conclusions of the scientists and engineers, this paper presented the design method of the combining shaped metamaterial invisible cloak, and deduced the electromagnetic parameter tensors of the cloak.

2. Design of the Combining Shaped Cloak

This cloak is designed by coordinate transformation, and the schematic diagram of the cloak is showed as Fig. 1.

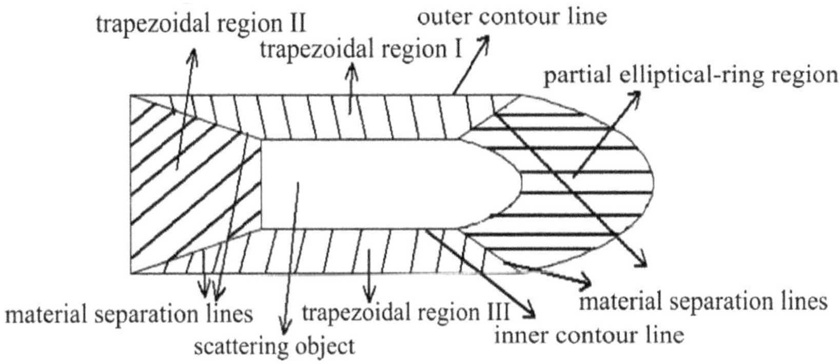

Fig. 1 The schematic diagram of the combining shaped cloak

In the research field of the two-dimension invisible cloak, scientists have accomplished a large amount of research works previously. However, most two-dimension shapes of the previous cloak have the characteristic of simplicity, including circular, elliptical, polygonal shapes and other shapes which are described by polar equations. And the cloak discussed in the paper is compose of three trapeziums and one partial elliptical-ring shape, which is showed in Fig. 1.

With regard to the partial elliptical-ring region, it assumes that the minor axis of the inner and outer ellipse is a and b respectively, and the axial ratio is k. Based on the conclusions of Jiang et al[4], the coordinate transformation expressions of this partial elliptical-ring region can by written as

$$r' = \frac{b-a}{b} r + \frac{kar}{\sqrt{x^2 + k^2 y^2}} \quad (1)$$

$$\theta = \theta' \quad (2)$$

$$z = z' \quad (3)$$

According to the coordinate transformation and the previous conclusions[4], the electromagnetic parameter tensors in the partial elliptical-ring shape can be deduced as following:

$$\varepsilon'_{xx} = \mu'_{xx} = \frac{r_1}{r_1 - ka} + \frac{k^2a^2R^2 - 2kar_1^3}{(r_1 - ka)r_1^5}x'^2 \tag{4}$$

$$\varepsilon'_{xy} = \varepsilon'_{yx} = \mu'_{xy} = \mu'_{yx} = \frac{k^2a^2R^2 - ka(1+k^2)r_1^3}{(r_1 - ka)r_1^5}x'y' \tag{5}$$

$$\varepsilon'_{yy} = \mu'_{yy} = \frac{r_1}{r_1 - ka} + \frac{k^2a^2R^2 - 2k^3ar_1^3}{(r_1 - ka)r_1^5}y'^2 \tag{6}$$

$$\varepsilon'_{zz} = \mu'_{zz} = \left(\frac{b}{b-a}\right)^2 \frac{r_1 - ka}{r_1} \tag{7}$$

As for the deduction of the electromagnetic parameter tensors in the trapezoidal regions, by virtue of the derived conclusions of Li et al[5], if setting the boundary curve to straight line, the tensors of these regions can be written as

$$\begin{aligned}\varepsilon'_{xx} = \mu'_{xx} = \{&[r' - \tau R(\theta')]^2\cos^2\theta' + \tau^2[\frac{dR(\theta')}{d\theta'}]^2\cos^2\theta' \\ &- 2\tau r'\frac{dR(\theta')}{d\theta'}\sin\theta'\cos\theta' + r'^2\sin^2\theta'\} \\ &/\{r'[r' - \tau R(\theta')]\}\end{aligned} \tag{8}$$

$$\begin{aligned}\varepsilon'_{xy} = \varepsilon'_{yx} = \mu'_{xy} = \mu'_{yx} = \{&-\tau R(\theta')[2r' - \tau R(\theta')]\sin\theta'\cos\theta' \\ &+ \tau^2[\frac{dR(\theta')}{d\theta'}]^2\sin\theta'\cos\theta' \\ &+ \tau r'\frac{dR(\theta')}{d\theta'}(\cos^2\theta' - \sin^2\theta')\} \\ &/\{r'[r' - \tau R(\theta')]\}\end{aligned} \tag{9}$$

$$\begin{aligned}\varepsilon'_{yy} = \mu'_{yy} = \{&[r' - \tau R(\theta')]^2\sin^2\theta' + \tau^2[\frac{dR(\theta')}{d\theta'}]^2\sin^2\theta' \\ &+ 2\tau r'\frac{dR(\theta')}{d\theta'}\sin\theta'\cos\theta' + r'^2\cos^2\theta'\} \\ &/\{r'[r' - \tau R(\theta')]\}\end{aligned} \tag{10}$$

$$\varepsilon'_{zz} = \mu'_{zz} = (\frac{1}{1-\tau})^2\frac{r' - \tau R(\theta')}{r'} \tag{11}$$

In above expressions, $r' = \sqrt{x'^2 + y'^2}$, τ is the similarity ratio of the inner and outer boundary line segment in each trapezoidal regions.

3. Validation of the Parameter Tensors of the Combining Shaped Cloak

In above expressions of the electromagnetic parameter tensors of the cloak, setting a=0.5m, b=1m, k=4, τ =0.5, the outer boundary line segments of the three trapezoidal regions of the cloak are $R_1(\theta') = \sqrt{3}/(2\sin\theta')$, $R_2(\theta') = -4/\cos\theta'$, $R_3(\theta') = -\sqrt{3}/(2\sin\theta')$ respectively. And the coordinate of the intersection points between the hypotenuses of the three trapeziums and the outer contour line are (-4, $\sqrt{3}/2$), (-4,-$\sqrt{3}/2$), (2, $\sqrt{3}/2$), (2,-$\sqrt{3}/2$) respectively.

COMSOL Multiphysics software based on the finite element method is used to finish the simulation of the combining shaped cloak. And the simulation model of this cloak is showed in Fig. 2. On the simulation conditions, the source of the incoming wave is TE wave, which frequency is 200MHz and the intensity is 1V/m. And the wave propagates from the left boundary of the rectangle region with $16m \times 12m$. The boundary condition of the solving region is perfectly matched layers (PML). The inner boundary of the cloak is set to be perfect electric boundary, and the outer boundary of the cloak and the material separation lines are set to be continuum boundary. Besides, the space outside the cloak region is considered to be free space.

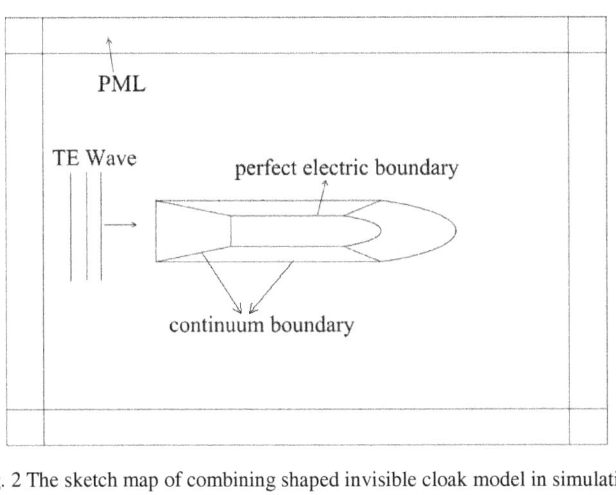

Fig. 2 The sketch map of combining shaped invisible cloak model in simulation

Fig. 3 shows the simulation results of the cloak on the above conditions:

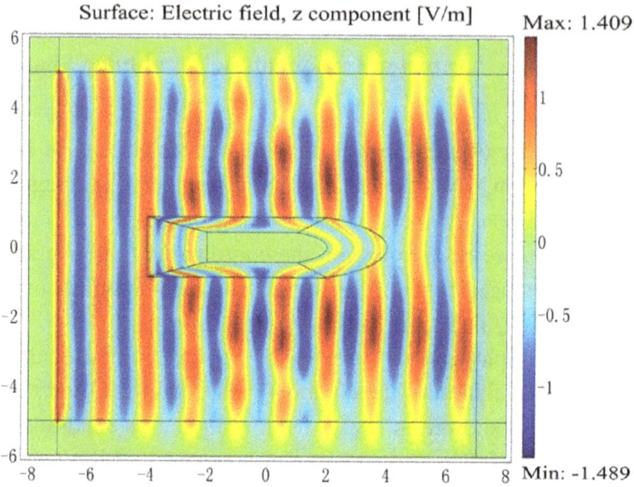

Fig. 3 The distribution of the electric field in the resolving region in the simulation

From Fig. 3, it is found that the distribution of the electric field in the resolving region shows the combining shaped cloak presents the excellent invisible performance to the incoming TE wave. When the TE wave propagated into the cloak region, it propagated around the inner boundary of the cloak with no scattering. And the wave almost returned to be the original characteristics when it transmitted to the free space from the cloak. In the free space outside the cloak, it is showed that there was little electromagnetic scattering in this region, and the intensity and wave form of the incoming TE wave almost keep invariance. Therefore, the cloak region is nearly transparent to the incoming TE wave for observer in the free space. It seemed for the observer to be consider that there is no obstacle in the whole solving region and the incoming TE wave propagated in the free space all the time.

In order to further exhibit the invisible performance of the combining shaped cloak, the numerical simulation with the same related condition was used to analyze the case of a scattering object without cloak. The simulation about the scattering object is showed by Fig. 4.

From Fig. 4, it is clearly exhibited that there is obvious electromagnetic scattering all over the free space, including forward scattering and backward scattering, leading to the incoming TE wave not keeping to be a plane wave. Besides, the intensity of the incoming TE wave has been changed in the free space, and there is no regularity in the whole free space.

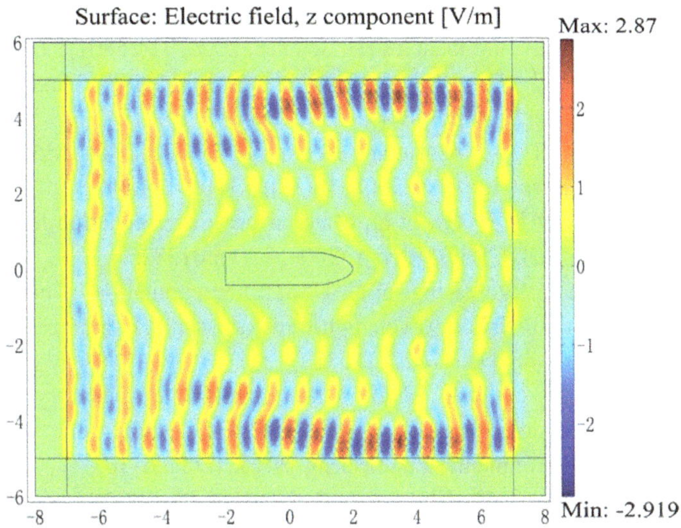

Fig. 4 The distribution of the electric field in the resolving region without the proposed cloak in the simulation

4. Conclusion

Based on the coordinate transformation, combining shaped metamaterial invisible cloak which is more approach to physical object is proposed. The simulation results show that the cloak is nearly invisible to the incoming electromagnetic wave. However, from the electromagnetic parameter tensor expressions of the cloak, it is found that there will be still a tremendous challenge which is to accomplish the intensive anisotropy of the parameter tensors of the cloak. Therefore, it will be a relative long way to realize the physical combining shaped cloak.

Acknowledgments

This work was supported in part by the National Defense Pre-research Foundation of China under Grant No. 9140A270304, in part by the Natural Science Foundation of Hubei Province in China under Grant No. 2013CFB439.

Reference

1. J.B. Pendry, D. Schurig, D.R. Smith. Controlling electromagnetic fields [J]. Science, 2006, 312(5781): 1780–1782.
2. Leonhardt U. Optical conformal mapping[J]. Science, 2006, 312(5781): 1777–1780.

3. D. Schurig, J.J. Mock, B.J. Justice, et al. Metamaterial electromagnetic cloak at microwave frequencies [J]. Science, 2006, 314(5801): 977–980.
4. W.X. Jiang, T.J. Cui, X.M. Yang, et al. Invisibility cloak elliptical cylindrical cloaks [J]. Appl. Phys. Lett., 2008, 93(19): 194102.
5. Li C, Li F. Two-dimensional electromagnetic cloaks with arbitrary geometries [J]. Opt. Express, 2008, 16(17): 13414–13420.
6. H. MA, S. B. QU, Z. XU, et al. Material parameter equation for elliptical cylindrical cloaks[J]. Phys. Rev. A, 2008, 77(1): 013825.
7. Y. Liu, Z.Y. Li, W.Z. Zhang, et al. Design and simulation of metamaterial electromagnetic invisible wavelength-transformer [J]. Journal of Functional Materials, 2014, 45(11): 11027–11031.
8. Y. Liu, Z.Y. Li, L. Zhao, et al. A design fundamental theory of elliptic cylindrical invisible cloak with non-singular electromagnetic tensors [J]. Equipment Environmental Engineering, 2015, 12(1): 6–10.
9. Wang X H, Qu S B, Xia S, et al. Nonsingularity in two-dimensional cylindrical invisible cloaks [J]. Chin. Phys. B, 2010, 19(6): 064102.

Design and development of self-circulating evaporative cooling system for 1600W transmitter

Rui Cao, Lin Ruan[†], Peng Zhang, Shu-Qin Guo, Hai-Hong Dong

Institute of Electrical Engineering,
Chinese Academy of Sciences, Beijing, 100190
[†]*E-mail: cr304@mail.iee.ac.cn*

This paper introduces a self-circulating evaporative cooling system for TV broadcast transmitters. Through practical testing, it was shown that the evaporative cooling technology can be used to cool the high power density electronic equipment. This system not only breaks the thermal bottleneck of air cooling, but also has the advantages of a better cooling effect, low noise output, a simple structure and high reliability. The system thus shows great potential in its cooling capability.

Keywords: TV broadcast transmitter; Self-circulating evaporative cooling system; High power density electronic equipment.

1. Introduction

In recent years, the design of power amplifier unit is integrated highly and air cooling mode left much to be desired[1], so the cooling has become bottlenecks to the transmitter's design and development. The most commonly used air-cooled technology has almost gradually reached the limit of its heat dissipation. Currently the most popular is the water cooling technology, but it also has many weaknesses[2,3]. In addition, there also has some hot and advance researches, heat pipe, semiconductor refrigeration mode and so on[4,5], but these researches are not widely used now because the process is complicated and the mechanism is being studying.

The evaporative cooling technology takes advantage of phase change to realize the cooling effect of the object. Its cooling capability is far larger than the other heat transfer method using specific heat. This cooling technology has been successfully used to large electrical machines and electronic equipments in china. The recent application of this technology to the large electrical machine is the world famous Three Gorges hydropower project. The characteristic of this cooling technology include: high efficiency for the heat transfer, high safety and most important low energy cost or zero energy cost from the cooling system itself. All of these characteristics show cases it will be very suitable for the electronic components such as transmitter etc.

2. Demo of Applying the Evaporative Cooling System to the 800W Unit

2.1 *Cooling structure of 800W unit*

With the development of the digital TV broadcasting, the integration of transmitter unit is higher and higher. Fig.1 shows that design trends of the power amplifier tube. The most commonly used air-cooled technology has almost gradually reached the limit of its heat dissipation. In view of this situation, we propose using adherent self-circulating evaporative cooling structure to cool the transmitter.

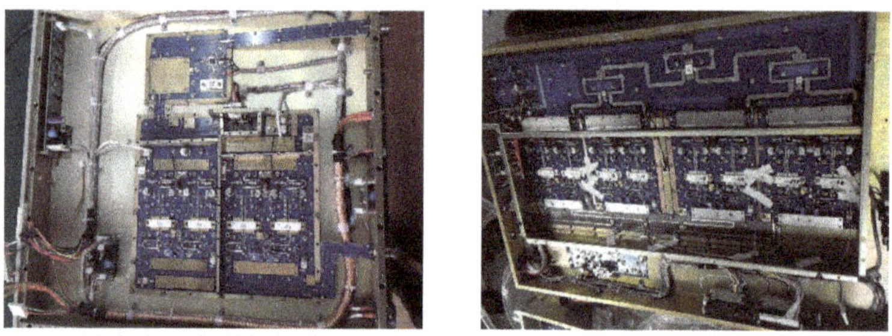

(a) 4tube unit(400W unit) (b) 8tube unit(800W unit)

Fig. 1. Trends of the power amplifier tube

The 1600W TV broadcast transmitter consists of two 800W units, so we design the cooling system of 800W unit first. We design the cooling structure according to the distribution of heat source of 800W power amplifier unit: There are two liquid-boxes on the unit, and one box is charge of cooling 4 tubes. Fig. 2 shows the assemblage of liquid boxes.

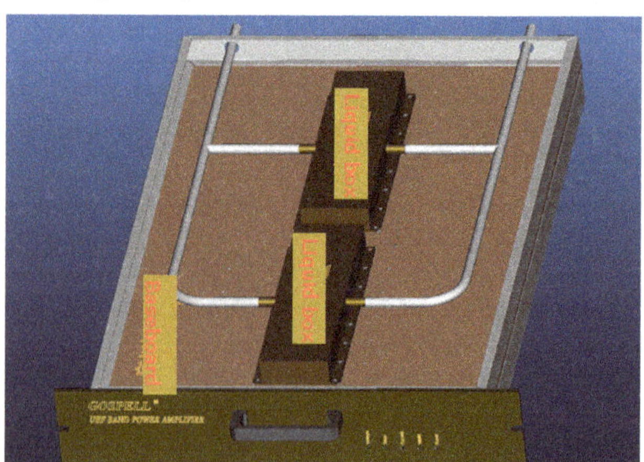

Fig. 2. The assemblage of liquid boxes on the unit

2.2 *Determination of cooling medium and material of liquid-box*

We established the experimental platform of the cooling system for the 800W transmitter unit as requested. We use 2 kinds of cooling medium (ZXB-1and ZXB-2) and 2 materials of liquid box (A and B), so there are 4 kinds of working condition, and we need to monitor the tube, the substrate and the baseboard's temperature in the experiment. When the temperature of the environment is 27.1°C. The Tab.1 shows the results of the cooling experiment.

Table 1. The temperature of components of the unit

	ZXB-1			ZXB-2		
A	T_t: 78°C	T_s: 65°C	T_b: 56°C	T_t: 72°C	T_s: 60°C	T_b: 49°C
B	T_t: 73°C	T_s: 59°C	T_b: 48°C	T_t: 67°C	T_s: 55°C	T_b: 43°C

The temperature of the surface of power amplifier tube is required to keep below 80°C,so we can take this temperature to the boundary condition in simulated calculation, in addition, the material of the liquid box is set to B, and the property of medium is set to ZXB-1.By the simulation we can get the distribution of the temperature of baseboard and substrate which is shown in Fig.3.

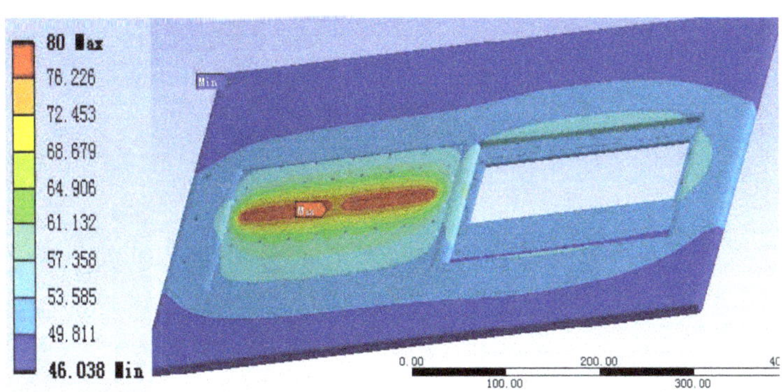

Fig. 3. The temperature of baseboard and substrate

From the simulation, we can get the results: 1. The temperature of the surface of power amplifier tube is kept below 80°C; 2. The temperature of the surface of substrate which is near the liquid box is about60°C; 3. The temperature of the surface of substrate which is far away from the liquid box is about 50°C. The experimental results and the simulation results are consistent which is proving that the heat capacity of liquid box can meet the cooling demand of power amplifier unit.

3. The Evaporative Cooling System for the 1600W Transmitter

On the base of the demo of 800W cooling unit, we established the final cooling system. We use the air cooling condenser in this cooling system as the second cooling part. The capacity of air cooler depends on the total heat of transmitter. We can hang the condenser on the wall of outdoor when the transmitter is placed in the chamber. The sketch map of circulation system is shown in Fig.4. According to the structure of the 1600W transmitter, we optimized the system, which makes the cooling structure more compact. The cooling structure of the unit is shown in Fig.5 and the structure of the loop is shown in Fig.6. The temperature of the surface on the tubes can maintain below 70°C.

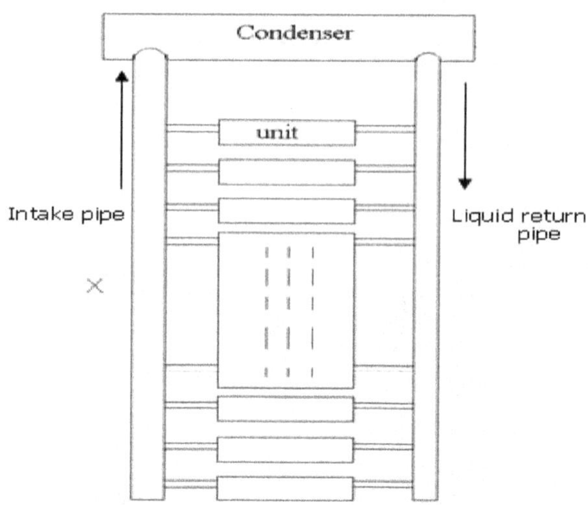

Fig. 4. Schematic diagram of evaporative cooling system

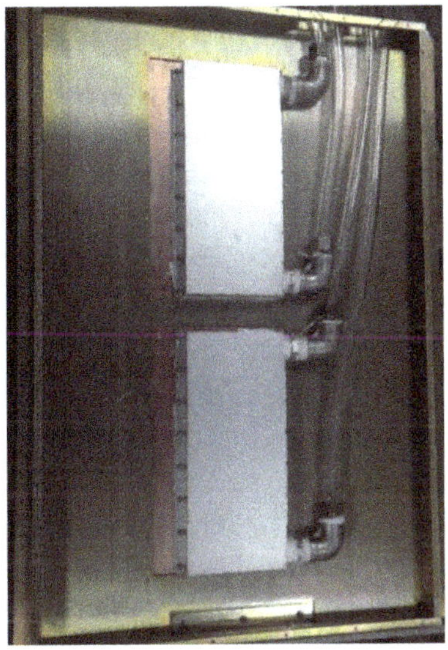

Fig. 5. Cooling structure of the unit

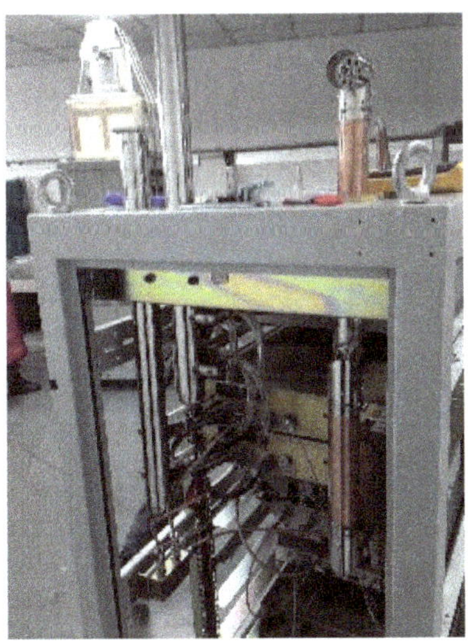

Fig. 6. Structure of the loop

4. Conclusion

In this paper, we design and development a self-circulation evaporative cooling system for the TV broadcast transmitter, which can solve the problem of cooling high power density electronic equipment. The technology has a better cooling effect, low noise, simple structure and high reliability, besides, this application shows great cooling capability potential, so it can be extended to various electronic and magnetic equipment.

References

1. Chen Deng Ke. New technologies of electronics cooling [J]. Chinese Journal of Low Temperature Physics, 2005, 27(3): 255–262.
2. Pastukhov V.G., Maidanik Yu F., Vershinin C.V. Miniature loop heat pipes for electronics cooling [J]. Applied Thermal Engineering. 2003, 23(9): 1125–1135.
3. Cooper G, Herskovits E.A Bayesian Method for the Induction of Probabilistic Networks from Data [J]. Machine Learning, 1992, 9(4): 309–347.
4. Chen Haisheng, Yang Wei, He Yurong, et al. Heat transfer and flow behaviour of aqueous suspensions of titanate nano-tubes(nanofluids) [J]. Powder Techn, 2008, 183: 63.
5. Leonard L1Vasiliev1Heat pipes in modern heat exchangers. Applied Thermal Engineering, 2005 (25): 18–20.

System design and experimental testing of humidity sensor based on condensation method*

Zhong-Hua Dai, Shi-Jun Zhao†,
Tai-Chang Gao, Zheng-Yu Zhang

Institute of Meteorology, PLA University of Science and Technology, Nanjing 211101, China Corresponding author
Phone: +8615250958223,
†Email: 15250958223@163.com

Air humidity is a basic element of ground and upper air meteorological observation. Measurement of air humidity in the whole range of air temperatures and humidity has been a worldwide problem in the field of atmospheric sounding. By relying on the principle of the condensation method, as well as the rapid developments in microscopic imaging, image recognition and embedded system technology, this article solves key problems such as the technology of condensation mirror design, mirror temperature control and condensate identification. A prototype condensation-type humidity sensor is also successfully developed. Synchronous observations with a humidity standard device and a number of preliminary tests are carried out in different conditions of environment temperature and humidity. The test results show that the deviation of relative humidity between the proposed sensor and the standard device falls within the range of [-2.86, -0.95] %RH, which meets the requirements of business applications for ground meteorological observation.

Keywords: Humidity sensor; Condensation-type; Image recognition.

1. Introduction

The air humidity is one of basic elements of ground and upper air meteorological observation, which is usually represented by parameters of mixing ratio, specific humidity, vapor pressure, relative humidity, absolute humidity and dew-point/frost point temperature, etc. As the humidity is strongly influenced by air temperature and pressure, the air humidity measurement in the whole range of temperature and air humidity has still been one of the main challenges in worldwide atmospheric sounding field[1-5].

* This research is supported by special industry project of meteorology GYHY201206078.

There has been a great variety of hygrometers, which can be classified by weighing method, thermodynamic method, hygroscopicity method, optics method and condensation method in principle[6-8]. The condensation-type hygrometer is the only kind of hygrometers that has a good performance in the whole range of measurement and can reach a high level of accuracy even in low temperature conditions[9-10]. They are mainly applied in the laboratory and used as reference calibration standards for measuring humidity of gases. There are two main problems associated with dew point hygrometers. First, the condensation mirror is very easy to be polluted. Second, the sub-cooled water error is difficult to eliminate. These problems strongly affect the dew point temperature measurement. So, it can hardly adapt the complex environment of outdoor and unattended operation. The stabilization and long-term reliability of measurement is of poor guarantee, which limits its promotion and application in surface meteorological observation[11-13].

The rapid development of microscopic imaging, image recognition, embedded system and automatic control technology provides the possibility to develop small condensation-type humidity sensor for applications. In this article, system structure of small condensation-type humidity sensor, key technologies of condensation mirror, controlling of mirror surface temperature and identification of condensate substance have been studied. And a prototype of condensation-type humidity sensor is fabricated successfully and a number of preliminary tests are carried out. The test results show that the relative humidity between the prototype and standard device is close, which meet the business application of ground meteorological observation.

2. Sensor Structure and Working Principle

The whole structure of the sensor consists of three parts, sensing and protection unit, microscopic imaging unit, acquisition and image processing unit (see Fig.1).

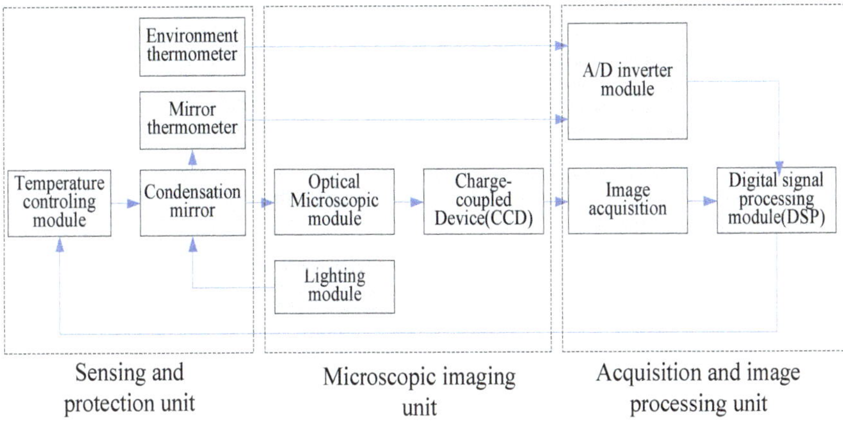

Fig. 1 Whole scheme of the sensor

The sensing and protection unit is designed to sense the environment temperature and air moisture conditions. This unit includes a semiconductor refrigeration device, a condensation mirror and two thermometers. One thermometer is to measure the temperature of the mirror and the other is to monitor the environmental temperature. Gases will be cooled by semiconductor refrigeration device constantly when it get into the sensor, waiting for the moment that gases temperature is below dew point temperature, water vapor in the gases will reach saturated state and drops of water will condense on mirror surface. The two thermometers will record the mirror temperature and environmental temperature separately.

The microscopic image unit is to realize the rapid and clear imaging of condensed substance. For this purpose, a set of optical magnifying glasses is fixed on top of condensation mirror to magnify the image of the mirror surface. The magnifying glass module is well designed that it can magnify mirror image without distortion and difficult to be polluted. Because light has a great influence on picture quality and external light is out of control, we shield the external light and add internal light with a group of monochrome LEDs. High resolution CCD gathers the mirror images instantaneity and transfer them to image processing module.

The main purpose of acquisition and image processing unit is to gather temperature data and process large of mirror image. This work mainly depends on a digital signal processing (DSP) chip. Image detection and recognition algorithm has been put into DSP in advance. When open the sensor, there will be mirror images transfer from CCD to DSP, the DSP preprocesses the images and classify the them into three kinds with recognition algorithm. Usually, there is nothing in mirror surface, these images are first kind. If dewdrops have been condensed on the mirror surface, the images will change, those are second kind, called dew-image. When the environment temperature is below the triple point temperature of the water (0 °C), water vapour is much possible to condense into frost on mirror surface, these images are third kind. Then using mirror thermometer to get the temperature when the second or third kind image first appeared. This temperature is considered to be dew point or frost point temperature. Basing on Goff-Gratch formula (see Eq. (1) and (2)), the dew point temperature is one-to-one relationship with vapor pressure. So, we can get the real water vapor content from dew point or frost point temperature.

To flat water surface, temperature between 223.25 K to 373.15K, the formula express as

$$\lg e_w = 10.79574(1 - T_1/T) - 5.028\lg(T/T_1) + 1.50475\times10^{-4}[1 - 10^{-8.2969(T/T_1 - 1)}]$$
$$+0.42873\times10^{-3}[10^{4.76955(1 - T_1/T)} - 1] + 0.78614 \quad (1)$$

To flat ice surface, temperature between 173.15 K to 273.15K, the formula express as

$$\lg e_{si} = -9.09685(T_1/T - 1) - 3.56654\lg(T_1/T) + 0.87682[1 - T/T_1] + 0.78614 \quad (2)$$

Depending on the whole structure of the sensor, we have designed the specific structure chart and prototype sample of humidity sensor (see Fig. 2).

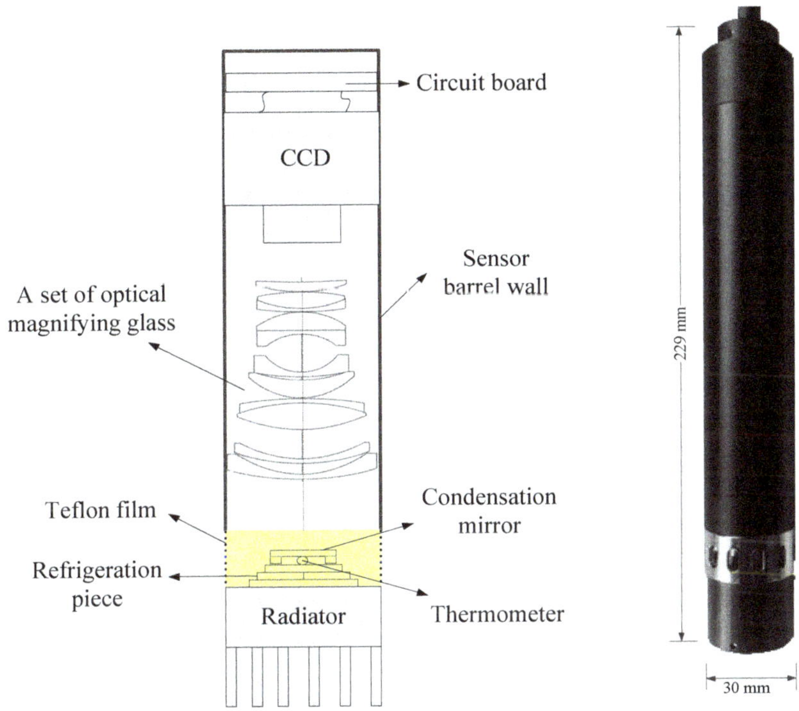

Fig. 2 Specific structure chart and sample appearance of the sensor

3. Key Technology

3.1. *Choosing of condensation mirror*

The condensed of water vapor on mirror surface is the important physical process in humidity sensor. Timing of the dewdrops condensed decides the accuracy of the temperature measurement of dew point. The basic condition of dewdrops' formation is to have a hydrophobic and polished surface and a certain number of nuclei (the small particles on polished surface), or tiny crack on the mirror surface. So, choosing good mirror material has a great influence on measurement of dew point temperature. If mirror material is not selected appropriately, dewdrops will condense in advance and have a bigger volume than the usual. The bigger volume of dewdrops, the stronger of their dissolving capacity, thus more soluble impurities will dissolve in them. Water drops will become salt solution with soluble impurities dissolved in. As is known to all, salt solution has lower saturated vapor pressure than pure water in same temperature, the water molecules evaporation from the mirror is less than pure water in per unit time. This kind of phenomenon will cause the measured dew point temperature higher than real dew point temperature (called Raoult effect). Besides, Mirror material must have good thermal performance to guarantee thermometer is consistent with mirror surface temperature.

In this work, stainless steel plated with rhodium was used as condensation mirror. It is corrosion resistant, high hardness, wear-resisting and hydrophobic.

3.2. *Rapid control of mirror surface temperature*

The change of mirror surface temperature is a second-order response system, it will approach dew or frost point temperature oscillatorily and gradually. At the beginning, cooling rate must be fast, when congelation is condensed on mirror surface, slow the heating or cooling rate until the condensation process reaching the steady state.

The system requires that the thermostat not only responses fast but also approach target temperature steadily. So, we have a high demand for temperature control algorithm and sensitivity of charged substance. In recent years, semiconductor refrigeration technology provides a good condition for miniaturization of dew point hygrometers, for it can reach different levels of low temperature accurately by multistage superposition method. PID (Proportion Integration Differentiation) controling algorithm is a method which is widely used in industry control and has been proved to be stable and reliable. PID controller is a linear controller, it gets control deviation according to the given value $y_d(t)$ and the actual output $y(t)$ (see Eq. (3)). Equation (4) shows the law of PID, where k_p is scale factor, T_I is integral time constant, T_D is differential time constant.

$$error(t) = y_d(t) - y(t) \tag{3}$$

$$u(t) = k_p[error(t) + \frac{1}{T_I}\int_0^t error(t)dt + T_D \frac{derror(t)}{dt}] \tag{4}$$

So, multistage semiconductor refrigeration and PID controling algorithm are hanged together to control mirror surface temperature for they not only have the advantage of small volume from the former but also get the flexibility and accuracy from the latter .

3.3. *Detection of dew and frost*

Detecting of dew and frost plays a very important role in hygrometers. The common way is photoelectric method which is used in most dew point hygrometers. There are many other methods include radio sounding, surface acoustic wave, quartz crystal and image-processing. In those ways, image-processing is a better method to reach the miniaturization goal and has an advantage in sub-cooled water detection. Dew images are different from frost images through human eyes (see Fig.3).

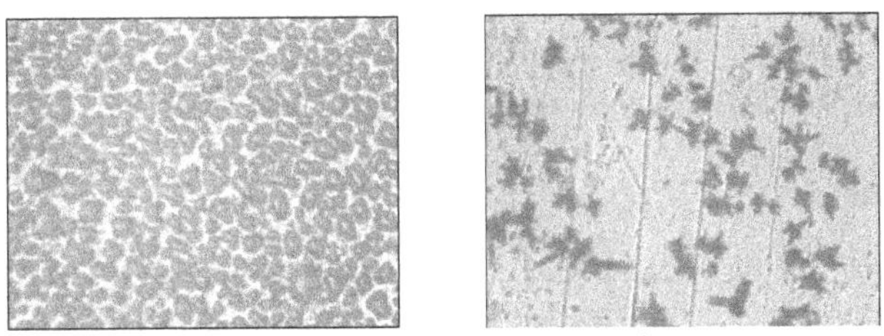

a)Dew image b)Frost image

Fig. 3 Dew and frost image

For detecting them from each other, many image features must be analyzed. The image entropy is one of texture feature which reflect the change of image gray level and is very easy to calculate (see Eq. (5)).

$$Ent = -\sum_{i=0}^{N} p(i) * \log p(i) \tag{5}$$

Where $p(i)$ is the probability of each gray pixels of whole image, N is the grayscale. In this work, 1,080 images including 360 mirror images, 360 dew images and 360 frost images have been picked out to count the image entropy. The result has been proved very useful in image detection (see Fig.4).

It is obvious from the figure that image entropy is different and have a boundary between each other. The mean entropy of mirror images is 3.2253, and is less than mean entropy 5.8122 of dew images, the mean entropy 9.1597 of frost images is biggest. The mean entropy shows the different gray scale distribution between different kinds of images that is same to human eyes observation. So, image entropy has been used in the humidity sensor as a parameter to detect the condensed substance.

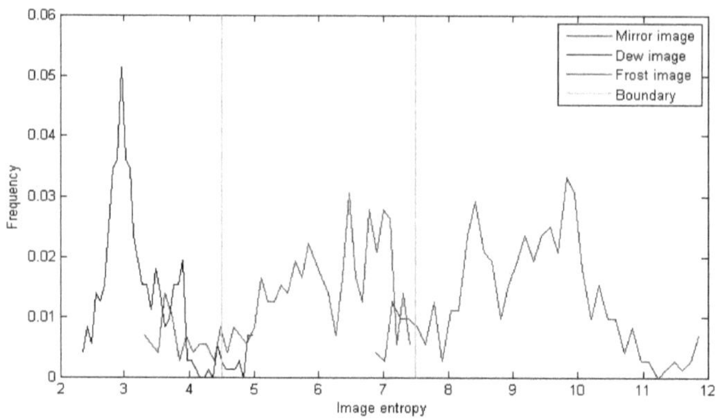

Fig. 4 The statistics of entropy of 1,080 images

4. Experiment

Depending on all the technologies and algorithm above, we have fabricated a prototype successfully. For testing the performance of prototype, a static and field test have been conducted. The static test use a two-pressure humidity generator and a digital intelligent dew point hygrometer MBW373 (dew point measurement precision is 0.1°C) which have been confirmed as reference calibration standard of first-class metering station, the former is used to form the target humidity environment and the latter is used as standard instrument.

Basing on the working temperature (-10~+50°C) of MBW373, we chose 3 temperature points (50°C, 20°C and -10°C) at different relative humidities (30%, 60% and 95%) in the static test. As instruments will be covered with frost in low temperature and high humidity, which is harmful to hygrometers, the test in -10°C at 95% is canceled. In the test, the goal temperature is set first, when humidity generator reach the target point, 30 minutes of waiting is needed before recording the data. To ensure accuracy, read 3 times data on each point and the interval is 30 seconds, then calculate the average. The finally result show good consistent with MBW373 hygrometer (see Table.1 and Fig.5).

Table.1 Humidity static test with hygrometer MBW373

Test temperature point (°C)	Test humidity point (%RH)	Condensate type (dew/frost)	MBW373 DPT (°C)	MBW373 H (%RH)	Prototype sensor DPT (°C)	Prototype sensor H (%RH)	DPT error (°C)	Humidity error (%RH)
50	95	dew	48.60	93.26	48.21	91.45	-0.39	-1.81
	60	dew	39.98	59.74	39.52	58.29	-0.46	-1.45
	30	dew	27.66	30.04	26.97	28.85	-0.69	-1.19
20	95	dew	19.24	95.39	19.00	93.97	-0.24	-1.42
	60	dew	11.93	59.71	11.63	58.54	-0.30	-1.17
	30	dew	1.98	30.15	1.53	29.20	-0.45	-0.95
-10	60	frost	-15.63	54.44	-16.21	51.58	-0.58	-2.86
	30	frost	-22.83	27.62	-23.69	25.17	-0.86	-2.45

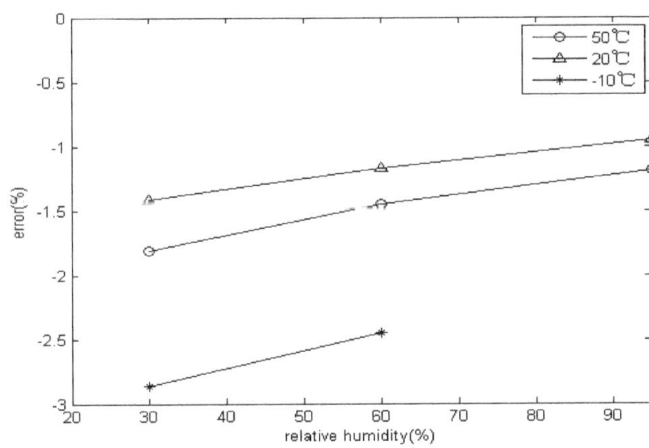

Fig. 5 The error of prototype sensor

Table 1 and Fig.7 shows the error of prototype sensor in different relative humidities. The condensate type detected by prototype sensor is same to MBW373, and the average error is only -1.66%, which prove the feasibility of the sensor design and methods on above. The deviation of relative humidity with prototype sensor and the MBW373 is in the range of [-2.86, -0.95] %RH, which meet the business application of ground meteorological observation.

On the other hand, the result shows some problems. First, all the error value is below 0, thus reflects a common phenomenon that the measured dew-point temperature is lower than real value. The main reason can explain this phenomenon is that the image entropy used in *detection of dew and frost* is not very sensitive to condensate. When dew or frost begin to condense on mirror surface, the mirror image entropy has little change compare to original image because of the condensate is very little, it can't recognize them. Second, the performance in different temperature is different, the mean error in 20°C is -1.18% and is good than the mean error -1.48% in 50°C, both of them is good than the mean error -2.66% in 50°C. This phenomenon is caused by the accuracy of thermometer below the mirror and the difference of condense speed of dew and frost. The thermometer is made by thermistor, thermistor has different accuracy in different temperature. Besides, in 20°C and 50°C, the condensate is dew, but in -10°C the condensate is frost, frost has slower condense speed than dew, which slows the response time of multistage semiconductor refrigeration and increase the measurement error. All those problems will be disposed in next step work.

For testing the dynamic characteristics of the prototype sensor, field comparison experiment has been conducted with HMP155a humidity sensor, which have been widely used in meteorological station around the world. The result show good consistent (see Fig.6).

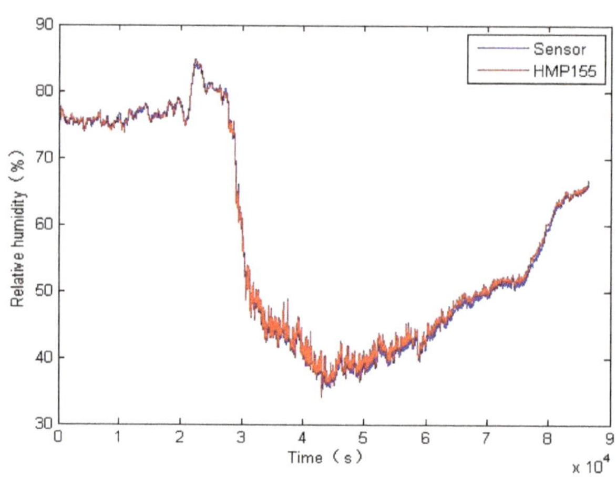

Fig. 6 Field comparison with HMP155a

5. Conclusions

The humidity sensor designed based on condensation method has adopted microscopic imaging modalities, using small planar array CCD to collect and recognize condensate on the mirror, which achieved miniaturization, low power consumption and low cost of hygrometer. Meanwhile, multistage semiconductor refrigeration and PID controlling algorithm are used to control mirror surface temperature. At last, image entropy is used as feature to detected dew and frost. The performance of sensor has been examined through a static and field test, the result shows good consistent with standard device MBW373 hygrometer and common humidity sensor HMP155a. However, more tests should be carried out in complex environmental conditions to ensure the reliability and long-term stability of sensor before its application in ground meteorological observations business. Meanwhile, the problems of this sensor require continual improvement to achieve the business application as soon as possible.

References

1. Kevin A.Vetelino, Patrick R. Story, Russell D. Mileham, David W.Galipeau. Improved dew point measurements based on a SAW sensor[J], Sensors and Actuators B: Chemical. 1996, 35–36: 91–98.
2. Jerzy Weremczuk. Distinct Optical Properties of Supercooled Water and Ice Covering Dew-Point Hygrometer Mirror, IEEE Transactions on Instrumentation and Measurement[J]. 2006, 55(3): 995–999.
3. Jerzy Weremczuk. Dew/Frost Point Recognition With Impedance Matrix of Fingerprint Sensor, IEEE Transactions on Instrumentation and Measurement[J], 57(8), 2008.08.
4. Mridul Sarkar. A new theoretical formulation of dew point temperatures applicable for comfort air-cooling systems[J], Energy and Buildings. 2015, 86: 243–256.
5. Jing Nie, Xiaofeng Meng. Dew point and relative humidity Measurement using a quartz resonant sensor[J], Microsyst Technol.2014, 20:1311–1315.
6. J. Weremczuk, G. Tarapata, D. Paczesny, R. Jachowicz. Fast dew point hygrometer with silicon integrated detector–Optimization of dynamic properties, Sensors and Actuators A. 2006, 132: 195–198.
7. Joung O J, Kim Y H. Dew point measurement for organic vapor mixture using a quartz crystal sensor[J]. Sensors and Actuators B: Chemical. 2006, 113(1): 335–340.
8. R. Jachowicz, J. Weremczuk. Sub-cooled water detection in silicon dew point hygrometer[J], Sensors and Actuators A. 2000, 85:75–83.
9. R.S. Jachowicz, Dew point hygrometer with heat injection—principle of construction and operation[J], Sensors and Actuators B, 1992,7: 455–459.
10. Sun Xuejin, Wang Xiaolei, Li Hao, et al. Atmospheric detection[M]. Beijing: Meteorological Press, 2009: 123–151.

11. Su Xiaoyong, Zhao Shijun, Gao Taichang, et al. Present Situation and Analysis of Key Technique of Chilled-Mirror Dew-Point Hygrometer[J]. Journal of Atmospheric and Environmental Optics, 2011, 6(05): 323–328.
12. LV Xueqin, AO Zhenlang. Experiment and Analysis of Chilled mirror dew point instrument for dynamic measurement errors[J]. Chinese Journal of Scientific Instrument. 2013, 34(12): 14–18.
13. Ingleby B, Moore D, Sloan C. Evolution and Accuracy of Surface Humidity Reports [J]. Journal of Atmospheric & Oceanic Technology. 2013, 30(9).

Fog/Low stratus detection during night-time: Application of a multi-channel threshold algorithm*

Jun Jiang and Wei Yan† and Shuo Ma and Xia-Rong Zhang
Institute of Meteorology and Oceanography,
PLA University of Science and Technology,
Nanjing, 211101, China
†E-mail: 18913979082@163.com

The Day/Night Band (DNB) low-light visible sensor, mounted on the Suomi National Polar-orbiting Partnership (S-NPP) satellite, can measure visible radiances from the earth and atmosphere (solar/lunar reflection, natural/anthropogenic nighttime light emissions) during both the day and night. In particular, it has achieved unprecedented nighttime low-light-level imaging with its accurate radiometric calibration and splendid spatio-temporal resolution. Based on the superior characteristics of DNB, a multi-channel threshold algorithm combining DNB with other VIIRS channels was proposed to monitor nighttime fog/low stratus. Through a gradual separation of underlying surface (land, vegetation, water bodies), snow, and medium/high clouds, a fog/low stratus region could ultimately be extracted by the algorithm. Algorithmic feasibility then was verified by a typical case of heavy fog/low stratus in China, 2012. The experimental results demonstrate that the outcomes of the algorithm approximately coincide with the ground measured results.

Keywords: NPP/DNB; Multi-channel threshold algorithm; Remote sensing; Cloud detection; Fog.

1. Introduction

Fog and low stratus are meteorological phenomena which cause bad visibility and air quality, seriously threatening navigation, aviation and land transportation. Their accurate description in time and space therefore is of significant importance in societal, economic and health aspects. While ground meteorological stations provide information about fog and low stratus episodes, the data acquired by point observation are spatially discontinuous and temporally dispersed. In comparison, weather satellite data have the advantage of continuous spatial coverage with providing reliable near-real time information on the spatio-temporal distribution of fog and low stratus.

* This work was supported by the National Natural Science Foundation of China (Grant No. 41375029).

Common techniques rely on the particular emissive properties of fog and low stratus droplets at the 3.9μm and 10.8μm channels: the small droplets found in fog and stratus are less emissive at the 3.9μm channel than those at 10.8μm, whereas both emissivities are roughly the same for larger droplets (Hunt, 1973). According to Hunt's theory, Eyre et al. (1984) firstly implemented the dual channel difference (DCD) algorithm on the National Oceanic and Atmospheric Administration (NOAA) Advanced Very High Resolution Radiometer (AVHRR). Subsequently, many researchers applied similar technique to various other sensors, promoting the development of the DCD algorithm (e.g. Dybbroe,1993; Bendix, 2002; Chaurasia et al, 2011). However, Ellrod (1995) pointed out some shortcomings of the DCD algorithm, such as poor performance on layers of very thin fog or stratus (less than 100m), occasional confusion between medium/high stratus and fog, and inapplicability to certain types of soils which produce similar signal at the 3.9μm and 10.8μm channels.

Meanwhile, the rapid development of low-light-level sensor technology has made it a reality to get visible imagery under low illumination conditions at twilight or night, providing a fresh approach to the issue. Zhou et al. (2011) presented a bi-channel threshold algorithm by a combination of visible and infrared data of the Defense Meteorological Satellite Program (DMSP) Operational Linescan System (OLS), and the feasibility of the algorithm was verified by validation experiments. However, visible data available on OLS are partitioned into a limited 64 gray levels and IR data into 256 levels. The lack of calibration and coarse radiometric resolution limits its utility in quantitative environmental applications. The VIIRS first launched onboard S-NPP on 28 October 2011 solves the problem above. Especially, the DNB low- light-level sensor, with accurate radiometric calibration and favorable nighttime imaging capacity, shows its great potential for quantitative remote sensing at twilight or night (Lee et al, 2006; Miller et al, 2012).

Based on the superior characteristics of DNB, we propose a multi-channel threshold algorithm combining DNB with other VIIRS channels to monitor the nighttime fog and low stratus. The data used in the algorithm are introduced in Section 2. The applied techniques are described in detail in Section 3. The validation experiments are completed and discussed in Section 4. And several corresponding conclusions are summarized at the end of the paper.

2. Data

Nighttime fog and low stratus detection has been carried out using S-NPP/VIIRS data. NPP/VIIRS is sensitive to 22 wavelength channels ranging in central wavelength from 0.412 μm to 12.01 μm in the electromagnetic spectrum (Seaman et al, 2013). The distinct radiation characteristics in certain spectral channels are presented among the fog/low stratus, underlying surface, snow, and medium/high clouds, causing the radiation difference received by satellite sensors. In the light of this principle, 5spectral channels are used in the algorithm. Table 1 summarizes the selected channels.

Table 1. Channels used in the algorithm

VIIRS Channels	Wavelength Range (μm)	Central Wavelength (μm)	Spatial Resolution at nadir (m)	Application
DNB	0.500-0.900	0.700	750	Removal of the underlying surface
I1	0.600-0.680	0.640	375	Removal of snow
I2	0.846-0.885	0.865	375	Removal of snow
I3	1.580-1.640	1.61	375	Removal of snow
I5	10.50-12.40	11.45	375	Removal of medium/high clouds

3. Algorithm

The nighttime fog and low stratus detection is executed by a chain of process for the background removal and surface homogeneity test. The overall structure of the multi-channel threshold algorithm is outlined in Figure 1.

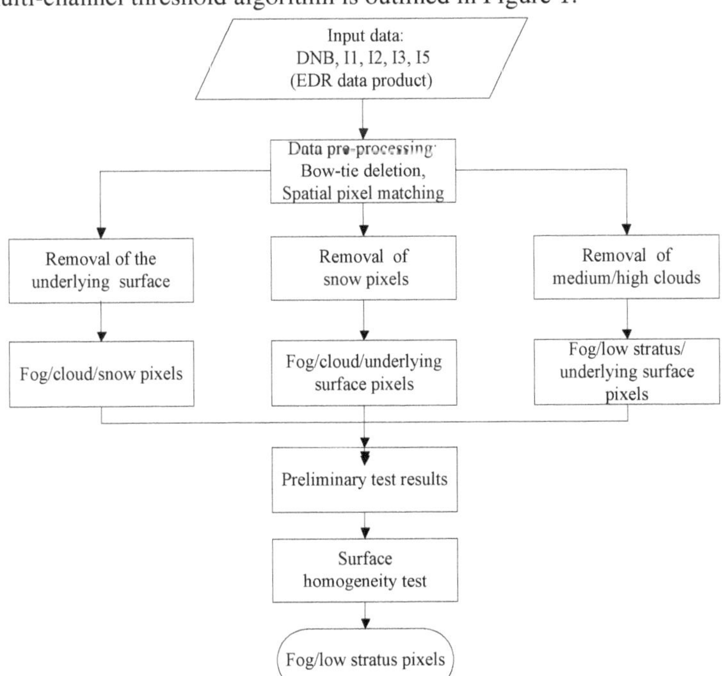

Figure 1. Overall structure of the multi-channel threshold algorithm

3.1. *Removal of the underlying surface*

During nighttime, the radiation energy received by the satellite sensors in visible channels depends on the reflectance of targets. According to the Mie scattering theory (Mie, 1908), the scattering of clouds is fairly obvious. The reflectance of the fog, clouds, and snow therefore is much higher than that of the underlying surface including land, vegetation, and water bodies. So, the underlying surface pixels can be removed by setting a proper reflectance threshold. However, the threshold varies with time and regions. To determinate a dynamic threshold is undoubtedly the key to accomplish the procedure. Otsu (1979) proposed an adaptive method to compute a threshold for grayscale image segmentation. The method recognizes an image as a composition of two types: backgrounds and objectives. The level of difference is measured by the variance between the two types. And a gray value which causes the maximum variance is considered to be the best threshold. For the reflectance values in the DNB-EDR data files (recorded as 0-65535), the data can be regarded as a grayscale image of 16 bits. Assuming that the image has a size of $M \times N$, a gray level of L, and N_i pixels totally where the gray value is i ($0 \leq i \leq L - 1$), the occurrence probability of the gray value i is $P_i = \frac{N_i}{M \times N}$. Accordingly, the best threshold of gray value is given by:

$$G = \arg\max_{0 \leq g \leq L-1} \left[P_a \left(w_a - w_0 \right)^2 + P_b \left(w_b - w_0 \right)^2 \right] \quad (1)$$

Where the parameters are defined as follows:

$$P_a = \sum_{i=0}^{g} P_i, \; P_b = \sum_{i=g+1}^{L-1} P_i, \; w_a = \sum_{i=0}^{g} i \frac{P_i}{P_a}$$
$$w_b = \sum_{i=g+1}^{L-1} i \frac{P_i}{P_b}, \; w_0 = \sum_{i=0}^{L-1} i P_i \quad (2)$$

3.2. *Removal of snow pixels*

Clouds and snow are both bright across the visible and near-infrared region (0.4-1.3μm), but clouds are much brighter than snow in the short wave infrared (1.3-3μm),especially near 1.61um spectrum, because of the smaller size of the scatterers in clouds decreasing the probability of absorption in this spectral region where snow is moderately absorptive (Crane and Anderson, 1984). On the basis of the difference, the Normalized Difference Snow Index(NDSI) and Normalized Difference Vegetation Index (NDVI) are used to remove snow pixels, referring to the snow classification techniques of the MODIS algorithm,

SNOMAP (Hall et al, 1998). For the VIIRS EDR data, the NDSI and NDVI are calculated as:

$$NDSI = \frac{R_{I1} - R_{I3}}{R_{I1} + R_{I3}} \quad (3)$$

$$NDVI = \frac{R_{I2} - R_{I1}}{R_{I2} + R_{I1}} \quad (4)$$

Where R_{I1}, R_{I2} and R_{I3} respectively mark the reflectance in 0.64um (I1), 0.865um (I2), and 1.61um (I3) VIIRS channels. Generally, the NDSI of pure snow pixels keeps high. A Pixel which meets the criteria: $NDSI \geq 0.4$ and $R_{I2} \geq 0.11$ is classified as snow. When the snow pixel locates under forest canopy, the NDSI decreases. On the premise of $0.1 < NDSI < 0.4$, a pixel can still be classified as snow if it fits the thresholds: $NDVI < NDVI\_1$ and $NDVI > NDVI\_2$. The thresholds are determined as a function of NDSI:

$$NDVI\_1 = a_1 + a_2 * NDSI \quad (5)$$

$$NDVI\_2 = b_1 + b_2 * NDSI + b_3 * NDSI^2 + b_4 * NDSI^3 \quad (6)$$

where the values of coefficients a_1, a_2, b_1, b_2, b_3, and b_4 are defined following Klein et al (1998).

3.3. *Removal of medium/high clouds*

At night, the emitting radiation from objects themselves is the main energy source received by the satellite sensors in long wave infrared region (8-14μm). Compared with the fog/low stratus and underlying surface (except snow), the medium/high clouds maintain a greater vertical height. Accordingly, the brightness temperature (BT) of the medium/high clouds is much lower than that of the fog/low stratus and underlying surface. Based on the BT difference, the removal of medium/high clouds can be executed by determining an appropriate BT threshold. For the BT values in the I5-EDR data files, to describe the statistical distribution helps determinate the threshold dynamically.

3.4. *Surface homogeneity test*

In terms of each preliminary test result, the fog/low stratus entity is expected to be fairly continuous and homogeneous. As a matter of fact, there exist a lot of isolated fog/low stratus pixels influenced by noise. In order to find and further eliminate the isolated pixels, we tested the surface homogeneity by calculating the standard deviation of gray value in every 3×3 pixel area where the central pixel appeared to be fog or low stratus in the preliminary test result. Only if it falls below a certain threshold (0.5), the central pixel is eventually identified as a fog/low stratus pixel. The threshold was determined empirically by continuous monitoring of operational processing.

4. Validation Experiments

To validate the feasibility of the algorithm, an experiment has been performed on a typical case of heavy fog/low stratus occurred in China, 2012. The satellite data collected at 1904 UTC 2 December 2012 are employed. For the ground measured data which are recorded every 3 hours, we applied the data of 1800UTC 2 December 2012. The ground meteorological stations discriminate fog and low stratus by the criteria of horiziontal visibility (HV), relative humidity (RH), and cloud base height (CBH): $HV \leq 10000\,m$ (fog: $HV \leq 1000\,m$, light fog: $1000\,m < HV \leq 10000\,m$), $RH \geq 95\%$, and $CBH \leq 2500\,m$. As is shown in Figure 2, the fog/low stratus results are obtained by the algorithm and verified by the ground measured results.

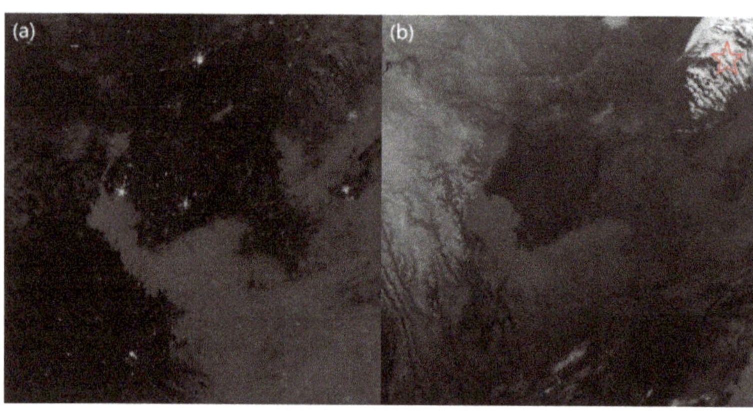

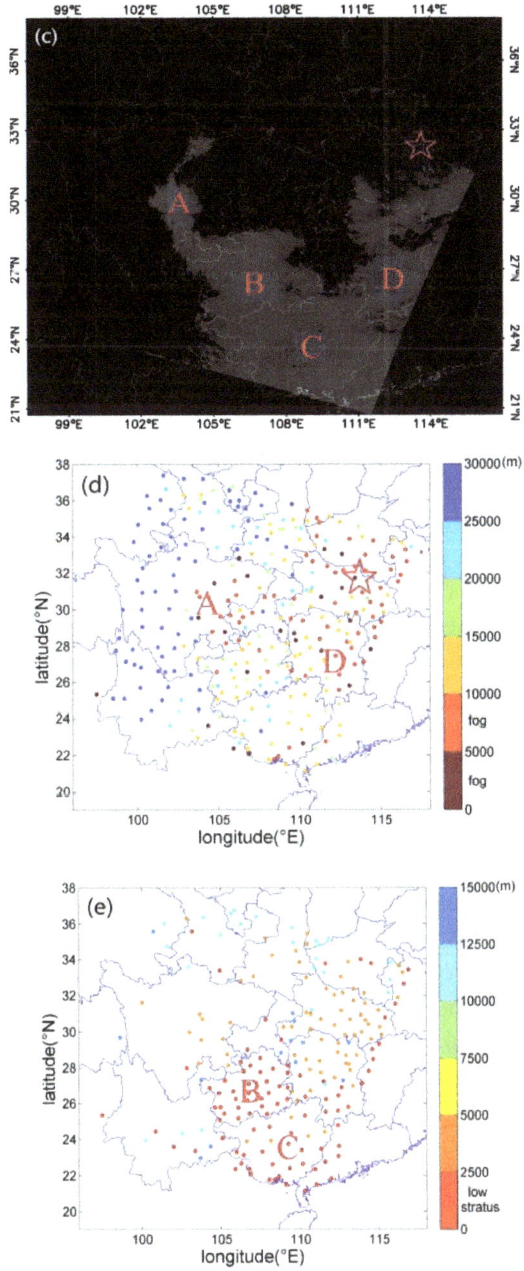

Figure 2. Experimental results of case 1 (a) Low-light-level image of DNB; (b) Infrared image of I5; (c) Monitoring result of the algorithm; (d)Ground measured result of HV; (e) Ground measured result of CBH

In Figure 2 (c), the gray area represents the fog/low stratus monitored by satellite. In Figure 2 (d, e), the ground mesured results are displayed as points with colors. For case 1, Figure 2 (c) shows that the fog/low stratus distributes in Northern China and adjacent North Korea, marked with letters: ‘A’ (central region of Liaoning Province), ‘B’ (the border between China and North Korea), and ‘C’ (East of Shandong Province). The corresponding regions are verified in Figure 2 (d).

The experimental results show that the fog/low stratus area monitored by the algorithm almost coincides with the ground measured result. However, the algorithm is found invalid in cases of overlapping clouds (marked with pentagrams), for only the uppermost layer can be detected reliably.

5. Conclusions

The DNB low-light-level sensor with accurate radiometric calibration and splendid spatio-temporal resolution has shown its vast prospect in promoting the development of nighttime sounding techniques. However, the research on its quantitative environmental applications has been still at the initial stage. Aiming at the topic of nighttime fog/low stratus detection, a muti-channel threshold algorithm is proposed based on DNB and other VIIRS channels. The algorithm regards the fog/low stratus as targets, and the underlying surface, snow, and medium/high clouds as backgrounds. Firstly, by analyzing the reflectance or radiation difference between the targets and backgrounds in DNB, I1, I2, I3, and I5 channels, a series of thresholds for removing the backgrounds are determined and implemented to get a preliminary test result. Secondly, according to the continuous and homogeneous characteristic of the targets, the surface homogeneity test is completed to further eliminate the noises in the preliminary test result, and then work out the result of nighttime fog/low stratus detection. Lastly, a typical cases of heavy fog/low stratus are analyzed with certain experiments to validate the effect of the algorithm. The results verify the high accuracy of the algorithm from both qualitative and quantitative aspects.

Although the present algorithm has achieved a satisfactory performance in the validation experiments, several problems remain to be settled in future research. For one thing, the Otsu method identifies the best threshold of gray value precisely under the conditions that a marked difference exists between targets and backgrounds. The practical gray level histograms, however, occasionally present multiple inconspicuous troughs due to noise interference, which imposes a bad influence on the best threshold acquiring. A timely manual identification therefore may be needed in such situations. For another, the algorithm proves to be invalid in cases of overlapping clouds, for the satellite can not obtain radiation information through the whole atmospheric layer. Additionally, the fog/low stratus are treated as one type in this paper, therefore the feasibility of discrimination between fog and low stratus using NPP data requires further study.

Acknowledgements

The authors would like to thank the National Oceanic and Atmospheric Administration (NOAA) for making the S-NPP/VIIRS data products publicly available. We acknowledge the S-NPP/VIIRS science team for the high quality products. The S-NPP/VIIRS data products were downloaded from http://www.class.ncdc.noaa.gov/saa/products. We also thank the China Meteorological Administration (CMA) for providing ground-based meteorological data. This work was supported by the National Natural Science Foundation of China (Grant No. 41375029).

References

1. G. E. Hunt, Radiative properties of terrestrial clouds at visible and infra-red thermal window wavelengths, *Quart. J. Roy. Meteor. Soc.***99**, 420 (1973).
2. J. R. Eyre, J. L. Brownscombe and R. J. Allam, Detection of fog at night using advanced very high resolution radiometer (AVHRR) imagery, *Mete. Mag.***113**, 1346 (1984).
3. A. Dybbroe, Automatic detection of fog at night using AVHRRdata, *Proceedings 6th AVHRR Data Users' Meeting,* (Belgirate, Italy, 1993).
4. J. Bendix, A satellite-based climatology of fog and low-levelstratus in Germany and adjacent areas, *Atmos.Res.***64**, 1 (2002).
5. S. Chaurasia, V.Sathiyamoorthy, B.Paul Shukla, et al, Night time fog detection using MODIS data over Northern India,*Meteor. Appl.***18**, 4 (2011).
6. G. P. Ellrod, Advances in the detection and analysis of fog at night using goes multispectral infrared imagery,*Wea. Forecasting.***10**, 3 (1995).
7. X. K. Zhou, W. Yan, H. Bai, et al, Detection of heavy fogs and low clouds during nighttime using DMSP-OLS data, *Remote Sens. Inf.***27**, 6 (2012).
8. T. E. Lee, S. D. Miller, F. J. Turk, et al, The NPOESS VIIRS day/night visible sensor, Bull. Amer. Meteor. Soc. **87**, 2 (2006).
9. S. D. Miller, S. P. Mills, C. D. Elvidge, et al, Suomi satellite brings to light a unique frontier of nighttime environmental sensing capabilities, *Proc. Natl. Acad. Sci.U.S.A.* **109**, 39 (2012).

10. C. Seaman, D. Hillger, T. Kopp, et al,Visible Infrared Imaging Radiometer Suite (VIIRS) ImageryEnvironmental Data Record (EDR) User's Guide, (2014), http://www.start.nesdis.noaa.gov/jpss/VIIRS.
11. G. Mie, Contributions to the optics of turbid media, particularly of colloidal metal solutions, *Ann.Phys.***25**,3 (1908).
12. N. Otsu, A threshold selection method from gray-level histograms, *Automatica.***11**, 285-296 (1975).
13. R. G. Crane and M. R. Anderson, Satellite discrimination of snow/cloud surfaces,*Int. J. Remote Sens.***5**, 1 (1984).
14. D. K. Hall, G. A. Riggs, V. V. Salomonson, et al, Development of methods for mapping global snow cover using moderate resolution imaging spectroradiometer data, *Remote Sens. Environ.* **54**, 2(1995).
15. A. G. Klein, D. K. Halland G. Riggs, Improving snow-cover mapping in forests through the use of a canopy reflectance model, *Hydrol. Processes.* **12**, 10-11 (1998).

Design of a tobacco leaf maturity degree detection device based on image processing technology*

Qiang Wang, Hai-Ping Si and Lei Xi, Guang Zheng†
College of Information & Management Science, Henan Agricultural University, Zhengzhou, 450002, China
†*E-mail: ggzan@163.com*

To develop a means for the rapid quantization and noninvasive measurement of the maturity degree of tobacco leaves in the fields, a tobacco leaf maturity degree detection device was designed based on image processing technology. Through the collection of image data from mature tobacco leaves, the leaves' chlorophyll content and the SPAD values were measured. The color pattern of the tobacco leaf image was converted from the RGB color model to the HSV color model by transforming and processing the tobacco leaf image. Adopting the linear regression analysis method, the functional relationship between the HSV color value, chlorophyll content value and SPAD value of the tobacco leaf image was established. Based on the functional relationship between the maturity degree and SPAD value, the relation model between color values of the tobacco leaf image and the tobacco leaf maturity degree was set up. The hardware of the tobacco leaf maturity degree detection device was then designed, followed by the device's application software. The results obtained from application testing showed that the device has the advantages of strong operability, non-destructive determination, and on-site measurement. The employing of this device could allow for rapid quantitative determination of the maturity degree of tobacco leaves in the field and provide a basis for deciding the optimal harvesting time.

Keywords: Image processing; tobacco leaf maturity degree; detecting device; HSV color model.

1. Introduction

The tobacco leaf maturity degree has great influence to the output and quality of the tobacco leaf in the harvesting period. The maturity degree of tobacco leaf was closely related with the color, aroma and taste of tobacco leaf and the premise of the guarantee and improvement of the product quality and appearance quality of the tobacco leaf. It was also the first element of the grade tanking in the grading standard of the flue-cured tobacco[1–4]. In the area of the

* This study has been funded by China National 863 Plans Projects (Grant No. 2008AA10Z220), also supported by special program of Henan Province (Grant No. 131100110400)

† Corresponding author: Guang ZHENG

mature tobacco leaf harvesting, the foreign researcher had carried out systematic research. For example, the American researcher adopted the method of conducting the chemical composition analysis to the picked tobacco leaf sample a week in advance to decide whether the tobacco leaf was mature; The Zimbabwe researcher employed the color of the tobacco leaf color picture and the quantitative index of the curing barn test and the drawer test to measure the maturity of the tobacco leaf; The Japanese researcher used the colorimetric compare by the color atlas to decide the maturity of the tobacco leaf and so on [5–8]. In China, the maturity of the tobacco leaf was judged by the appearance and shape features such as the leaf color, leaf vein, leaf nap, leaf morphology and so on. In Henan Agricultural University, the researcher employed the colorimetric method to identify the maturity of the tobacco leaf through the color card produced by the painted stiff paper. Moreover, some other methods were still used to identify the maturity of the tobacco leaf through the appearance quality, inherent quality, chemical component, physical characteristics and other aspects[9–11]. All the research about the judgment of the maturity of the tobacco leaf had the ambiguity and the subjectivity for the description of appearance of the tobacco leaf. Likewise, these studies also lacked the instantaneity and veracity of the on-the-spot judgment in the field. The lack of the systematic and accurate quantitative research for the evaluation of the tobacco leaf maturity became the bottleneck of the tobacco leaf quality improvement in our country[5,7–8,12–13]. Therefore, adopting the image processing techniques and the linear-regression analysis methods, through the construction of the quantitative relation model between the tobacco leaf image color and the tobacco leaf maturity, the tobacco leaf maturity detector using the tobacco leaf image as the input data was designed in this paper.

2. System Structure

As shown in Figure 1, the tobacco leaf maturity detector was composed of the image capture unit, the data input unit, the processing control unit and the data output unit or the data storage unit. The processing control unit controlled the image capture unit to collect the image data of the detected tobacco leaves. Through the processing of the collected image data, the processing control unit could obtain the H, S color component values. After gaining the phyllotaxis values of the detected tobacco leaves through the data input unit, the processing control unit would work out the MD, SPAD values of the grade of maturity and the chlorophyll content of the detected tobacco leaves on the basis of the tobacco leaf maturity model. Finally, these data would be sent to the data output unit for on screen display or the data storage unit to be saved.

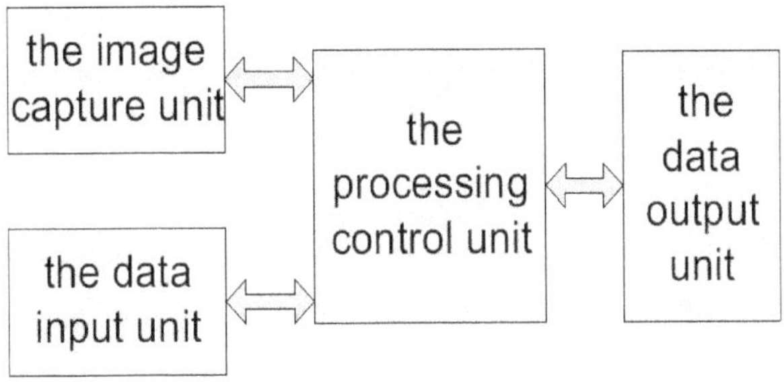

Fig. 1. The functional structure of the detector.

The circuitous philosophy of the tobacco leaf maturity detector was shown in Figure 2. The image capture unit was made by the camera or the vidicon. The data output unit was the keyboard, touch screen or the light pen. The processing control unit could use one of the 51 series microcontroller, AVR microcontroller, PIC microcontroller, ARM processor and the matched intelligent terminal (for example: smart phone). The data output unit was the viewing screen or the miniprinter[14].

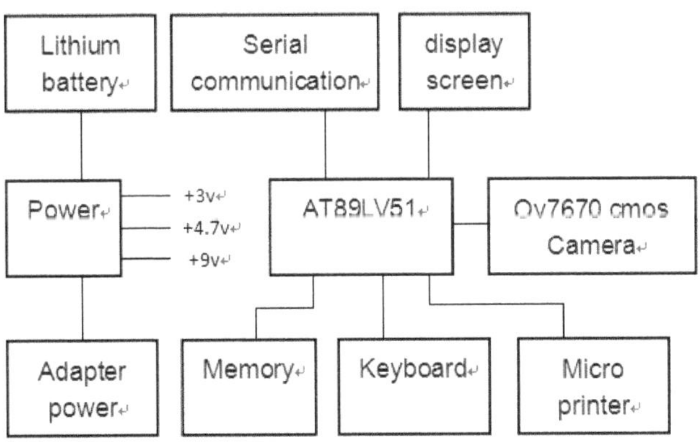

Fig. 2. The circuitous philosophy of the tobacco leaf maturity detector.

3. Technology Implementation

3.1. *The model building of the detecting algorithm*

The Sony DSC-F707 digital camera was used to shoot the front image of the tobacco leaves. Moreover, the RGB image of the tobacco leaves was gained for obtaining the data of the collected tobacco leaves sample. Using the acetone extraction spectrophotometry, the chlorophyll a and chlorophyll b values of the samples was tested. The SPAD values of the tobacco leaves were also measured using the SPAD-502 chlorophyll meter. In the process of the determination, the averagely distributed 8 positions of the sample blade were selected to be tested. The average value of the SPAD values of these 8 positions was validated the SPAD values of the tobacco leaf.

3.1.1. *The relation model of the color feature and the chlorophyll content*

Using the PHOTOSHOP CS3 and VISUAL BASIC 6.0, the tobacco leaf images were processed. The RGB color model of the tobacco leaf images was transformed to the HSV color model. The obtained HSV color values were contrasted with the tested chlorophyll content. After that, the linear regression analysis method was used to build up the relational model between the HSV values of the tobacco leaf images and the chlorophyll content values (a+b). The specific computational formula was shown in Eq. (1) [15–16].

$$CHL = 10.498\left(\frac{H}{H+S}\right) - 4.8807 \tag{1}$$

In the formula, the CHL means the content values of the chlorophyll (a+b); the H, S were the value H and the value E in the HSV color model.

3.1.2. *The relational model between the chlorophyll content and the SAPD values*

The tested chlorophyll content values of the tobacco leaves and the SPAD values were imported in EXCEL 2003 and analyzed. Furthermore, the Relational Model between the Chlorophyll Content and the SAPD values was built, as shown in Eq. (2).

$$SPADNA = 38.393(CHL) - 2.8236 \tag{2}$$

In Eq. (2), the SPADNV means the standardized SPAD values of the tobacco leaf; the CHL was the chlorophyll content value (a+b).

3.1.3. *The relational model between the color features of the tobacco leaves and maturity of the tobacco leaves*

As shown in Eq. (3)-(5), the criteria of the tobacco leaf maturity could refer to the Meter Reading Model of the Chlorophyll and the Maturity of Tobacco Leaf—TMDSPADV which get from the Li folin[7].

$$MD = \begin{cases} 0, SPADNV > 40 \\ 1, SPADNV \in (30,40] \\ 2, SPADNV \in (25,30] \\ 3, SPADNV \in [15,25] \\ 4, SPADNV < 15 \end{cases} \tag{3}$$

$$\text{SPADNV= f SPADV} \tag{4}$$

$$f = \begin{cases} 0.737, PHY \in [1,7] \\ 1, PHY \in [8,14] \\ 1.331, PHY \in [15,21] \end{cases} \tag{5}$$

In these equations, the MD means the grade of maturity; the SPADV was the SPAD values of the tobacco leaf; the SPADNV was the standardized value of the SPAD of the tobacco leaf; f was the coefficient; the PHY was the phyllotaxis from top to bottom of the tobacco leaf. The meaning of the MD values evaluation was shown in Table 1.

Table 1. The Grading Standard of the Tobacco Leaf Maturity

Grade of Maturity	Maturity Feature
Immature Leaf(M0)	dark green leaf color, not yellowing, main vein all green, fuzz not fall off completely.
Undercuring Leaf (M1)	light green leaf color, just yellowing, 2/3 of main vein whitening, green branch vein, a few fuzz fall off.
Maturity in a Mellow Range (M2)	olivine leaf color, main vein all white, 1/3 of branch vein whitening, part of fuzz fall off, leaf opex down-turned slightly.
Maturity (M3)	yellow leaf color, green less than yellow, appear white color in yellow leaf surface, main vein whole whitening, 2/3 of branch vein whitening, most of fuzz fall off, blade surface were covered with yellow spots all over, leaf opex and leaf margin whitening, , leaf opex down-turned, slight wither leaf tip and sallow leaf side.
Postmaturity (M4)	main vein whole whitening and shining, appear white color in yellow leaf Surface, most of fuzz fall off, leaf auricle all yellow, many Gymnosporangium haraeanum-liked plaques, Wither leaf tip and sallow leaf side.

As shown in Figure 3, on the basic of the function relationship of the formula1-5, the relational model between the HSV color feature values of the tobacco leaves and the maturity of the tobacco leaves was established, as shown in Formula 6[15–16].

In the Eq.(6), the MD means the grade of the maturity; the evaluation meaning was shown in the Table 1; the meaning of the other variable was same with Eq. (1)-(5).

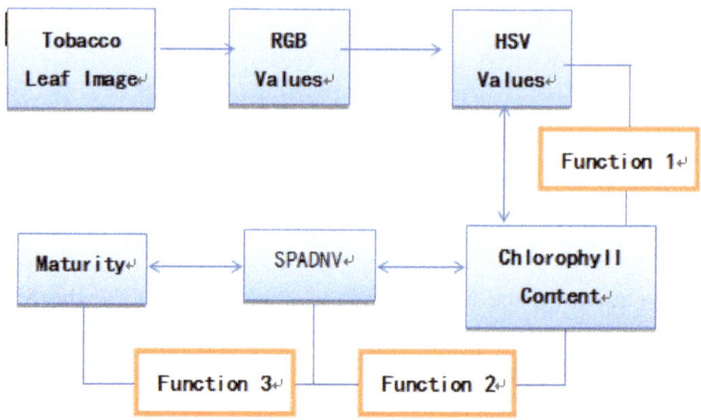

Fig. 3. The function relationship between each data of the tobacco leaf.

$$MD = \begin{cases} 0, H/(H+S) > 0.0992f + 0.4579 \\ 1, H/(H+S) \in (0.0744f + 0.4579, 0.0992f + 0.4579] \\ 2, H/(H+S) \in (0.0620f + 0.4579, 0.0744f + 0.4579] \\ 3, H/(H+S) \in [0.0372f + 0.4579, 0.0620f + 0.4579] \\ 4, H/(H+S) < 0.0372f + 0.4579 \end{cases} \tag{6}$$

3.2. *The design of the system software*

On the basis of the design of the system structure principle and the circuitous philosophy, this tobacco leaf maturity detector could choose corresponding hardware configuration to conduct the assembling and integration according to the design requirements and get integrated tobacco leaf maturity detection hardware equipment. Based on the algorithm model of the tobacco leaf maturity detection, following the design principle of the computer software engineering, the software of the tobacco leaf maturity detection was designed and realized. The utility software not only could be solidified the storage of the checkout equipment and provide the operation-interface to the users directly, but also be installed on the hardware equipments according with the hardware requirements as the form of the third-party software. In this paper, the smart phone was adopted as the hardware platform for the detector. Following the detective algorithm model, adopting the VISUAL STUDIO 2005 as the software development platform, using the VISUAL C# as the development language, employing the WINDOWSCE as the operating environment, the corresponding software for the tobacco leaf maturity detection was developed. At the same time, the specific application test was also conducted. The system contains 3 main functions including data inputting, data processing and operation and data outputting. The data inputting included the inputting of the tobacco leaf images and the corresponding tobacco leaf phyllotaxy. The image inputting obtained the tobacco leaf images through the camera of the smart phone. Moreover, the images inputting also could choose the obtained tobacco leaf images from the storage equipment directly, as shown in Figure 4.

The data processing and operation included the extraction of the image color, the conversion of the image color mode, the importing of the detection model and the calculation of the detection results. Through the methods of the GetPixel(x,y).R, GetPixel(x,y).G and GetPixel(x,y).B, the R, G, B color values of the pixels of the tobacco leaf regional pictures get from the inputted images were extracted. The RGB color values were converted to the corresponding HSV values of the HSV color model. These HSV values were transmitted to the system and conducted the operation by the core algorithm including Ylab( ), SPAD( ) and MD( ). The obtained chlorophyll content values of the tobacco leaf, the calculated SPAD values and the grades of maturity were exported to the display screen, as shown in Figure 5.

Fig. 4. The inputting of the tobacco leaf images and the phyllotaxy

Fig. 5. The outputting of detection results

4. Application Effect

The grade of maturity of the randomly collected 80 K 326 tobacco leaf samples in the Scientific & Educational Park of Henan Agricultural University were tested by the designed checkout equipment in this paper. The tested results were compared with the determinate maturity grade results by the color identification method. The comparison result was shown in Table 2. The coincidence rate of the measured tobacco leaf samples maturity was 89%.

Table 2. The comparison of results between using the detector and the colorimetric method

Category Judgment by Colorimetric Method	Category Calculation by TMDHSV Model					Total	Same Quantity	Coincidence Rate(%)
	M0	M1	M2	M3	M4			
M0	5	1	0	0	0	6	5	83
M 1	0	8	1	0	0	9	8	89
M 2	0	1	16	1	0	18	16	89
M 3	0	0	2	35	2	39	35	90
M 4	0	0	0	1	7	8	7	88
Total	5	10	19	37	9	80	71	89

Figure 6 was the relational graph of the tested results used the detector in this paper and the colorimetric method. The dependency between the prior verified maturity category and the calculated maturity category by the model was very good; the correlation coefficient was 0.9013; the straight slope reached up to 0.9394; the intercept closed to 0. All the results showed that the tobacco leaf maturity detector had better application effect and higher accuracy rate.

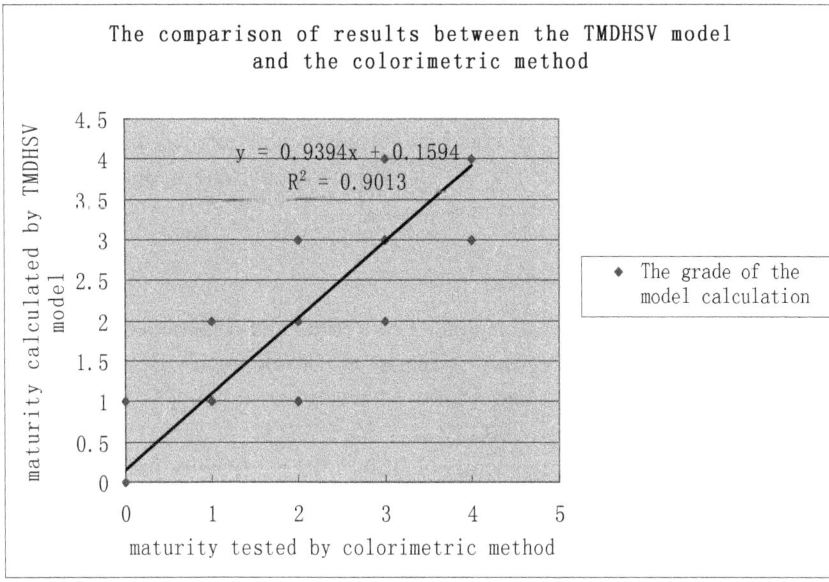

Fig. 6. The comparison of results between using the detector and the colorimetric method

5. Conclusions

Rapidly quantitative determination of the tobacco leaf maturity could provide the decision basis for the matured harvest of the tobacco leaf. In this paper, the relational model among the image data of tobacco leaves, chlorophyll content, SPAD values and grades of maturity was constructed by the linear regression analysis method; the hardware structure and the circuit configuration of the tobacco leaf maturity detector was designed; the detection software of the tobacco leaf maturity was developed. The designed detector was also tested and applied. The application results showed that the detector had higher testing results and better application effect. The characteristics of the detector as follows:

(1) This tobacco leaf maturity detector could conduct rapid operation in the fields and provide decision basis for the timely harvesting of the tobacco grower. The detector had the advantages such as the on-site detection, living body lossless testing and quantitatively intuitive detection result.

(2) During the detecting algorithm model establishment, through abandoning the V (luminance) component in the HSV color components, the influence

of the luminance to the algorithm model was excluded. The influence of the factors such as weather or illumination to the detection results was also shielded during field test in the farm. Thus, the detection could be conducted in the fields directly and not need move back to the laboratory with specific light source after harvesting.

(3) The tobacco leaf maturity detector could be designed and produced. The detection software of it could be installed and applied in third party platform such as smart phone. The popularization and application of the total system was convenient and flexible.

References

1. Zhao Yuankuan, Chen Jianghua. The Results and Experience of the High Quality Tobacco Leaves Cooperative Development Between China Tobacco Corporation and Philip Morris Corporation (2) [J]. Tobacco Science & Technology, 2000, (8): 34–37.
2. Zhu Zunquan. The Key Factors of Restricting the Quality Improvement of the Tobacco Leaf and Cigarette[J]. Tobacco Science & Technology, 1998, (4): 3–4.
3. Zhou Jie, Yang Hongqi, Zhou Jiheng, et al. Difference of Flue-cured Tobacco's Chloroplast of Main Nerve in Different Maturing[J]. Journal of Hunan Agricultural University: Natural Sciences: 2007, 33(2): 174–176. (in Chinese)
4. Lu Tiansheng, Liu Hanchuan, Shao Fuwen, et al. The Research between the Different Maturity and Internal and Appearance quality in Flue-cured Tobacco 40 Level Standard[J]. Journal of Anhui Agricultural Sciences, 1994, 22(3): 256–258.
5. Xuan Xiaoquan, Bo Yunchuan, Xu Ruyan, et al. Studies on analysis of flavor constituents of different maturity tobacco leaves[J]. Chinese Agricultural Science Bulletin, 2007, 23(2): 98–102. (in Chinese)
6. Gao Hanjie, Chen Hanxin, Peng Shiyang, et al. Distinguishing of Tobacco Leaves Maturity and Application of Five-parted Curing Technology[J]. Chinese Tobacco Science, 2002(4): 39–41. (in Chinese)
7. Peedin G F. Effects of Nitrogen Rate and Ripeness at Harvest on Some Agronomic and Chemical Characteristics of Flue-cured Tobacco[C]// Coresta Congress, Agro-Photo Groups, 1995, 7.
8. Li Folin, Zhao Chunjiang, Liu Liangyun, et al. Quantitation of Maturity of Fresh Flue-cured Tobacco Leaf[J]. Tobacco Science and Technology, 2007 (1): 54–58. (in Chinese)

9. Xu Guanghui, Hu Xiaohong, Xiong Shuping, et al. Relationship Between Chlorophyll Concentration and Color Feature of Tobacco Leaf[J]. Journal of Henan Agricultural University, 2007, 41(6): 600–604. (in Chinese)
10. Yan Keyu. The Grading of Tobacco Leaf[M]. China Agriculture Press, Beijing, 2003.
11. Yang Shuxun. The Primary Exploration of the Accurate Judgment for Harvested Tobacco Leaves Maturity[J]. Chinese Tobacco Science, 2003(4): 34–36.
12. Cun Guomin. The Standard of Tobacco Leaves Maturity[J]. Yunnan Aagriculture, 2006(9): 16–17.
13. Li Folin, Chen Fenglei. The Theory and Practice of Tobacco Commodity Science[M]. Yunnan Science and Technology Publishing Press, Kunming, 2004:901.
14. Li Folin, Zhao Chunjiang, Wang Jihua, et al. Spectral Characteristic of Flue-cured Virginia Tobacco Leaves of Different Maturity Grades and Spectra Discriminant Classification[J]. Journal of Fujian Agriculture and Forestry University: Natural Science Edition, 2008, 37(6): 565–569. (in Chinese).

Design of an experiment rig for traffic light control system based on freescale MCU

Miao-Zhong Sun†, Shi-Hai Cui, Bing-Fa Tan

College of Mechanical Engineering, Tianjin University of Science & Technology, No.1038 Dagu South Road, Hexi District, Tianjin City, Tianjin 300222, PR China

Tianjin Key Laboratory of Integrated Design and On-line Monitoring for Light Industry & Food Machinery and Equipment

†E-mail: sunmzh66@sina.com, shihaicui@tust.edu.cn, tanbf@tust.edu.cn

This paper presents hardware and software designs of an experiment rig for traffic light control system based on Freescale MC9S12D64 MCU. In the design process, time counting of lights is controlled by comparison of adjusted output pulse signal and interrupt subroutines with the help of the enhanced capture timer (ECT). The circuits of traffic lights linking to MCU and the circuits of five nixie tube display are designed. The control programs are performed and the main codes are analyzed. The efficiency of the system is verified using the designed experiment rig. The system can be able to simulate the control process of actual traffic light.

Keywords: MC9S12D64; Traffic Light; Hardware Design; Software Design.

1. Introduction

It has been reported much that Intel 8051 series MCU (Microcontroller Unit) was applied to the design of traffic light control system[1–3]. But recently Freescale MCU has been developed rapidly and has been widely used in the areas of automobile electron and industry control etc. We owe to it the strength of its performance such as the easy programming of C language and the function of BDM (Background Debug Module)[5]. In this paper, we use the Freescale MC9S12D64 MCU to design imitatively a control system of traffic light on an experiment rig. In the hardware design, seventeen I/O pins are used to control respectively twelve LED traffic lights and five little passing yellow lights. Five nixie tubes display passing time, warning time and forbidding time. MCU power circuit and oscillation circuit are also designed. In the software design, C language is employed to program for the system. The programming flow chart and the main programming codes are presented.

2. Control Project of Traffic Light

The traffic light control system must simulate the actual city traffic light system. The project should be performed as follows:

(1) The system must illuminate normally red lights, yellow lights and green lights in the east-west crossing and south-north crossing.
(2) When the east-west is permitted, the east-west green lights and the south-north red lights are illuminated simultaneously.
(3) When the south-north is permitted, the east-west red lights and the south-north green lights are illuminated simultaneously.
(4) The passing time is set to be 40 second according to the south-north road as the main one, and the forbidding time is 30 second due to the east-west road as the secondary main one.
(5) When the green light time of the main road decreases to 3 second, the yellow lights of the main road start to be illuminated to warn vehicles and people. Three second later, the red lights of the main road are illuminated. At the same time, the green lights of the secondary main road turn to be lighted.
(6) When the green light time of the secondary main road counts to 3 second, the yellow lights of the secondary main road start to be illuminated to warn vehicles and people. Three second later, the red lights of the secondary main road are illuminated. The green lights of the main road turn to be lighted meanwhile.
(7) Five nixie tubes display passing time, warning time and forbidding time.
(8) Five little passing yellow lights are set to show that the vehicles are going through which the road is.

3. System Hardware Design

The MC9S12D64 MCU is a 16-bit device composed of standard on-chip peripherals including a 16-bit central processing unit (HCS12 CPU), 64K bytes of Flash EEPROM, 4Kbytes of RAM, 1K bytes of EEPROM, an 8-channel IC/OC enhanced capture timer, two 8-channel, 10-bitanalog-to-digital converters (ADC), an 8-channel pulse-width modulator (PWM), 11 I/O ports including 91 I/O pins etc. The MC9S12DJ64 has a function of background debug module. With the help of this function a compiled program of C language can be downloaded easily to the MCU flash memory. The MCU PLL circuit allows power consumption and performance to be adjusted to suit operational requirements [4]. Figure 1 indicates the framework of traffic light control system.

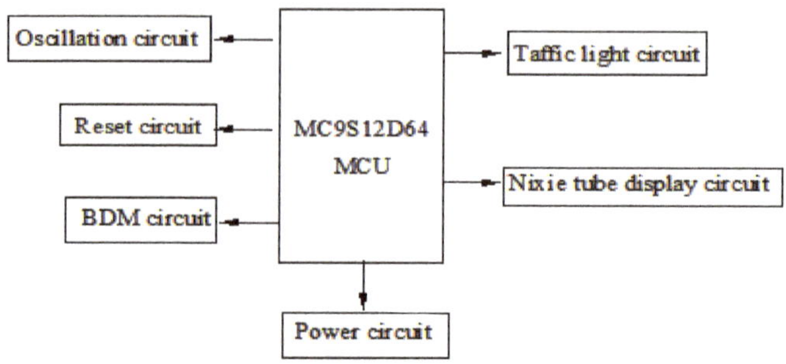

Fig. 1. Framework of traffic light control system.

3.1. *Minimum system circuit*

Minimum system circuit consists of reset circuit, oscillation circuit and BDM circuit etc. Figure 2 displays the minimum system circuit.

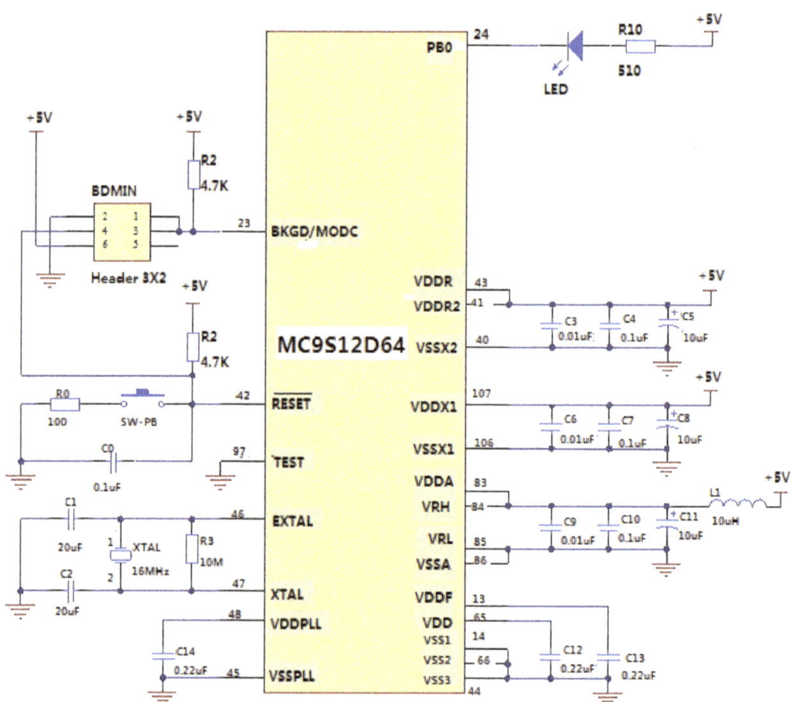

Fig. 2. Minimum system circuit.

3.2. *Design of LED traffic light and nixie tube circuit*

The traffic light circuit uses twelve I/O pins to be connected with red, green and yellow LED lights of the main road and secondary main road. The another five pins link to the five little passing yellow lights. Table 1 denotes the condition of MC9S12D64 I/O pins connected with LED traffic lights.

Table 1. MC9S12D64 I/O pins are connected with LED traffic lights

No.	I/O pin	Location	Light	No.	I/O pin	Location	Light
1	PA7(64)	north	red	5	PA4(61)	east	red
2	PA6(63)	north	yellow	6	PA3(60)	east	yellow
3	PA5(62)	north	green	7	PA2(59)	east	green
4	PM5(100)	south-north passing	Little yellow	8	PA1(58)	east-west passing	little yellow
9	PM7(87)	south	red	13	PK7(108)	west	red
10	PB2(26)	south	yellow	14	PE2(54)	west	yellow
11	PB0(24)	south	green	15	PE3(53)	west	green
12	PM6(88)	south-north passing	little yellow	16	PM4(101)	east-west passing	little yellow
17	PB7(31)	center passing	little yellow				

The traffic light time is displayed with the five nixie tubes that assemble the TEC9607 chip. The TEC9607 uses a driver circuit of serial common-anode CMOS5×8.It consists of clear circuit, oscillation circuit, bit scanning driver circuit, 5×8 bit shifting register circuit, 5×8 bit data latch circuit, multi-channel selector, common-anode driver circuit. It can connect with a serial port or a parallel port of a lot types MCU and offer specially five common-anode LED nixie tubes to display decimal data [6-7].

MC9S12D64 MCU transfers data to TEC9607 through the SPI communication. In order to perform the process, TEC9607 serial bit shifting pin(SCP) must link to MC9S12D64 pin PWM2/KWP2/PP2(2), data latching pin(LCP) to XADDR15/PK1(7) and serial input data pin(SI) to PWM1/KWP1/PP1(3). Figure 3 shows the circuit of the MCU linking to the LED traffic lights and five nixie tubes.

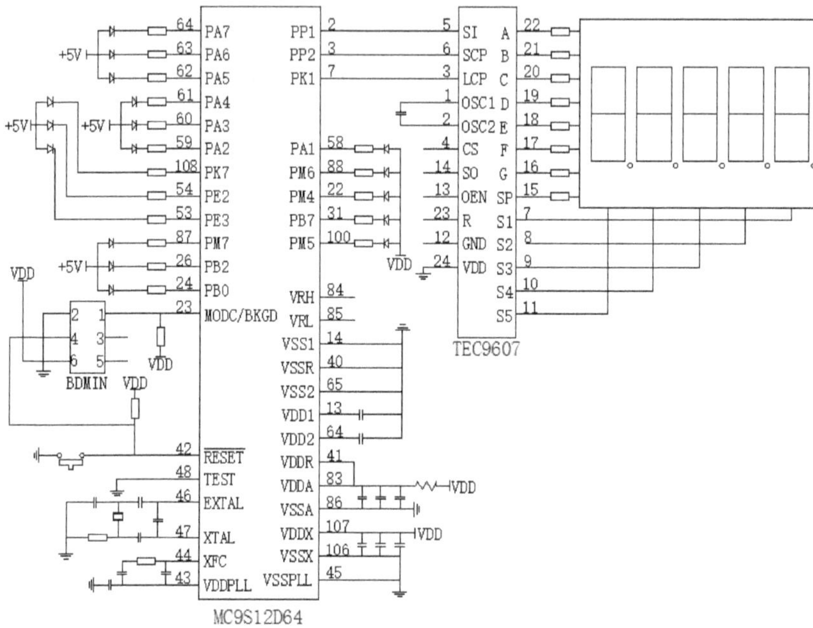

Fig. 3. MC9S12D64 I/O pins link to the LED traffic lights and five nixie tubes.

4. System Software Design

4.1. *Programming flow chart*

The Freescale MCU has more superiority than the type 8051 MCU in aspect of programming. For example, I/O ports can easily control the LED lights through DDRx register and PORTx register. The C language program compiled by PC can be easily downloaded to the MCU flash memory and erased with more than one hundred thousand times. We use CodeWarrior5.1 to compile the C language program of the traffic light control system. The programming flow chart is presented in Figure 4.

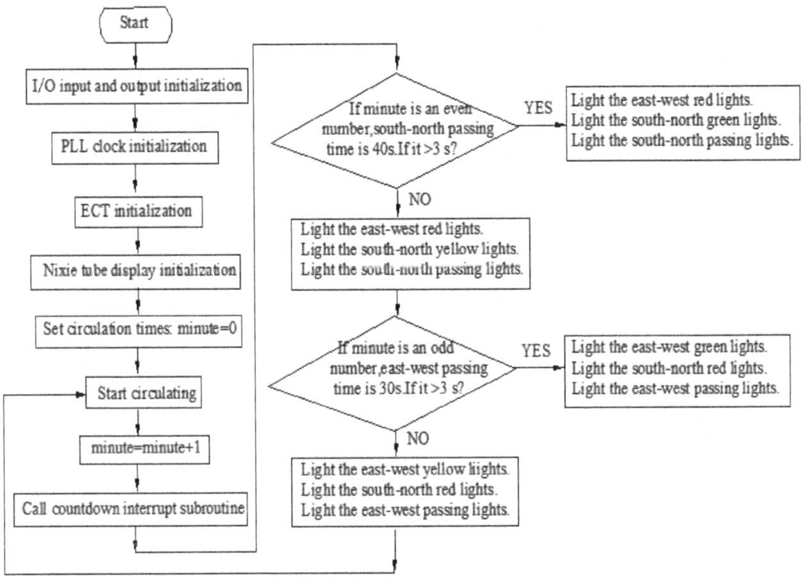

Fig. 4. Programming flow chart

4.2. *Programming codes and explanation*

MC9S12D64 MCU can control high or low voltage for an output I/O pin to illuminate or not illuminate the corresponding traffic light accordance with requirements. The main programming codes are introduced and explained as follows:

```
for(;;)// Circulate all the time
{ EnableInterrupts;    //Enable interrupt.
if(!counter)   //If Counter is not equal to zero.
ledCONTROL(second);// Display passing time according to interrupt.
if(minute%2==0&ew1==1&second>>3)//If circulation times minute is an even number  the south-north passing time is 40s, and the time>3?
{PTM_PTM5=1;PTM_PTM6=1;PORTA_BIT1=0;PORTB_BIT7=0;PTM_PTM4=0; PORTA_BIT7 = 0;PORTA_BIT2 = 0;PTM_PTM7 = 0;PORTE_BIT3 = 0;  }//Yes, Light the south-north green lights. Light the east-west red lights. Light the south-north passing lights. Don’t light the other lights.
if(minute%2==0&ew2==1&second<=3)// If circulation times minute is an even number, the south-north passing time is 40s, and the time<=3?
{ PORTA_BIT7 =0;PORTA_BIT2=1;PTM_PTM7= 0;PORTE_BIT3 =1;
```

PORTA_BIT3=0;PORTE_BIT2=0; }// Yes, Light the south-north yellow lights. Light the east-west red lights. Light the south-north passing lights. Don't light the other lights.

if(minute%2==1&sn1==1&second>>3) //If circulation times minute is an odd number, the east-west passing time is 30s, and the time>3?

{PORTA_BIT1=1; PTM_PTM4=1; PTM_PTM5=0; PORTB_BIT7=0;

PTM_PTM6=0; PORTA_BIT7= 1; PORTA_BIT2 = 1; PTM_PTM7 = 1;PORTE_BIT3 = 1;

PORTA_BIT3=1; PORTE_BIT2=1; PORTA_BIT5 = 0;PORTA_BIT4 = 0;

PORTB_BIT0 = 0;PORTK_BIT7= 0; }//Yes, Light the south-north red lights. Light the east-west green lights. Light the east-west passing lights. Don't light the other lights.

if(minute%2==1&sn2==1&second<=3)// If circulation times minute is an odd number, the east-west passing time is 30s, and the time<=3?

{PORTA_BIT5 = 1;PORTA_BIT4 = 0;PORTA_BIT6=0; PORTB_BIT0 = 1;PORTK_BIT7 = 0; PORTA_BIT2=1;PORTB_BIT2=0; PORTA_BIT5 = 1;PORTB_BIT0 = 1;PORTA_BIT4 = 1;PORTK_BIT7 = 1; PORTA_BIT6=1; PORTB_BIT2=1; }//Yes, Light the south-north red lights. Light the east-west yellow lights. Light the east-west passing lights. Don't light the other lights.

}

When programming for the traffic light passing time, we first initialize PLL (Phase locked loop) clock and set the bus clock 24 MHz, then use the ECT output compare function to compare the clock pulses of ECT channel 6, set the ECT clock frequency division 32 and channel 6 (TC6) 37500.So the clock circle=(24000000/32)/37500= 20Hz=0.05s, 20 circulating times will be needed to form one second. We can set total second number to be decreased circularly and perform the countdown function. The programming codes of ECT countdown interrupt subroutine are listed as follows:

#pragma CODE_SEG __NEAR_SEG NON_BANKED//Minimum starting code, small memory mode.

void interrupt VectorNumber_Vtimch6 TimerCh6(void)// Interrupt subroutine name.

{ TFLG1_C6F = 1;//Set channel 6 as interrupt.

TC6+= 37500;//the clock pulse=(24000000/32)/37500=20Hz=0.05s.

counter += 1; //Count.

if (counter == 20)//Accumulate 20 times to 20*0.05=1 second.

{ second-=1;//Countdown.

```
counter=0; // Count again.
ew1=1;//Label east-west direction and passing time >37s.
sn1=1;// Label south-north direction and passing time>27s.
if(second = = 3){ew2=1;sn2=1;}//Label respectively south-north and
east-west yellow lights are lighted 3 s.
if((second==0)&(minute%2= =0) //If second is zero and circulation time
is an even number?
{ second=40;}//Set south-north passing time 40s.
if((second==0)&(minute%2= =1) //If second is zero and circulation
time is an odd number?
{second=30; }// Set east-west passing time 30s.
minute+=1; //Increase circulation time.
}
}
```

5. Experiment Testing

Finally we have developed an experiment rig assembling the traffic light control system. On the rig, we perform the testing to obtain Figure 5 that shows the running case of the traffic lights. We can see that two east-west red lights, two south-north green lights and three little passing yellow lights are lighting, and five nixie tubes are displaying time 27 second to explain that is permitted,

Fig. 5. The running case of the traffic lights on an experiment rig.

On the contrary east-west is forbidden. The three lighting passing yellow lights indicate that the south-north vehicles are going across the road. Through the testing, the system is proved to be successful, and it can simulate running case of actual city traffic light.

5. Conclusions

This paper utilizes the MC9S12D64 MCU to carry out a simulation design for the traffic light control system. Hardware is designed including MC9S12D64 minimum system, linking circuit of seventeen traffic LED lights and displaying circuit of five nixie tubes etc. C language is used to program for the traffic light control with the help of CodeWarrior5.1.After the program is tested repeatedly to be success, it is downloaded to the MCU flash memory using a BDM debugger. Finally the designed system is tested on an experiment rig, and it can simulate the running case of the actual traffic light.

References

1. D.M. Wang, J.Q. Zhang and J.Y. Lu, Design and Realization of Traffic Light Control System based on Micro Controller Unit, Journal of Jiamusi University (Natural Science Edition),Vol.27,(2009), pp.94–96.
2. L.P. Xiong, C.Y. Zhang and G.Y. Ma, The Control System Design of the Traffic Light Based on Single Chip Microcomputer, Electromechanical Engineering Techology,Vol.42, (2013), PP.80–82.
3. G.W. Wu, Design and simulation of traffic light system based on AT89C51 single-chip processor, Modern Technology, Vol.35, (2012), PP.144–146.
4. Motorola, Inc, *MC9S12DJ64 Device User Guide*, (2004).
5. Y. Ren, Y.D. Wang and W. He, *Application and Principle of MCU* (Tsinghua Press, 2012).
6. Y.X. Wang, Application of serial LED Decoder/Driver TEC9607 in Microcomputer PIC16C5X, Foreign Electronics and Electrical Appliance, Vol. 6,(2000), PP. 31–34.
7. J.K. Zheng and T. Pei, New LED Driver TEC9607 and Its Application, Electronic Component and Device Applications, Vol.5, (2003), PP.15–16, 20.

On-board measurements of CO/CO_2 emissions from city buses with different emission standards*

Kun You†, Yu-Jun Zhang, Yan-Wei Gao,
Chen Chen, Ying He and Wen-Qing Liu

*Key Laboratory of Environmental Optics and Technology,
Anhui Institute of Optics and Fine Mechanics,
Chinese Academy of Sciences, Hefei 230031, China*
†*Email: kunyou@aiofm.ac.cn*

In this paper, non-dispersive infrared analyzer (NDIR) methods were used to detect pollutant CO/CO_2 from vehicle exhaust. On-board emissions tests of four city buses in Macao were conducted using SEMTECH–FEM, an emissions measuring instrument provided by Sensors, Inc. We selected three diesel city buses, each of which complies with a different national emission standard (GB3, GB4 or GB5), and a new type of compressed natural gas bus for testing. We set up the four city buses to travel the same route with identical speed. The test results reflect that CO emissions decrease significantly when bus emission standards are enhanced. For the compressed natural gas bus, CO emission is several times lower than that of the diesel buses, at about only 200ppm, but its CO_2 emission is about 10%, which is twice that of the diesel buses. This test also verifies the feasibility and reliability of on-board CO/CO_2 emissions measurement based on NDIR.

Keywords: NDIR; On-board emissions tests; SEMTECH – FEM; Macao; Diesel; Compressed natural gas; CO/CO_2 emissions.

1. Introduction

Vehicle exhaust pollutants mainly include CO, CO_2, NO_X and total hydrocarbons (THC), which are great harmful to the survival of human and environment. CO which is colorless, tasteless and toxic is the intermediate of hydrocarbon fuel combustion, and it is exhausted by Vehicle caused by the incomplete combustion of hydrocarbon [1–3]. Hydrocarbon fuel combustion completely with air generated greenhouse gas of CO_2, as in equation (1).

* Work partially supported by grant 2-4570.5 of the Swiss National Science Foundation.
† This work is supported by National High Technology Research and Development Program of China (863 Program), No. 2014AA06A503

$$CmHn + (m + \frac{n}{4})O_2 = mCO_2 + \frac{n}{2}H_2O \tag{1}$$

However, when the air is insufficient, part of the fuel will not be able to complete combustion. H2 and CO will be produced in that way, as show in equation (2).

$$CmHn + \frac{m}{2}O_2 = mCO + \frac{n}{2}H_2 \tag{2}$$

Many countries include China have made more stringent vehicle emission standards in order to enforce manufacturers to improve product to reduce exhaust pollution and to prevent the vehicle emissions of harmful pollutants, and scrap vehicles which are overweight emissions. Therefore strengthen automobile exhaust emission measurement is great significance.

We used SEMTECH–FEM which is provided by Sensors Inc for on-board measurements CO/CO_2 emissions of four city buses in Macao, and these city buses comply with different emission standards. SEMTECH–FEM is an all-new fuel economy meter with unsurpassed accuracy and ease of use based on non-dispersive infrared analyzer (NDIR). Compared with the conventional electrochemical method, electric method or Chromatography, NDIR has lots of advantages, such as high sensitivity, high precision, high reliability, fast response, long life and wide measuring range, in addition, NDIR also can measure multi-component gas simultaneously and can continuous measurement. So SEMTECH–FEM is suitable for measuring CO/CO_2 emissions of vehicle exhaust.

2. Measuring Principle and Devices

2.1. *Measuring principle of operation*

Gas molecules absorb their infrared energy of corresponding wavelength and convert it into molecular internal energy when asymmetric double atoms and polyatomic molecule (such as CO or CO_2) irradiated by infrared beam. In another way, when the infrared light through the measured gas, gas molecules will absorb energy of infrared light, as shown in figure 1:

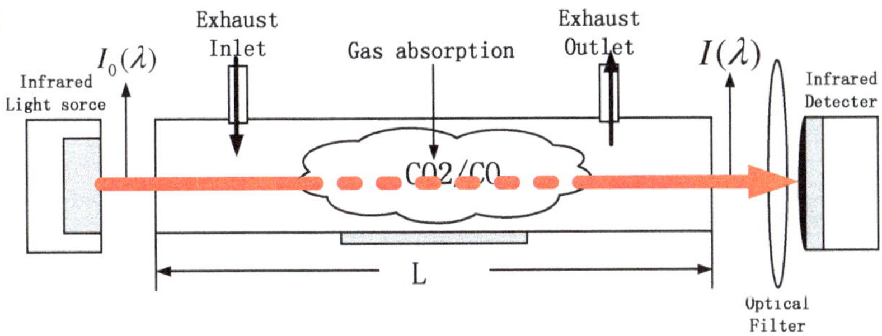

Figure 1. NDIR measuring principle

Absorption relations follow the Lambert- Beer law [4]:

$$I(\lambda) = I_0(\lambda)e^{-K(\lambda)CL} \tag{3}$$

$I(\lambda)$ is the light intensity after infrared light through the gas(such as CO/CO_2) , otherwise $I_0(\lambda)$ is the light intensity before infrared light through the gas, K is the absorption coefficient of the gas, C is the gas concentration; L is the length of the gas absorption, λ is the wavelength of infrared light. We can deduce equation (4) from equation (3):

$$E(\lambda) = \ln\frac{I_0(\lambda)}{I(\lambda)} = K(\lambda)CL \tag{4}$$

Formula (4) shows that $E(\lambda)$ (infrared absorbance) is proportional to the C (gas concentration) of the absorbed radiation when the L (length of the gas absorption) fixed. Because equation (3) and equation (4) are only apply to certain wavelengths of infrared light, so we need to use a specific filter to eliminate other infrared wave band during measurement.CO/CO_2 all have their own strong absorption band spectrum and the absorption band non-overlapping each other, so we can measure CO/CO_2 simultaneously by NDIR[5–6].

2.2. *Measuring Devices*

We used SEMTECH- FEM(Sensors, Inc) to measure vehicle exhaust emissions of CO/CO_2.The SEMTECH - FEM includes a sampling flow meter and a CO/CO_2 gas analyzer. The exhaust sample is filtered and dried directly at the exhaust flow tube, with an additional dryer located in the control module. The intake manifold on the control module is custom built of stainless steel, to mitigate leaks. Sample handling is integrated into the system, with an internal sample probe, heated filter, water trap, and Nafion dryer. The dried and filtered sample then passes through CO/CO_2 gas analyzer for measurement of CO/CO_2.

The CO/CO_2 gas analyzer uses NDIR technology, incorporating a heated sample cell and relative humidity sensor, along with the interface electronics and control electronics. The technical indicators of instrument are shown in table 1 below:

Table 1. FEM Analytical specifications

Parameter	CO	CO2
Max Range(Full Scale)	85 vol.	18 % vol.
Resolution	10 ppm	0.01 % vol.
Accuracy	< ± 2% of reading or < ± 0.3% of full scale , whichever is larger	
Repeatability	< ± 1% of point (span) or < ± 1% of measured, whichever is larger	
Precision	< 1% Full Scale if range is > 155 ppm or < 2% of range if range < 155 ppm	
Noise	< 1 % of max	
1-hr Zero Drift	< ± 50 ppm	< ± 0.1 % vol.
8-hr Span Drift	< ± 20 ppm	< ± 0.1 % vol.
Rise Time	$T_{10\text{-}90}$ < 2.5 seconds	
System Response Time	$T_{0\text{-}90}$< 10 sec. with rise time $T_{10\text{-}90}$ < 10 sec.	
Data Rate	Up to 5 Hz	
Sample Flow Rate	1.5 L/min	

3. Test Experiment

We participate in the project of "Macau vehicle exhaust test experiment" organized by College of the Environment, Tsinghua and university of Macau in may 2013. The test includes a variety of tail gas analysis instruments. We are mainly responsible for the SEMTECH-FEM instrument to test CO/CO_2 emissions of vehicle exhaust. In the test we select three diesel city buses that comply with national emissions standards GB3, GB4, GB5 respectively and a new type of compressed natural gas bus for testing [7]. The specific information of test vehicle is shown in table 2 below:

Table 2. Test vehicle information

Test object	Weight (t)	Load (t)	Total Weight (t)	Fuel type	Emission standard	Manufacturer	Year of Manufacture	Trip Distance (km)
BUS_1	15	3	18	diesel	GB3	JinlongK 132	2012	107261
BUS_2	16	3	19	diesel	GB4	Wuzhoulong	2011	173691
BUS_3	13.5	4	17.5	diesel	GB5	Wuzhoulong	2013	1051
BUS_4	16	2	18	CNG	GB5	Dongfeng	2013	3313

The testing process is shown in Figure 2. First of all, we install sample handing connect vehicle exhaust pipe, then high temperature exhaust gases can be preprocessed by sample handing and tail gases' flow velocity are controlled by sampling flow meter. Secondly, processed exhaust gases are pumped into heated sample cell of NDIR gas analyzer by sampling tube, then CO/CO_2 gas absorption spectrum are detected by infrared detector and processed by signal processing module. Lastly, CO/CO_2 concentration is obtained by inversion of absorption spectra. In addition, we can operation and control SEMTECH- FEM conveniently through a PC. GPS is used to record the position and speed of the vehicle and the humidity meter is used to measure the humidity of the tail gas.

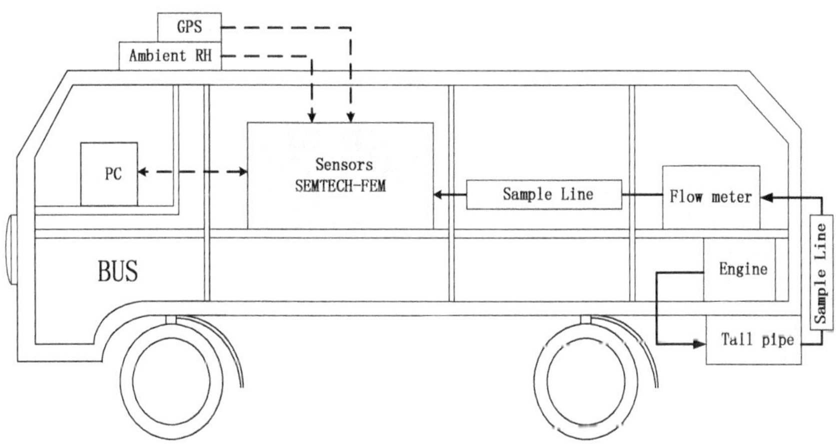

Figure 2. On-board emissions measurement CO/CO_2

We test four city buses in the four days continuously. Firstly, we installed equipment into the city bus and adjustment and calibration at New Welfare bus factory in the Island Typa Island of Macau. After the completion of the preliminary work, the bus travel on the city road to start testing, which route is starting from the New Welfare bus factory, then through the Friendship Bridge to the Macau peninsula, and back to the New Welfare bus factory from the West Bay Bridge, about 16.8km and about 1 hour.

4. Data Analysis

We sorted out the data of the CO/CO_2 emissions and exhaust gas flow velocity measured by the four city buses, as shown in Figure 3:

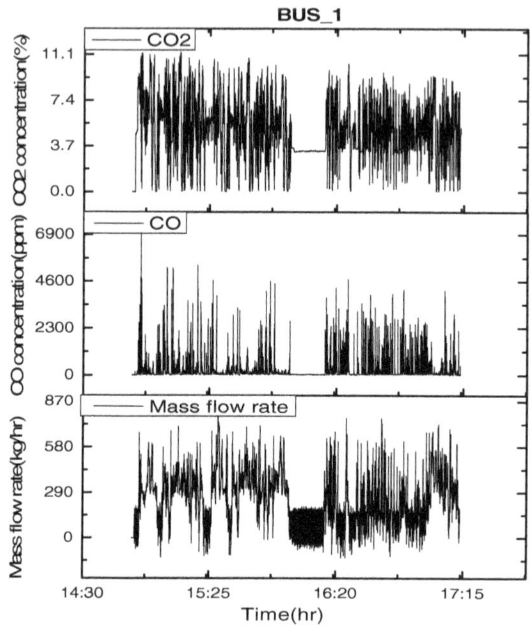
BUS_1
CO2
CO2 concentration(%)
11.1
7.4
3.7
0.0
CO
CO concentration(ppm)
6900
4600
2300
0
Mass flow rate
Mass flow rate(kg/hr)
870
580
290
0
14:30
15:25
16:20
17:15
Time(hr)

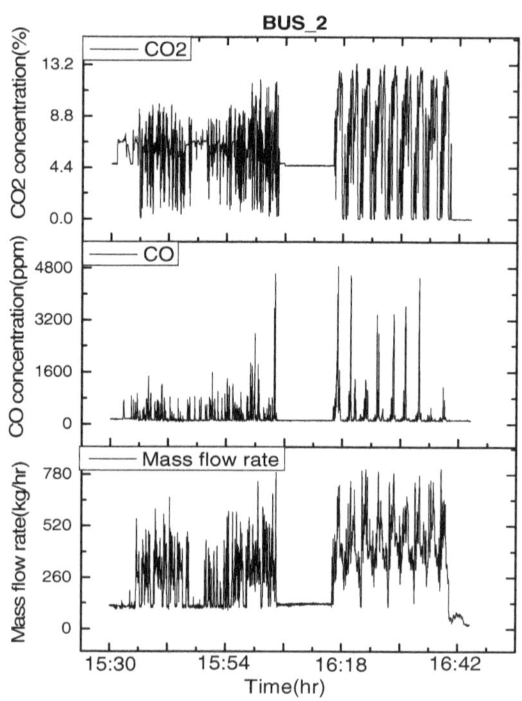
BUS_2
CO2
CO2 concentration(%)
13.2
8.8
4.4
0.0
CO
CO concentration(ppm)
4800
3200
1600
0
Mass flow rate
Mass flow rate(kg/hr)
780
520
260
0
15:30
15:54
16:18
16:42
Time(hr)

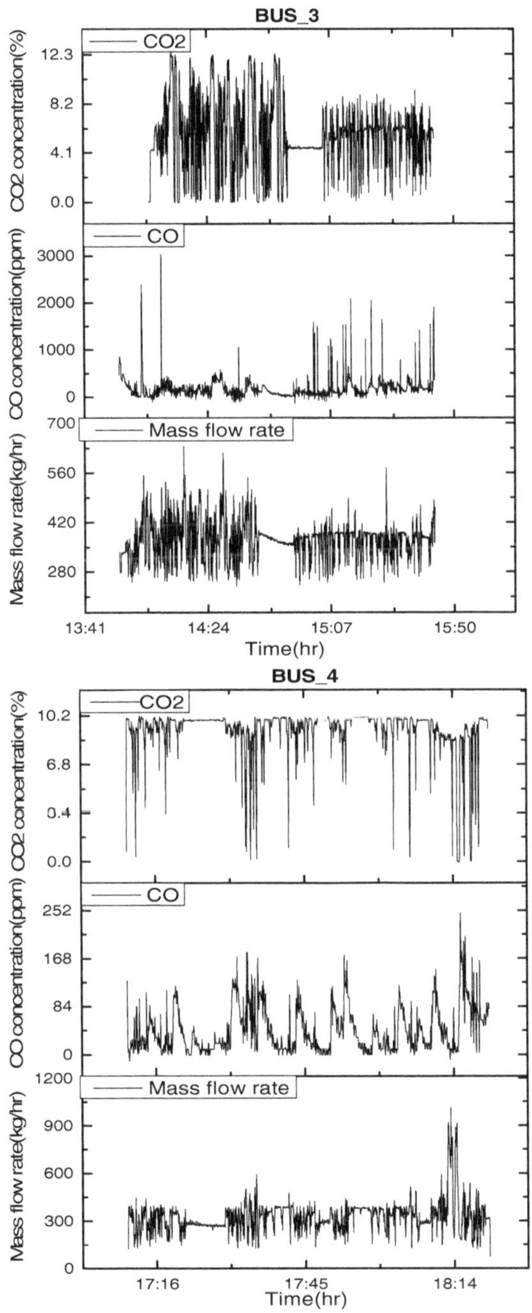

Figure 3. Original data of CO/CO2 emissions and exhaust gas flow velocity measured by SEMTECH-FEM

According to the original data of four city buses, CO/CO_2 exhaust is related to the exhaust gas flow velocity. The exhaust gas flow velocity of Bus_1, Bus_2 and Bus_3 are reduced during test because buses idled for a while, so there is no CO emissions exhaust.

Table 2 shows that the manufacturing period of four buses is different, so it is not the same as the rules of the emission standards, there are three national emissions standards of GB3, GB4, GB5. The emissions of CO and CO_2 mean are summarized according to the original data, as shown in Figure 4:

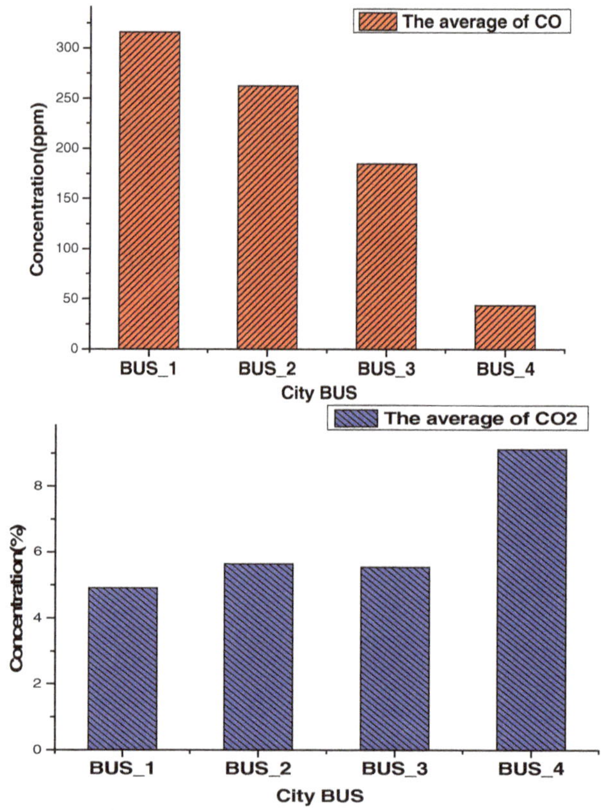

Figure 4. The mean of CO/CO2 emission for four buses with different emission standards

The national emission standards of BUS_1, BUS_2 and BUS_3 are GB3, GB4 and GB5 respectively, emission standards are increasingly stringent requirements. From above Figure 4 we realize that CO emission of buses reduced when the emissions standards of buses enhanced, and the maximum emission values are 0.69%, 0.485%, 0.303% of BUS_1, BUS_2 and BUS_3. Furthermore, CO emissions of compressed natural gas BUS_4 is several times lower than diesel buses. However, CO_2 emissions of compressed natural gas BUS_4 is twice than diesel buses, which also verifies the relationship between the CO and CO_2 emissions of equation (1) and equation (2).

5. Conclusion

In this paper, we study on-board CO/CO_2 emissions measurement based on NDIR. We use SEMTECH-FEM as the measuring instrument and carry out the test of city buses emissions of CO/CO_2 pollutants. This result verifies that with the increasing of the national emission standards, it can effectively reduce the control of emissions. The content of the test is mainly to test the CO/CO_2 emissions of buses with different emission standards and obtain lots of reliable data. Test results show that the introduction of new national emission standards is conducive to reducing the pollution of motor vehicles on the roads of the city, and new energy vehicles is more environmentally friendly than conventional fuel vehicles. So it is helpful to limit the emissions of motor vehicle emissions that the national emission standards are becoming increasingly stringent.

Reference

1. Octavio Armas, Magín Lapuerta and Carmen Mata. Methodology for the analysis of pollutant emissions from a city bus (Meas. Sci. Technol 2012).
2. Behrentz, E., Ling, R., Rieger, P, Winder, A.M. Measurements of nitrous oxide emissions from light-duty motor vehicles: a pilot study. Atmospheric Environment 38, 4291–4303. 2004.
3. California Air Resources Board (CARB), Methodology for Estimating Emissions from On-road Motor Vehicles, Vols. I–V. 1996.
4. Jane Hodgkinson and Ralph P Tatam. Optical gas sensing: A review (Meas. Sci. Technol. 24 (2013) 012004).
5. W.G. Fastie and A.H. Pfund. Selective Infra-Red Gas Analyzers (Received june 11, 1947).
6. Robert Frodl, and Thomas Tille. A High-Precision NDIR CO2 Gas Sensor for Automotive Applications (IEEE Sensors Journal, Vol. 6, No. 6, December 2006).
7. Limits and measurement methods for exhaust pollutants from compression ignition and gas fuelled positive ignition engines of vehicles (III, IV, V).

In-situ laser monitoring of ammonia concentration in dairy feedlot *

Ying He[12†], Yu-Jun Zhang[1], Kun You[1], Yan-Wei Gao[1],
Chen Chen[1], Min-Guang Gao[1], Wei Jia[3]

[1]*Key Laboratory of Optical Monitoring Technology for Environment, Anhui Province, Anhui Institute of Optics and Fine Mechanics, Chinese Academy of Sciences, Hefei, 230031, China*

[2]*University of Science and Technology of China, Hefei, 230026, China*

[3]*College of mechanical and electrical engineering, Anhui Jiangzhu University, Hefei, 230022, China*

[†]*E-mail: heying@aiofm.ac.cn*

High sensitivity sensing and in-situ monitoring of the ammonia concentration in feedlots could allow for quick acquisition of ammonia information and effective breeding. A laser monitoring system for ammonia was designed based on open-path Tunable Diode Laser Absorption Spectroscopy (TDLAS) technology. The system has a millisecond time resolution, and the detection limit is about 5ppm-m. The ammonia concentration of a dairy feedlot in Changshu, Jiangsu Province was monitored in summer. Based on the results, ammonia concentration had a diurnal trend such that it was higher in the daytime than at night. The ammonia concentration increased during the activities of cow feeding and feedlot arranging. It peaked at a value of 5.3ppm at 9:00. In sum, optical technology for ammonia detection with high sensitivity, high time resolution without sampling was used and found to be an effective method for ammonia monitoring in dairy feedlots which provides technical support for scientific breeding.

Keywords: Laser Absorption Spectroscopy; Ammonia; Dairy Feedlot; In-situ Monitoring.

1. Introduction

Ammonia (NH3) is one of the most abundant alkaline trace gases in the atmosphere. In recent years, intensive breeding in our country has developed rapidly which had great social and economic benefits. However, the ammonia concentration in large-scale breeding was significantly higher than in other developed countries in the world. Ammonia emission influences the performance of aquaculture production, and also affects the regional

* This work is supported by the National major scientific instruments and equipment development projects (No. 2012YQ22011902) and the "12th five-year" National science and technology plan of rural areas (No. 2012BAJ24B02).

atmospheric visibility and regional air quality [1–2]. Therefore, it's significant to master the ammonia emission concentration characteristics in dairy feedlot for the environment protection and scientific breeding.

Recently research shows that ammonia emission in China is mainly produced in animal husbandry and agriculture, for example, animal feedlot, sewage pool as well as manure management places are the main emission source of ammonia [3]. But, the study of ammonia monitoring in dairy feedlot monitoring is still weak. The traditional ammonia measurement methods, such as membrane method, diffusion tube sampling method and denuder method require gas sampling, so the measurement accuracy is low. At present, a variety of optical method for ammonia concentration on-line monitoring, such as chemical ionization mass spectrometry (CIMS) technique[4], photo-fragmentation laser-induced fluorescence (LIF) technique[5], Tunable Diode Laser Absorption Spectroscopy (TDLAS) technology, etc. TDLAS technology is widely used for trace gas concentration monitoring with its high sensitivity, high resolution and high time resolution[6–7].

An ammonia concentration laser monitoring system has been designed based on TDLAS technology with open path. Then, this system was used for ammonia concentration continuously monitoring experiment in a dairy feedlot in Changshu, Jiangsu Province in summer. The data analysis and methods validation were completed, and the system performance was verified during the experiment. Other more, the ammonia emission characteristics were studies in the paper.

2. Detection Principle

As the Beer-Lambert's law, the original intensity of the tunable diode laser is I_0, and the laser frequency is ν, so the intensity detected at the receiving detector through the absorbing medium whose length is l can be expressed as in Eq. (1).

$$I(\nu) = I_0(\nu)\exp(-\sigma(\nu)cl) \tag{1}$$

Where c is the gas concentration, $\sigma(\nu)$ is the absorption cross section[8].

As ammonia molecular absorption in near-infrared accord to the condition $|\sigma(\nu)cL| << 1$, so after the intensity normalized, the gas absorption signal intensity is only direct proportion to the gas concentration and the optical length, but entirely unrelated to the original laser intensity I_0.

The measurement concentration can be gained with least square fitting between the measurement absorption signal and the standard absorption signal, and be expressed as in Eq. (2).

$$c_d\, l_1 = a\, c_s l_2 \tag{2}$$

Where c_d and c_s are the measurement concentration and the standard concentration in the calibration cell, l_1 and l_2 are the length of measurement optical path and the calibration cell, a is the least square fitting coefficient. So, c_d can be expressed as in Eq. (3).

$$c_d = a\, c_s l_2/l_1 \tag{3}$$

Now, the emission factor of ammonia has big uncertainty in the current world, especially in breeding. The mainly method for ammonia emission measurement at present is chemical method and micrometeorology, but the measuring cycle of the former is long, and the emission factor maybe too high with the latter method. The ammonia emission measurement is rapid and high-sensitive with the open TDLAS technology combined with the inverse dispersion technique[9].

The backward Lagrangian stochastic (BLS) model is a typical inverse dispersion technique, the model simulates the emission rate as in Eq. (4), and the ratio $(c/Q)_{sim}$ of gas concentration is related with the number of particles released at each point N, the number of specific release points along the laser path P and the vertical touchdown velocity ω_0. The true concentration c can be expressed as in Eq. (5), where c_b is the background concentration[10–11].

$$Q = \frac{c}{(c/Q)_{sim}} \tag{4}$$

$$c = c_d - c_b \tag{5}$$

3. System Design

An ammonia concentration laser monitoring system has been designed based on TDLAS technology with long open-path technology, and the system structure is illustrated in Fig. 1. The rotational-vibrational absorption line of ammonia at 1.53μm was chose for detection. The DFB laser of NEL Company as the laser source was tuned to the absorption line of ammonia by a current controller and a temperature controller of Lightwave Company, and the laser wavelength was modulated by wavelength scanning signal of 200Hz. The modulated laser beam was coupled to a beam splitter to divide into two parts, 2% output laser passed through a calibration cell of 10cm containing 5% standard ammonia to get the standard absorption signal, and 98% output laser directed to the open optical path for measurement. The open optical path was mainly consisted of a telescope and a retro-reflector to realize open path signal detection. The two detection signals were connected to the signal processing module to amplify and filter, and then they were sent to the data acquisition processor to realize the concentration inversion.

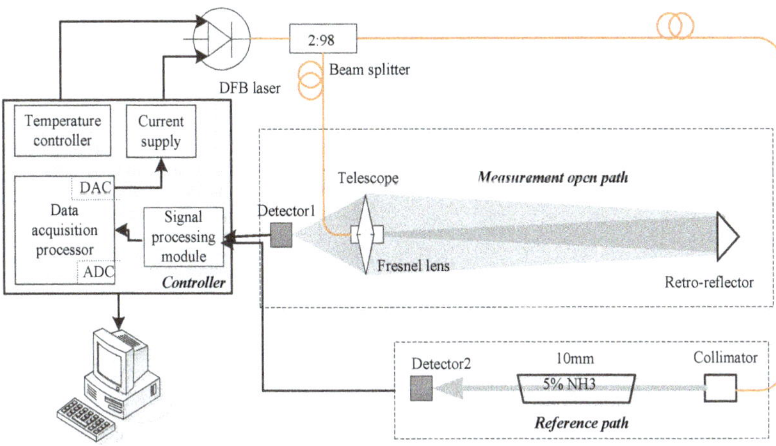

Fig. 1 This is the system construction.

As the wavelength scanning of the TDLAS system is 200Hz, the time resolution is 5ms here. In open path detection, the average time of the signal is 30, so the effective signal can be got with the time resolution of about 150ms. This time resolution is so shorter than the traditional ammonia measurement methods which maybe several hours even one day.

A multiple reflection cell of 24.3m was connected to the TDLAS system, and a set of ammonia standard gas were sent to the multiple reflection cell. The detection sensitivity is about 5ppm-m. So, the TDLAS system could obtain ammonia emission concentration characteristics quickly and accurately.

4. Experimental Site and Set-up

The experimental site for ammonia concentration monitoring was at Weizhong pasture, in Changshu, Jiangsu Province, China(120°43'E, 31°39'N) from May 21-23, 2015. Changshu is located in mid-latitude region in China where the climate is mild and the rainfall is abundant as its subtropical monsoon climate. Therefore, the feedlot in Changshu is the typical experimental site in a rapidly developed large-sized city in east of China.

There are more than 300 cows in Weizhong pasture feed with imported grass. There are three times for waste removal, feeding and milking at 5:00, 12:00 and 19:00 everyday, and it lasts more than two hours one time.

The experimental site is essentially flat with no significant obstructions for 10m south of the feedlot at the east-west corridor (the prevailing wind direction in summer during this experiment). There are several rows of wheat to the north of the feedlot approximately 30m away, and there is a canal to the east of the feedlot nearly 80m away. The experimental set-up is shown in Fig.2.

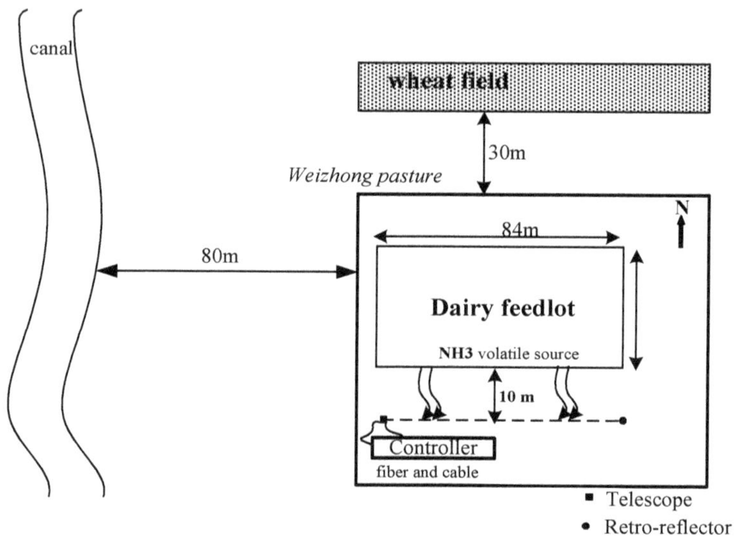

Fig. 2. This is the experimental set-up.

The open optical path for detection between the telescope and the retro-reflector is 84m, and the height of the telescope and the retro-reflector is 0.5m. The telescope system and retro-reflector were fixed on the adjusting bracket, and the controller placed nearby the telescope system connected with the fiber and cable whose lengths were 30m for signal detection. The system installation is shown in Fig.3.

5. Experiment Results

From May 21 to 23, the observation was carried out at Weizhong pasture. The concentration were chose as that the peak to peak of signal intensity was higher than 5% data acquisition range of signals after amplification and filter, and the linear correlation with the standard absorption signal was higher than 90% at the same time. The monitoring results are obtained as shown in Fig.4. The ammonia concentration peak was 5.3ppm during the period.

Fig. 3. This is the system installation situation.

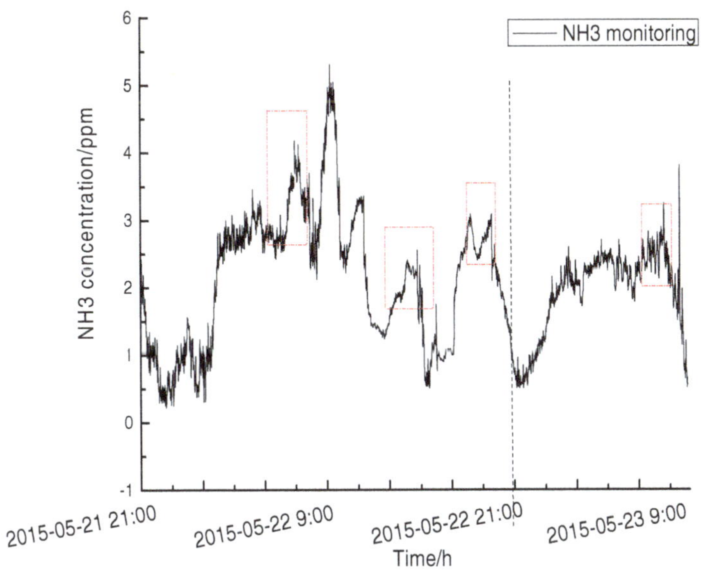

Fig. 4. There was the continuously monitoring result of NH_3 at Weizhong pasture.

The dairy feedlot is an ammonia emission source. From the monitoring results, ammonia concentration decreased after 3:00, and appeared the maximum value at about 9:00, then decreased again. The ammonia concentration increased again at about 12:00 and 19:00, and decreased to the minimum value of a day at 21:00. Comparing the results of the two days, it is obvious that the concentration increased after 5:00, 12:00 and 19:00, when the dairy cows were feed and the breeding area was arranged. In addition, the meteorological conditions such as wind speed, wind direction and atmospheric stability are also the important factors that affect the change of ammonia concentration. Due to the temperature increase, the ammonia volatilization rate was accelerated, so the ammonia concentration increased obviously.

The monitoring data of average hours were analyzed as follows: the diurnal variation of the ammonia concentration has a trend that the concentration was high in the daytime than it at night. The main reason may be that there are many rivers in Changshu, such as the canal nearby, so the relative humidity is high at night. Because ammonia is easily soluble in water, so the concentration may decreased at night. During the daytime, the temperature increased and the humidity reduced, so the ammonia concentration increased as the result, the daily trend is shown in Fig.5.

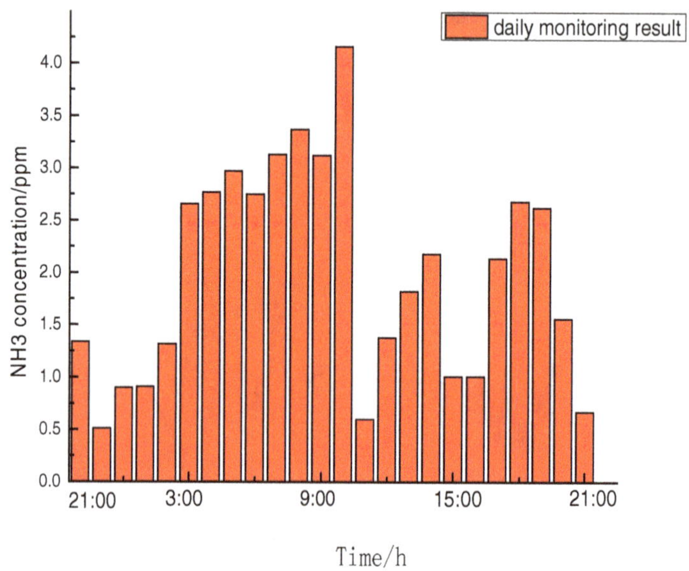

Fig. 5. There was the daily trend of NH_3 concentration at Weizhong pasture.

6. Conclusions

(1) An ammonia concentration laser monitoring system has been designed based on TDLAS technology with long open-path technology with high sensitivity, high time resolution without gas sampling, so the system could implement in-situ dynamic monitoring.

(2) The ammonia monitoring experiment has been carried out in a dairy feedlot in Changshu, Jiangsu Province in summer. The continuous results show that the variation of ammonia concentration had a diurnal trend that the concentration was higher in the daytime than it at night. The ammonia

concentration increased when the dairy cows were feed and the breeding area was arranged, and it reached the peak value of 5.3ppm. When the temperature increased, the ammonia volatilization rate was accelerated.

(3) It was verified that the open TDLAS system is an effective method for ammonia monitoring in dairy feedlot environment. If the meteorological data and ammonia concentrations were calculated for both 30-min intervals, the ammonia emission can be calculated with continuous modeled 30-min BLS. Thus, the study provides technical support for scientific breeding.

References

1. Yong Hou, Zhiling Gao, Wenqi Ma, et al. ACTA ECOLOGICA SINICA, 32, 4(2012).
2. P. Michopoulos, G. Baloutsos, A. Economou, et al. Science of the Total Environment, 323(2004).
3. A.B. Leytem, R.S. Dungan, D.L. Bjorneberg, et al. J. Environ. Qual, 40, 5 (2011).
4. S. S. John, E. S. Robert, A.V. Cornelis, et al. Applied Optics, 29, 33, (1990).
5. F.C. Fehsenfeld, L.G. Huey, E. Leibrock, et al. Journal of Geophysical Research Atmosphere, 107, (2002).
6. H. Riris, C. B. Carlisle, L.W. Carr, et al. Applied Optics, 33, 30,(1994).
7. Ruifeng Kan, Wenqing Liu, Yujun Zhang, et al. Chinese Optics Letters, 4,2, (2006).
8. J. Reid and D. Labrie, Appl. Phys. B., 26, 5(1981).
9. T. K. Flesch, L. A. Harper, J. M. Powell, et al. Transactions of the American Society of Agricultural and Biological Engineers, 52,1(2009).
10. L. A. Harper, T. K. Flesch, J. D. Wilson, et al. Poultry Science, 89(2010).
11. Zhiling Gao, R. L. Desjardins, T. K. Flesch. Atmospheric Environ, 44, 26(2010).

Oxygen sensor based on laser absorption spectroscopy for combustion process*

Yan-Wei Gao, Yu-Jun Zhang†, Ying He, Kun You,
Chen Chen, and Wen-Qing Liu

Key Laboratory of Environmental Optics and Technology,
Anhui Institute of Optics and Fine Mechanics,
Chinese Academy of Sciences, Hefei 230031, China
†*E-mail: ywgao@aiofm.ac.cn*

Oxygen concentration is the key to optimizing combustion and improving efficiency. Aimed at meeting demands for quickly and accurately detecting oxygen concentration for the combustion process, an oxygen sensor based on wavelength modulation spectroscopy is designed using the 760nm oxygen absorption line. Single circuit board based on FPGA is developed to implement sawtooth wave and sine wave generation, transimpedance amplification, analog-digital conversion and other functions. A finite impulse response (FIR) digital filter based on FPGA is used. The simulation results show that the algorithm can achieve the signal filtering, which then serves as a basis to achieve the digital phase lock, following which the second harmonic signal is extracted. This system is used to measure different concentrations of oxygen. The detection limit is 0.2%, and the experimental results show that the sensor meets the requirements for industrial application and vehicle exhaust emission tests.

Keywords: TDLAS; Oxygen Detection System; Finite impulse response filter; FPGA circuit design.

1. Introduction

Oxygen is the most important combustion-supporting gas for combustion process. The fast, sensitive and cost-effective on-line monitoring of oxygen concentration is meaningful for combustion diagnostics and process control to improve combustion efficiency, for example, in the roasting process of gold concentrate containing arsenic, oxygen has important influence on gold production rate and quality[1]; Now vehicle exhaust emissions have become a major source of urban gaseous pollutants, real-time monitoring of oxygen concentration of vehicle exhaust is the key for combustion diagnostics of vehicle

* This work is supported by National High Technology Research and Development Program of China (863 Program), No. 2014AA06A503

combustion engines and optimize combustion process to reducing emissions of pollutant[2]. So the research on-line monitoring oxygen concentration technology has great significance.

Tunable diode laser absorption spectroscopy (Tunable Diode Laser Absorption Spectroscopy, TDLAS) technology[3–6], using a tunable diode laser with output wavelength continuous tuning characteristics, by changing the laser injection current to scan absorption lines of target gases, to achieve accurate measurement of gases. TDLAS technology is non-invasive, high selectivity, fast response and high sensitivity, has been widely used in gas sensing applications. With requirements of high response speed, high precision, easy to carry and commissioning, oxygen sensors based on TDLAS technology is on the way of low power consumption, small size [10–12].

In this paper, oxygen absorption line near 760 nm is selected, an oxygen sensor based on WMS - TDLAS technology is developed to achieve low-power dissipation and miniaturization function, is suitable for the industrial applications and on-board monitoring applications.

2. Design of Oxygen Sensor

The basic principles of the oxygen sensor are the absorption law of Beer-Lambert and wavelength modulation spectroscopy (WMS) [7–9].

Sensor structure as shown in Fig.2, includes electronics components and optical components. Electronics components consist a FPGA chip (EP4CE15F17C8) as the control chip[10–11], two D/A conversion chips (LTC1668) generate sawtooth signal and sine signal to realize the wavelength modulation for laser, a 16-bits A/D conversion chip (LTC2380) to realize spectral data acquisition, a preamplifier carry on I-V transimpedance conversion for photocurrent signal of photodiode, as well as filter and amplification of voltage signal; FPGA logic include control procedures of DACs and ADC, as well as digital phase-locked amplifier algorithm. Optical component mainly is small gas absorption cell, effective optical path reach 1.2m.Temperature and current controller control DFB laser (nanoplus GmbH) center wavelength around 760.8 nm, the output power is 4 mw, the wavelength tuning rate is 0.023 nm/mA. Laser beam transferred through the absorption cell and reach the Pin Si photodiode (THORLABS) to realize the photoelectric signal conversion.

2.1. *The absorption lines of oxygen*

Fig.1 shows the absorption lines of oxygen molecules near 760 nm waveband, according to the HITRAN2008 database at T = 296k, P = 1atm. In this waveband oxygen absorption lines are not affected by common gas molecules such as water vapor molecules and CO_2 cross interference, so oxygen absorption lines around 760.8 nm chosed for measurement.

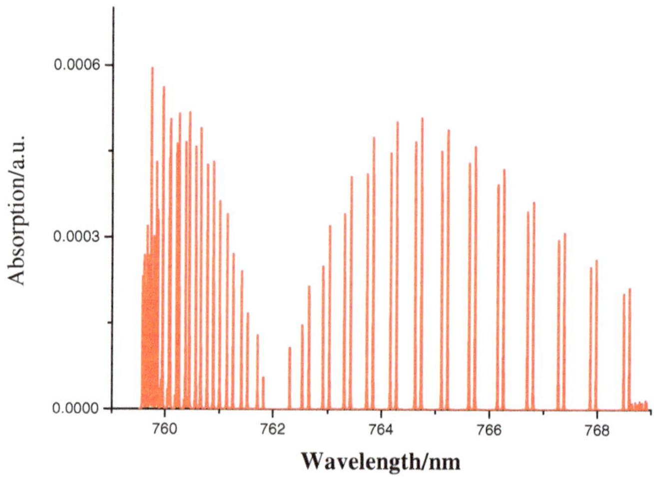

Fig. 1. Absorption lines of oxygen near 760nm

2.2. *Design of hardware circuit*

Sensor structure as shown in Fig. 2, includes electronics components and optical components. Electronics components consist a FPGA chip(EP4CE15F17C8) as the control chip[10–11], two D/A conversion chips(LTC1668) generate sawtooth signal and sine signal to realize the wavelength modulation for laser, a 16-bits A/D conversion chip(LTC2380) to realize spectral data acquisition, a preamplifier carry on I-V transimpedance conversion for photocurrent signal of photodiode, as well as filter and amplification of voltage signal; FPGA logic include control procedures of DACs and ADC, as well as digital phase-locked amplifier algorithm. Optical component mainly is small gas absorption cell, effective optical path reach 1.2m.Temperature and current controller control DFB laser (nanoplus GmbH) center wavelength around 760.8 nm, the output power is 4 mw, the wavelength tuning rate is 0.023 nm/mA. Laser beam transferred through the absorption cell and reach the Pin Si photodiode (THORLABS) to realize the photoelectric signal conversion.

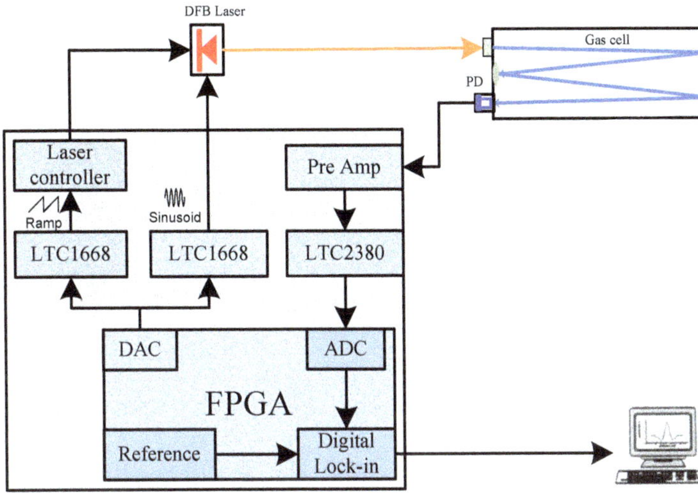

Fig. 2. Scheme diagram of oxygen sensor

2.2.1 *D/A conversion circuit*

Wavelength modulation technique needs two signals, respectively sawtooth scanning signal and sine modulation signal, are generated by two 16-bits D/A conversion chips controlled by FPGA.

The FPGA design logic of D/A is shown in Fig. 3. Sawtooth signal generate by linear count, and the generation of sine signal is composed of sine waveform ROM look-up table and address generator, the FPGA pre-store the sine waveform ROM look-up table as the sine amplitude information, address generator is equal to the phase information, sine signal is produced by phase increment lead to amplitude change, so frequency of sawtooth and sine signal can be changed by changing the clock signal to adapt different applications.

The frequency of clock is 1 MHz, each sawtooth signal cycle contains 8192 points, each sine signal cycle contains 64 points within ROM look-up table, so the sawtooth signal frequency is 1/8192 MHz≈122Hz, sine signal frequency is 1/64 MHz≈15.625 KHz, as shown in Fig.3.

2.2.2 *Low noise pre-amplifier circuit*

Photodiode abtains laser irradiation and converts light signal to current signal.in order to make the signal easy to process, current signal have to convert to voltage signal. Pre-amplifier determines the noise level of the whole circuit, so we choose low noise operational amplifier OPA627 to implement I/V conversion in this design, OPA627 maximum current noise density is 2.5 $fA/\sqrt{Hz}$, unit gain bandwidth is 16 MHZ, is suitful to the transimpedance amplifier applications. The pre-amplifier circuit is shown in Fig.4. If we choose resistance RF=R2=10K, R1=2.4K,the noise-voltage density is $15\,nV/\sqrt{Hz}$, meets the requirements of detection of weak photocurrent signal.

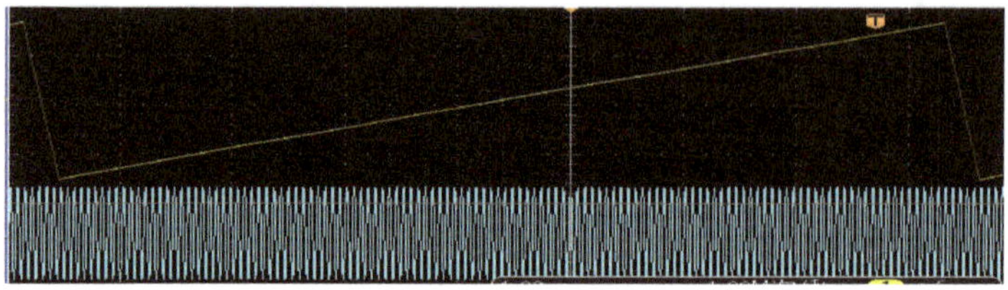

Fig. 3. Actual waveform produced by LTC1668.

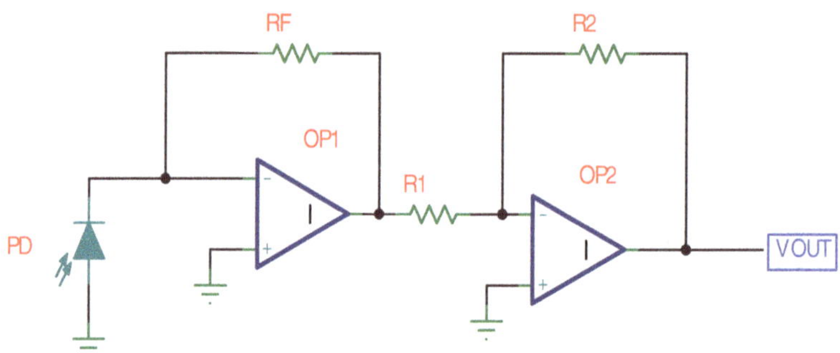

Fig. 4. Design of pre-amplifier circuit

Then signal is transferred to low pass filter to realize band-limited for Gaussian noise and filter out the high frequency component of the signal to improve signal to noise ratio, and provide anti-aliasing filter for subsequent analog-to-digital conversion. We choose analog integrated filter to realize low pass filter.

2.3. *Digital lock-in algorithm*

The extraction of second harmonic signal is realized by digital phase locked based on FPGA. According to the principle of phase sensitive detection, multiply spectral signal by double frequency reference signal, then after a low-pass filter to extract the DC component, we can get the second harmonic signal.

FPGA embedded firmware multipliers are used to realize multiplication. This paper mainly discusses the realization of the digital filter based on FPGA, We choose Finite Impulse Response (Finite Impulse Response, FIR) filter to realize low pass filtering, implementation way is simplify the Adder Graph (Reduced Adder Graph, RAG), the implementation steps are as follows:

(1) The filter initialization. According to the performance requirements of filter to design L order filter coefficients h(0), h(1)...h(L-1), and then quantitative treatment to 16-bits, the size of L is determined by filter performance requirements and the FPGA resources. FPGA distribute L consecutive storage space to store the filter coefficient, and then assign another set of L a 16-bits continuous storage space to store L digital quantitative spectral data sequence x(n), x(n-1)...x(n-L+1).

(2) Multiplication. Using the FPGA embedded firmware multiplier to realize Multiplication, in turn corresponding multiply L data of two sets storage space, namely $h(\mathrm{i}) \times \mathrm{x}(\mathrm{n}-\mathrm{i})(\mathrm{i}=0,1,2 \cdots L-1)$.

(3) Accumulative operation. Accumulation of L products in the step (2), accumulation is y(n).

When the neat value x(n+1) come, reset the second set of storage space and reload, then repeat steps (1) (2) (3) to get y(n+1).

According to the above steps to design FIR filter based on FPGA, filter order is 80, cut-off frequency FC is 100Hz, near the cut-off frequency FC, signal amplitude attenuate 40 dB, within the range of the passband, phase keep linear.

Using Modelsim simulation software to carry on the simulation of dewsigned FIR filter, the simulation result is shown in Fig.5, the "RST" is the reset signal, "CLK" is the clock signal, the frequency is 50 MHZ, "raw_data" is product signal of spectral signal and reference signal, "fir_out" is the second harmonic signal. From the Fig.7 we can get that in the beginning of each period there is a rush, the length of rush is 80 times of clock cycle, this is because in the beginning of filtering cycle there is 80 invalid datas (equal to the number of filter order), the rush can be abandoned, and will not affect the measurement result. By the simulation results show that the implementation of the FIR filter based on FPGA can realize the second harmonic signal extraction.

Filter consumes 3613 logical unit (LEs), 2648 registers, respectively accounted for 23%, 17% of total resources in FPGA. Filter order number is 80, due to the filter adopts symmetrical structure, so filter consumes only 40 18 x 18 bits multiplier resources, accounting for 71% of the total number of FPGA multiplier.

The oxygen sensor adopt small, compact design, power consumption of electrical component based on the FPGA is less than 10w, can use solar or battery as power supply, is suitable for the industrial environment, the on-board vehicle exhaust measuring.

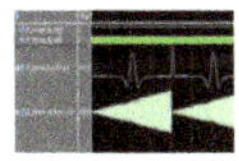

Fig. 5. Simulation diagram of FIR filter

3. Sensor Performance Test

The performance of the sensor was evaluated. Fig.6(a) is the second harmonic signal of 5 different oxygen concentrations within the range of 1-20%, and Fig.6(b) is the relationship between the measured concentrations and the actual concentrations. The linear fitting coefficient is 0.997, the results show that measured values have a good linear relationship with the actual values. We find the detection of the sensor is 0.2% by testing. The error sources include the uncertainty of different oxygen concentrations, the A/D quantization error, noise interference between optical mirror and influence of oxygen in the air environment etc.

4. Conclusion

In this paper, we use one oxygen absorption line near 760nm to design an oxygen sensor based on WMS–TDLAS. A control circuit based on FPGA is designed, including signal generating circuit, I/V transform circuit, analog-to-digital conversion circuit, etc. Based on the FPGA we realize the FIR digital filter, and on which basis to realize digital phase locking algorithm to extract the second harmonic signal, the whole circuit system is low power consumption is lower than 10W.The different concentration of oxygen is measured using this sensor, the linearity of 0.996, the detection limit is 0.2%, the results show that oxygen sensor designed in this paper can realize reliable measurement, high stability, is suitable for industrial combustion process and vehicle exhaust emission test application.

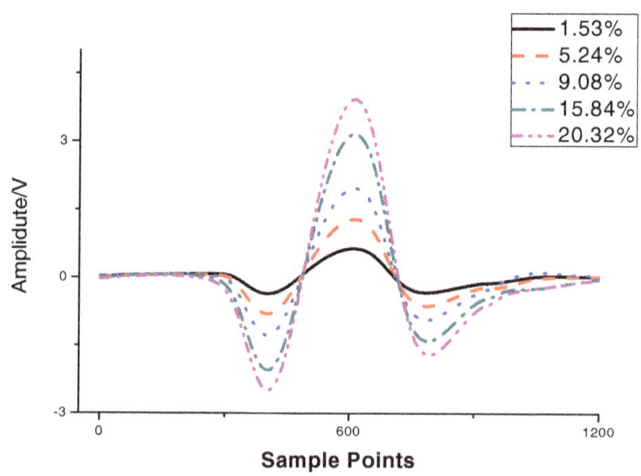

(a)

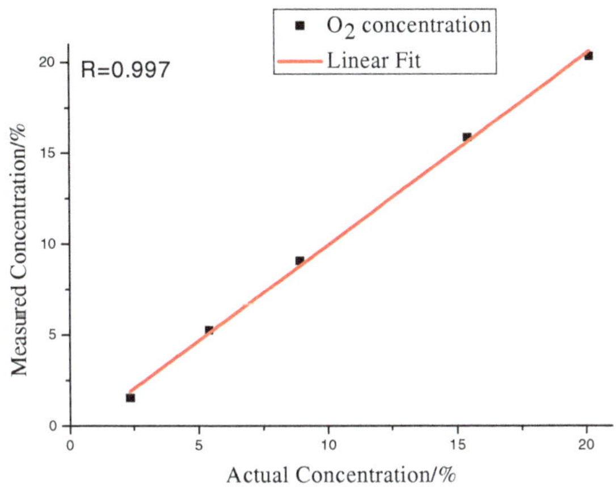

(b)

Fig. 6(a). The second harmonic signal of 5 different oxygen concentrations; Fig. 6(b). Comparison of the measured concentrations and the actual concentrations

References

1. Jieshu Xu. An approach to treatment of high-arsenic and high-sulphur gold concentrate using two-stage roasting technology[J]. SULPHURIC ACID INDUSTRY, 2009, 03: 40 41 (In Chinese).
2. Longhua Zheng. Production of the Planar Exhaust Gas Oxygen Sensor[D]. Wuhan: Huazhong University of Science and Technology, 2009:1–19. (In Chinese).
3. Ebert V, Pleban K U, Wolfrum J. In situ oxygen-monitoring using near-infrared diode lasers and wavelength modulation spectroscopy[J]. Laser Applications to Chemical and Environmental Analysis, 1998, 3: 206–209.
4. Nguyen Q V, Day T, Dibble R W. High-resolution oxygen absorption spectrum obtained with an external-cavity continuously tunable diode laser[J]. Optics letters, 1994, 19(24): 2134–2136.
5. Chen J, Hangauer A, Strzoda R, et al. Laser spectroscopic oxygen sensor using diffuse reflector based optical cell and advanced signal processing [J]. Applied Physics B, 2010, 100(2): 417–425.
6. Weldon V, O'Gorman J, Pérez-Camacho J J, Oxygen sensing using single frequency GaAs-AlGaAs DFB laser diodes and VCSELs [J]. Electronics Letters, 1996, 32(3): 219–221.
7. Chen J, Hangauer A, Strzoda R, et al. Laser spectroscopic oxygen sensor using diffuse reflector based optical cell and advanced signal processing [J]. Applied Physics B, 2010, 100(2): 417–425.

8. Wei H, Xiaoming G, Xiaoyun L, et al. Near-IR diode laser-based sensor for remote sensing of methane leakage [J]. Optica Applicata, 2005, 35(1): 23-32.
9. So S, Sani A A, Zhong L, et al. Laser spectroscopic trace-gas sensor networks for atmospheric monitoring applications[C]//ESSA Workshop. 2009, 9.
10. Frish M B, Wainner R T, Green B D, et al. Progress in reducing size and cost of trace gas analyzers based on Tunable Diode Laser Absorption Spectroscopy[C]//Proc. of SPIE Vol. 2004, 5586: 77.
11. Meyer-Baese U, Meyer-Baese U. Digital signal processing with field programmable gate arrays[M]. Heidelberg: Springer, 2007.
12. Spence T G, Calzada M E, Gardner H M, et al. Real-time FPGA data collection of pulsed-laser cavity ringdown signals[J]. Optics express, 2012, 20(8): 8804–8814.
13. Guangshu Hu. Digital Signal Processing: Theory, Arithmetic and Realization[M]. Tsinghua University Press. 2003. (In Chinese).

CHAPTER 4

Information Science

Study on a lightning flash detection algorithm based on VIIRS/DNB*

Xia-Rong Zhang and Yun-Xian Huang† and
Wei Yan and Shuo Ma and Jun Jiang

Institute of Meteorology and Oceanography,
PLA University of Science and Technology
Nanjing 211101, China
†E-mail: gsdxzxr@163.com

Lightning flash detection from space is an important way for gathering information on lightning and its distribution. Research shows that VIIRS/DNB, sensors with high gain performance, can detect light, fire and lightning flashes in low illumination conditions. In this paper, VIIRS/DNB data was first used to gather lightning information, following which a new method of lightning flash detection based on the reasonable gradient threshold and the threshold range obtained from the statistics was proposed. This method was then used to monitor lightning information, and the data obtained over half an hour was compared with those provided by the Chinese meteorological observation center. This new method was found to be accurate, especially with regards to the positional information.

Keywords: VIIRS; DNB; lightning flash detection.

1. Introduction

Lightning is a kind of severe natural phenomenon, it caused great harm to human life. According to statistics, about 8 million cloud-to-ground lightning occurs on the earth in one day, there were 100 lightning strikes per second, the loss caused by lightning has risen to a third of natural disasters (Chen et al, 2003; Xu et al, 2002). Therefore, how to monitor and forecast the lightning accurately is of great realistic significance.

Lightning flash detection methods is diverse, including ground detection, space-borne satellite detection, doppler weather radar and air electric field detection, etc (Ye et al, 2007; Zhang et al, 2010). Compared with the ground detection method, lightning detection by satellite can realize 24-hour continuous observations, and the method has the advantages of wide coverage. Using the data of LIS (Lightning Imaging Sensor, carrying on the TRMM satellite),

* This work was supported by the National Natural Science Foundation of China (Grant No. 41375029).

Williams (Williams E Ret al, 2000) and Boccippio (Boccippio D Jet al, 2000) analysis of the lightning activities and regional differences around the world. Orville and Henderson (Orville R E et al, 1986) calculated the number of lightning occurred in the land and sea each month and the ratio between them by DMSP (Defense Meteorological Satellite Program) photographic data. American Bankert (Bankert R Let al, 2011) used the data get from the OLS (Operational Linescan System) and proposed a lightning automated detection algorithm, however, due to the poor positioning accuracy and low radiation resolution of OLS, inevitably, it must exist error detection probability (Elvidge C Det al, 1997), which affecting the recognition effect of lightning detection.

The Visible–Infrared Imaging Radiometer Suite (VIIRS), which carried on the Suomi National Polar-Orbiting Partnership (NPP), is currently the world's strongest comprehensive detection ability of scanning imaging radiometer. VIIRS is a scanning radiometer that collects visible and infrared imagery and radiometric measurements of the land, atmosphere, cryosphere, and oceans. The data is used to measure cloud and aerosol properties, sea and land surface temperature, ice motion and temperature, fires as well as lightning flash (http://www.class.ncdc.noaa.gov/).

VIIRS has visible capabilities for nighttime imaging using the Daytime/nighttime Band (DNB), at much better sensitivity and better spatial resolution than previously seen from OLS (Bankert R Let al, 2011). The panchromatic DNB band measures night lights, reflected solar or moon lights with a large dynamic ange of 45000000:1, which allows the detection of reflected signals from as low as quarter-moon illumination to the brightest daylight (Liuet al, 2013). The DNB channel is mainly used to detect the earth's surface and atmosphere of visible radiation, including the sun or the moon's reflection radiation and emission radiation of man-made and natural.

This study focuses on the lightning flash detection through VIIRS/DNB data, inspired by Bankert's work (Bankert R Let al, 2011), the paper proposes a lightning detection algorithm based on luminance and scale gradient. Because of higher spatial resolution, more accurate geolocation and smaller instantaneous field of view (Zhou et al, 2012) of VIIRS/DNB compared with DMSP/OLS, the results obtained from our algorithm can be more consistent with the ground monitoring data.

2. Data

Visible Infrared Imaging Radiometer Suite (VIIRS) data records from Suomi NPP. This data family contains the raw (RDR), sensor (SDR) and environmental data records (EDR) (Maet al, 2011). Only VIIRS Day Night Band (DNB) SDR data products are utilized for the lightning flash detection. The DNB will measure VIS radiances from the Earth and atmosphere (solar/lunar reflection and both natural and anthropogenic nighttime light emissions) during both day and night portions of the orbit (Lee T E et al, 2006). We collected two months of the DNB data from August to September 2014, in order to verify the exactness of the algorithm, the corresponding ground lightning monitoring data downloaded from the China meteorological administration center meteorological observation were also utilized at the same time.

3. Algorithm

The DNB is implemented as a dedicated focal plane assembly (FPA) that shares the optics and scan mechanism of the other VIIRS spectral bands (Lee T E et al, 2006). It has 16 different sensors and every scan direction (along track) contains 16 pixels, so when there is lightning occurs, on the DNB imagery with 16 adjacent stripes, it shows up in DNB imagery as relatively large blocks of light in a nearly rectangular shape (Hillger D et al, 2013), as shown in Figure 1 in the red areas.

In the twilight or night, under the low light conditions, the urban lightning is the main noise of lightning flash monitoring through DNB imagery, as shown in Figure 1 in green area. When lightning occurs over the town, because of the humans light source, the identification of lightning will be more difficult than it occurs over other areas. Therefore, how to reduce the effect of urban lightning is our main research content.

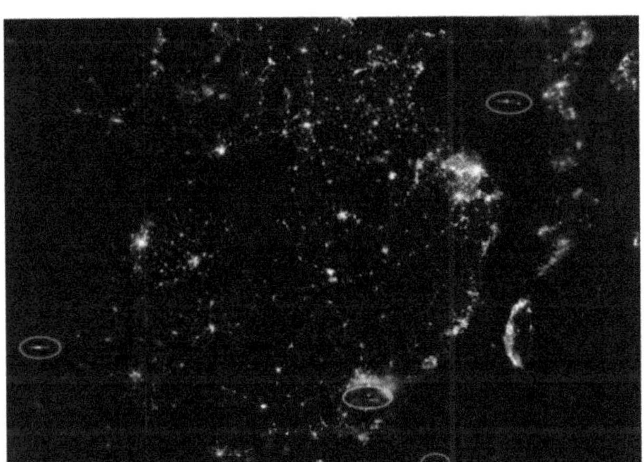

Fig. 1. The VIIRS/DNB image over China at 1808UTC 15Aug 2012

According to the characteristics of the lightning in DNB images, we propose the following algorithm:

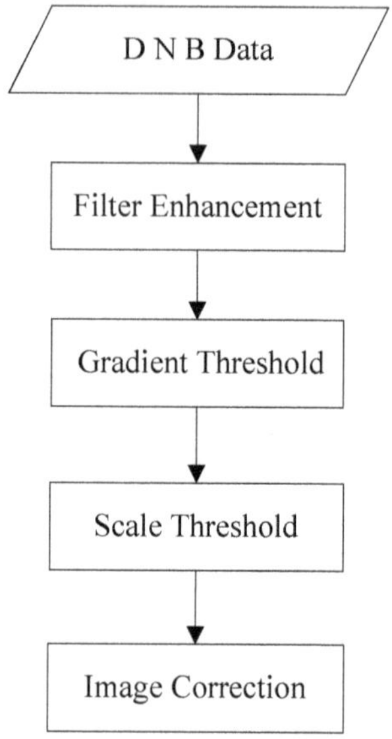

Fig. 2. Overall structure of the double threshold algorithm.

3.1. *Filter enhancement*

First, in order to identify lightning areas and urban lightning area much better, according to the characteristics of the DNB data, we adopt an 11×3 linear filter (Bankert R Let al, 2011) to the original data. The filter is used to enhance the lightning streaks of radiation intensity, convolving this filter to the DNB data results in an image on which lightning streaks stand out. It should be pointed out that the size and magnitude of this line-directional filter evolved through experimentation to reveal relatively faint and short lightning streaks that might otherwise go undetected, by a human analyst, as well as the more obvious (long, bright) streaks (Bankert R Let al, 2011).

Table 1. The 11×3 filterutilized in the algorithm.

-1	-1	-1	-1	-1	-1	-1	-1	-1	-1	-1
2	2	2	2	2	2	2	2	2	2	2
-1	-1	-1	-1	-1	-1	-1	-1	-1	-1	-1

Figure 3(a) shows the original DNB image, figure 3(b) shows the DNB image after filter treatment. We can found in figure 3(b), obviously, lightning streaks characteristics obtained enhancement.

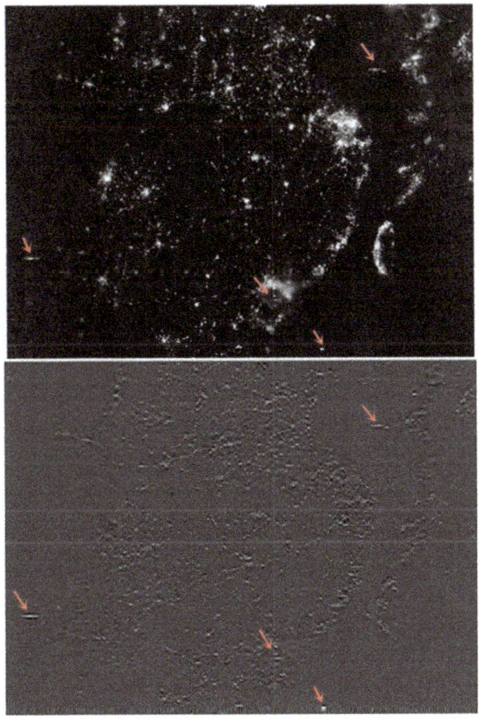

(a) The original image(b)The data created by the 11×3 convolution kernel

Fig. 3. TheVIIRS/DNB image over China at 1808UTC 15Aug 2012

3.2. *Gradient threshold processing*

In the area of lightning occurs, the brightness values of radiation will have a significant change, this is the equivalent that the radiation brightness value gradient greatly (including gradient above and gradient below), according to gradient difference of lightning regional and urban light regional, determine the gradient threshold of the lightning intensity.

$$\nabla f = \begin{bmatrix} G_{x(up)} \\ G_{y\ (down)} \end{bmatrix} = \begin{bmatrix} \dfrac{\partial f}{\partial x} \\ \dfrac{\partial f}{\partial y} \end{bmatrix} \tag{1}$$

Among them, $G_{x(up)}$ is at the top of the gradient of the region, $G_{y(\mathrm{down})}$ is at the bottom of the gradient of the region, then use the following formula for the first step processing:

$$\begin{aligned} &Data = Data, \quad if \ \frac{\partial f}{\partial x} > T_1 \\ &Data = 0, \qquad if \ \frac{\partial f}{\partial y} \le T_1 \end{aligned} \tag{2}$$

Among them, T_1 is lightning gradient threshold, it can be calculated according to the DNB data.

3.3 *Scale threshold processing*

In order to eliminate the influence of urban lightning further, combined with the feature of lightning scale and a certain number of samples of the DNB data, determine the horizontal scale of the lightning threshold for the next processing:

$$\begin{aligned} &Data = Data, \quad if \ \sum_{i=1}^{n} \theta_i > T_2 \\ &Data = 0, \qquad if \ \sum_{i=1}^{n} \theta_i \le T_2 \end{aligned} \tag{3}$$

Among them, θ is a particular pixel gray value, n is range from 65 to 85 (according to specific data statistics), T_2 is lightning horizontal scale threshold.

Finally, correct image and eliminate other noise.

4. Results

4.1. *Sample 1*

As shown in Figure 4(a), it is DNB imagery over China at 1919 UTC 7 Aug 2014. Some lightning flash in the figure are marked. Among them, A lies at the intersection of northeast of Inner Mongolia in China and Mongolia, B position along the border in the west of Chongqing and Sichuan province, C located in the junction of southeast Guizhou and northern Guangxi province.

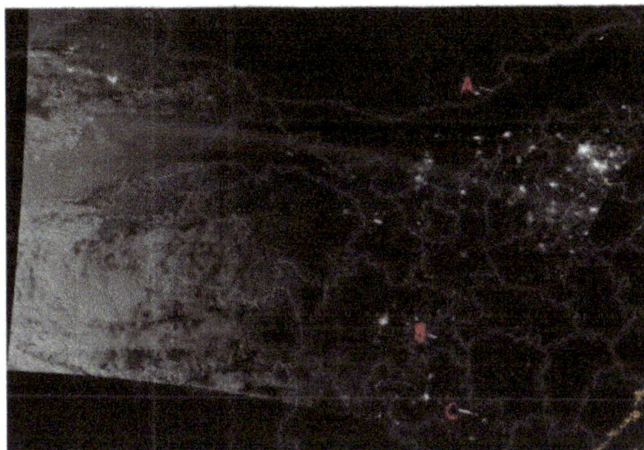

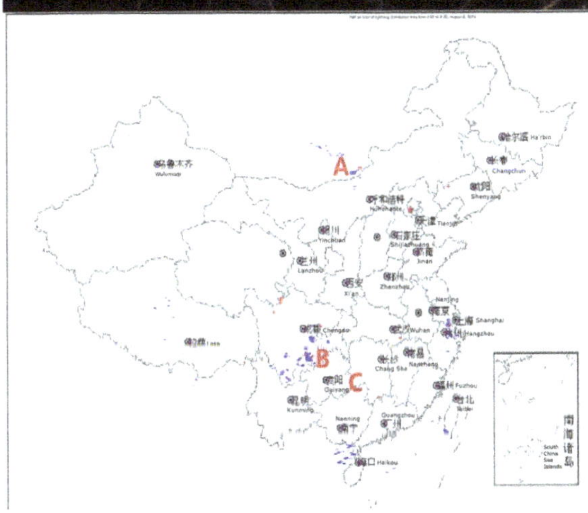

(a) The original image (b) The ground lightning data in half an hour

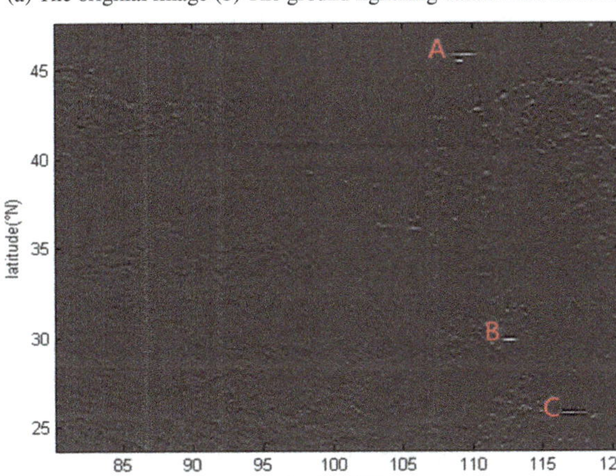

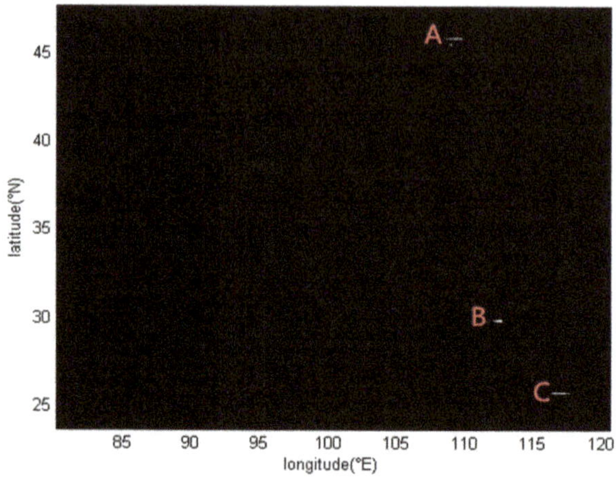

(c)The result after the gradient threshold processing (d)The result after the scale threshold processing

Fig. 4. The VIIRS/DNB image over China at 1919UTC 7 Aug 2014

First of all, making geometric correction with DNB image, then applying an 11×3 convolution kernel to the DNB data. Some lightning flashes are more apparent than in the original image, as is shown in Figure 4(c).

According to the features of the DNB data, we determine the values of T_1 and T_2, then make threshold processing through the formula (2), (3), the result is shown in Figure 4(d).

Finally, we made a comparison with the data of ground station, which downloaded from the China meteorological administration. The data provided the locations of cloud-to-ground lightning flashes over China in half an hour.

As shown in Figure 4 (b), this is lightning monitoring figure in half an hour from 3:00 to 3:30 Beijing time on 8 Aug 2014, corresponding to the UTC on August 7, 2014, from 19:00 to 19:30. We can see from the figure, lightning information is detected in A, B and C, the result is consistent with the algorithm, it confirms that lightning happened in these areas during a period of time.

4.2. *Sample 2*

In order to prove the algorithm, we find another data and make a comparison. As shown in figure 5(a), it is DNB imagery over China at 1805UTC 28 Sep 2014.

Some lightning flash in the figure are marked. Among them, A position in the northeast of Shandong province in China, lies at the yellow sea, B appeared in the west of Anhui province, C is located in the southwest of Guizhou province, there is some flash at this moment, D is located in the territory of Vietnam ,the south of Guangxi and Yunnan bungalows of China.

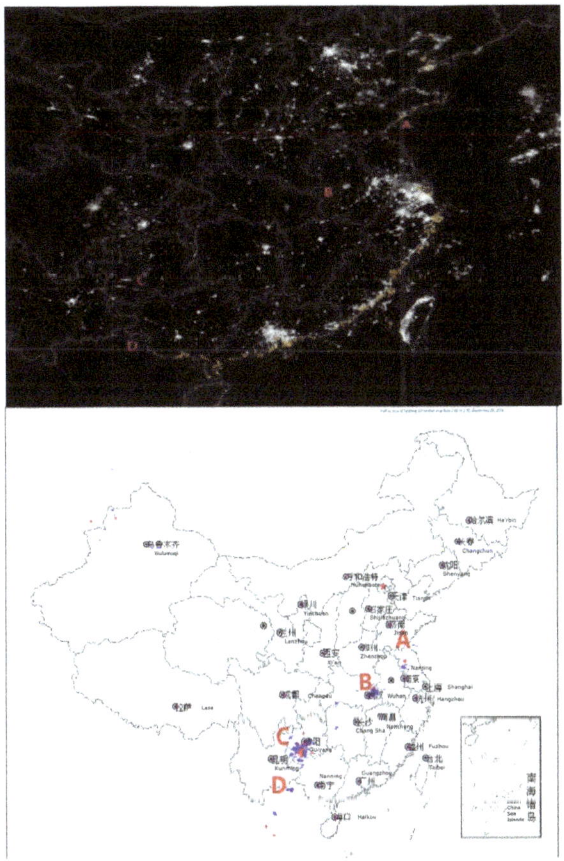

(a) The original image(b)The ground lightning data in half an hour

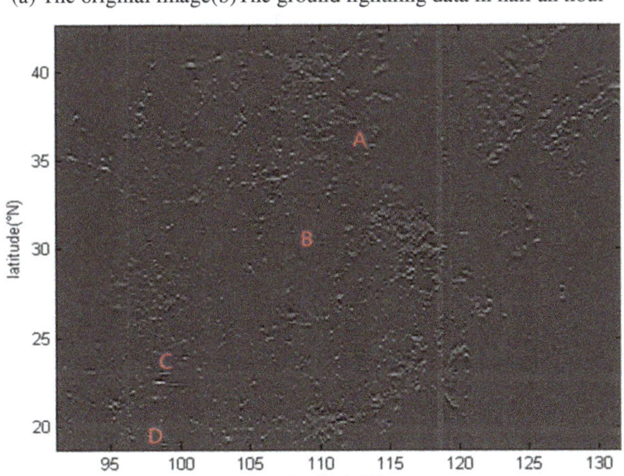

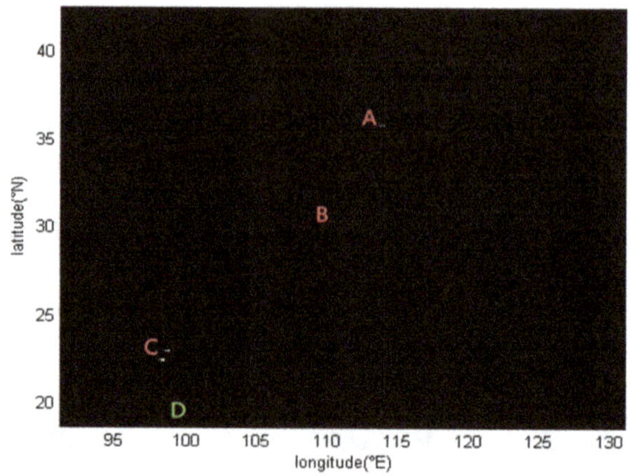

(c)The result after the gradient threshold processing (d)The result after the scale threshold processing

Fig. 5. The VIIRS/DNB image over China at 1805UTC 28Sep 2014

Similarly, applying an 11×3 convolution kernel to the DNB data, as shown in Figure 5(c).

Determine the values of T_1 and T_2, then make threshold processing through the formula (2), (3), the result is shown in Figure 5(d). It is important to note that lightning occurs in the D, after twice threshold processing, has not been detected. This is mainly because of its lateral dimension characteristic is different from others, this caused the selection of T_2 is difficult to achieve the best situation, therefore, lightning occurs in the D has not been identified.

Finally, we made a comparison with the data of ground station. As shown in figure 5 (b), this is lightning monitoring figure in half an hour from 2:00 to 2:30 Beijing time on 29 Sep 2014, corresponding to the UTC on September 28, 2014, from 18:00 to 18:30. Can see from the figure, lightning information is detected in A, B, C and D, which different from the results obtained through the algorithm, it proves that this algorithm still exist some limitations.

5. Conclusions

This paper proposes a lightning detection algorithm based on VIIRS/DNB data, the results prove the feasibility of using the VIIRS/DNB data monitoring of lightning flash in low light conditions. It provides a new train of thought for the lightning flash monitoring. Compared with the lightning figure in half an hour offered by China meteorological administration meteorological center, we found that the algorithm could reflect the lightning location information accurately.

However, due to instantaneity of lightning flash, the sensor can't manner to capture some of the lightning timely, also, the differences of the intensity of lightning makes some lower strength value lightning is missing, these leads to differences between the satellite data and ground data. In addition, how to realize the automation of threshold adjustment, further studies are needed.

Acknowledgments

The authors would like to thank the National Oceanic and Atmospheric Administration (NOAA) for making the S-NPP/VIIRS data products publicly available. Our S-NPP/VIIRS data products were downloaded from http://www.class.ncdc.noaa.gov/saa/products. We appreciate the work done by Bankert et al., their automated lightning flash detection algorithm based on DMSP/OLS provide us with a good idea and direction. We also thank the China Meteorological Administration (CMA) Meteorological Observation Center for providing ground-based monitoring data.

References

1. Chen Weilong, Lightning theory (China, 2003).
2. Xu Xiaofeng, National lightning monitoring network construction and technical analysis Engineering Science (China, 2002).
3. Ye Jiangrong, Thunder sound and light signal monitoring and positioning research (Zhejiang University, 2007).
4. Zhang Haowei, ShuYue Jiang, Li Jun, Summarization of lightning location technology (China, 2010).
5. Williams E R, Rothkin K, Stevenson D, et al, Global lightning variations caused by changes in thunderstorm flash rateand by changes in the number of thunderstorms (American, 2000).
6. Boccippio D J, Goodman S J. Regional differences in tropical lightning distributions (American, 2000).
7. Orville R E, Henderson R W, Global distribution of midnight lightning: September 1977 to August 1978 (American, 1986).
8. Bankert R L, Solbrig J E, and Lee T F, Automated Lightning Flash Detection in Nighttime Visible Satellite Data Forecasting (American, 2011).
9. Elvidge C D, Baugh K E, Kihn E A, et al, Mapping of city lights using DMSP Operational Linescan System data (American, 1997).
10. http://www.class.ncdc.noaa.gov/

11. Liu Q, Cao C, Weng F, Striping in the Suomi NPP VIIRS thermal bands through anisotropic surface reflection (China, 2013).
12. Zhou X K, Yan W, Bai H, et al, Detection of heavy fogs and low clouds during nighttime using DMSP-OLS data (China, 2012).
13. Ma S, Yan W, Huang Y X, et al, Vicarious calibration of S-NPP/VIIRS day–night band using deep convective clouds (China, 2015).
14. Lee T E, Miller S D, Turk F J, et al, The NPOESS VIIRS Day/Night Visible Sensor (American, 2006).
15. Hillger D, Kopp T, Lee T, et al, First-light imagery from Suomi NPP VIIRS (American, 2013).

Investigation and simulation of retrieving moisture independently along a LEO-LEO satellite link

Shen-Sen Hu, Hua-Ping Lv†, and Ying-Qiang Wang, Wei Yan
Institute of Meteorology and Oceanography,
PLA University of Science and Technology
Nanjing 211101, China
†*E-mail: hushensen@hotmail.com*

LEO-LEO radio occultation measurement is a new and efficient method of detecting Earth's atmosphere. Unlike the GNSS-LEO retrieval method, multi-frequency transmission technique is used in the LEO-LEO occultation detection, with signal frequencies located around a water vapor absorption line. This enables the measurement of water vapor independently from temperature. In this paper, the LEO-LEO occultation detection system and the ATOMMS algorithm for retrieving water vapor independently are first introduced. These provide a foundation for high precision water vapor inversion along a LEO-LEO satellite link. Compared with the standard atmosphere model, simulation results show that at the height of 2 to 35 km, the standard deviation of temperature is less than 0.65K, the standard deviation of atmosphere pressure is less than 1.95%, and the standard deviation of atmosphere density is less than 1.98%, while at the height of 2 to 8 km the standard deviation of humidity is less than 0.72g/kg. These are a great improvement from the precision of GNSS-LEO occultation observations. The simulation results make a valuable reference for the development of China's independent LEO-LEO occultation exploration program.

Keywords: Occultation detection; Atmospheric water vapor; Independent retrieval; Inversion error.

1. Introduction

Water vapor is one of the most active component in Earth atmosphere. It is a key factor that governs weather system. To grasp the variation trend of water vapor content in the atmosphere, is of great significance for weather and climate, especially for numerical weather prediction. Occultation observation at present mainly uses GNSS receiver installed on LEO (Low Earth Orbit) satellite to receive GNSS occultation signals, thus to retrieve global atmospheric and ionospheric information, including global refractive index profile, as well as the profiles of temperature, humidity and ionospheric electron density. However the inversion of lower atmosphere water vapor using GNSS-LEO occultation technique cannot avoid water vapor fuzzy phenomenon [1–2] and need outside data to separate humidity and temperature.

LEO-LEO occultation detection technology does not need to use the global satellite navigation and positioning system, but a LEO satellite launching signal and the other LEO satellite receiving occultation signal, which uses differential absorption principle to retrieve atmospheric water vapor profile independently[3]. Different from GNSS-LEO retrieval method, multi-frequency transmission technique is used in the LEO-LEO occultation detection, with signal frequencies located around a water vapor absorption line, which enables to measure water vapor independently from temperature.

Many researchers concentrates on LEO occultation events distribution. Zhao[4] used ray tracing method to study effect of LEO satellite's orbit parameter on GPS occultation event's number and distribution. Xu[5] simulated to study impacts of single LEO satellite orbit parameters on the distribution and number of occultation events. Du[6] studied simulation of the impacts of satellite orbit parameters on the number and distribution of LEO –LEO occultation events. Few researches retrieved atmospheric parameters from LEO-LEO occultation events.

2. ATOMMS

Here we introduce a LEO-LEO occultation system under development called the Active Temperature, Ozone Moisture Microwave Spectrometer (ATOMMS) that will probe the absorption lines of water vapor, ozone and other constituents[8]. The occultation detection system proposed by NASA is designed to include 12 LEO satellites and 2 orbital planes at the height of 650km and 800km respectively, which theoretically can observe 6000-8000 GNSS-LEO occultation events and more than 1000 LEO-LEO occultation events. As a result, it has the ability to probe atmospheric water vapor, pressure and temperature from surface to mesosphere, not influenced by clouds.

3. Algorithm

ATOMMS system launches three signals at the frequency of 9.7, 17.25 and 22.6GHz respectively to retrieve atmospheric water vapor, which are all located near water vapor absorption line of 22GHz. Similar with GNSS-LEO occultation inversion methods, ATOMMS needs to measure phase change of receiving signals thus to obtain refractive index and to measure amplitude change to obtain absorption coefficient[9–10].

We can obtain parameters such as temperature, pressure, humidity and so on by establishing a system of nonlinear equations. In this equations set, real refractive index is related to temperature, pressure and humidity by Smith-Weintraub Equation, and absorption coefficient is related to these parameters by Liebe Model[11]. Combined with the atmospheric static formula, the equations set can be written as

$$\begin{cases} k_1 - k_2 = F(p,e,T) \\ N = 77.6\dfrac{p}{T} + 71.7\dfrac{e}{T} + 3.75\times10^5\dfrac{e}{T^2} \\ \dfrac{d(p+e)}{p+e} = -\dfrac{gdz}{RT} \end{cases} \tag{1}$$

where k_1, k_2 are absorption coefficients near water vapor absorption line, T, p, e are atmospheric temperature, pressure and humidity respectively. At each height of earth atmosphere, temperature, pressure and humidity can be retrieved by solving these equations without outside data.

$$\begin{cases} k_1 - k_2 = F_{12}(p,e,T,L_c,T_c) \\ k_2 - k_3 = F_{23}(p,e,T,L_c,T_c) \\ k_3 - k_4 = F_{34}(p,e,T,L_c,T_c) \\ N = 77.6\dfrac{p}{T} + 71.7\dfrac{e}{T} + 3.75\times10^5\dfrac{e}{T^2} \\ \dfrac{d(p+e)}{p+e} = -\dfrac{gdz}{RT} \end{cases} \tag{2}$$

Due to the existence of clouds, the equations set in troposphere is expanded to where L_c is the liquid water content in clouds, T_c is the temperature in clouds. In the similar way, the five unknown parameters can be retrieved from the five closed equations.

The total LEO-LEO occultation inversion algorithm to retrieve atmospheric temperature, pressure and humidity is as follows.

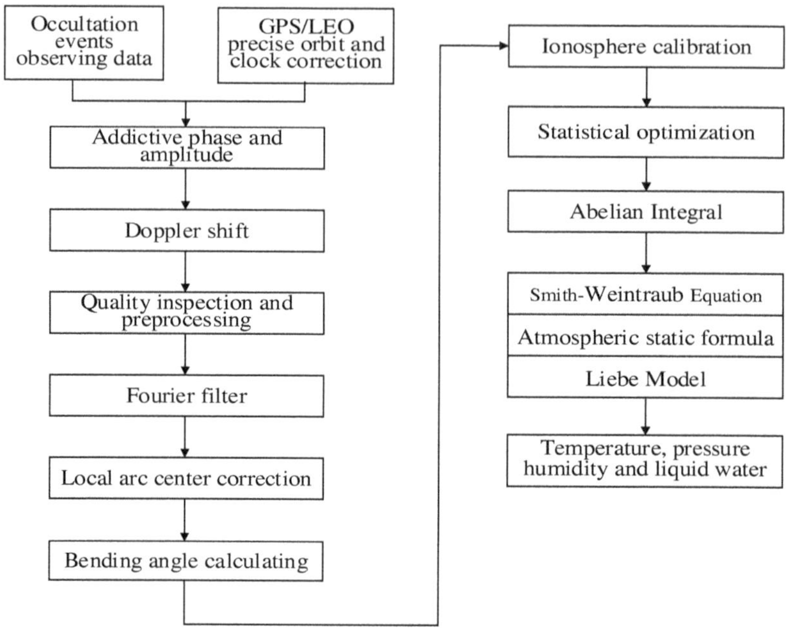

Fig. 1 Total LEO-LEO occultation inversion algorithm to retrieve atmospheric temperature, pressure and humidity.

4. Simulation

By far, study on LEO-LEO occultation detection is still in stage of demonstration, thus there is not a LEO-LEO occultation satellite on orbit around the world. Here we use the simulation platform EGOPS to conduct our study on high precise LEO-LEO occultation inversion.

We do this simulation in the following steps: 1) LEO-LEO satellite geometry relationship simulation; 2) LEO satellite signal propagating simulation; 3) occultation observation system error and observed value simulation; 4) bending angle calculating and inversion of atmospheric refractive index as well as parameters such as temperature, humidity and pressure.

In the simulation of LEO-LEO satellite geometry relationship, the occultation detection system includes 6 LEO satellites on 2 orbital planes, of which 3 satellites carries transmitters in X and K bands at the height of 800km and the other 3 satellites onboard receivers counter rotates at the height of 600km. As a result, 528 LEO-LEO occultation events can be observed every day. After the simulations of LEO satellite signal propagating in the atmosphere and of observation system error and observed value, amplitude and phase of electromagnetic wave can be obtained.

Relevant error in transmitter and receiver and in propagating process is of great effect on LEO-LEO occultation detection. In the simulation of error setting process, location error of transmitter and receiver are both 25cmRMS, relative velocity error in LEO-LEO attachment direction is 1mm/s, relative acceleration in LEO-LEO attachment direction is 0.5um/s^2, phase observation error of receiver carrier is 2mm, and clock error of receiver is $1e^{-12}$(1sAllan).

5. Results

We take 210 occultation events at random from aforementioned 528 occultation events into consideration to analyze relevant error of occultation detection. Absolute difference, relative difference, mean bias and standard deviation are defined as follows in the example of atmospheric pressure.

Absolute difference:

$$\varepsilon_a(h_i) = P(h_i) - P_{ref}(h_i) \tag{3}$$

where h_i is height, $P(h_i)$ is simulated atmospheric pressure profile and $P_{ref}(h_i)$ is referenced atmospheric pressure profile.

Relative difference:

$$\varepsilon_r(h_i) = 100 \times \varepsilon_a(h_i) / P_{ref}(h_i) = 100 \times [P(h_i) - P_{ref}(h_i)] / P_{ref}(h_i) \tag{4}$$

Mean bias:

$$\bar{\varepsilon} = \sum_{i=1}^{n} \varepsilon(h_i) / n \tag{5}$$

where n is the number of sampling points.

Standard deviation:

$$\sigma = \sqrt{\sum_{i=1}^{n} (\varepsilon(h_i) - \bar{\varepsilon})^2 / (n-1)} \tag{6}$$

To analyze relevant error of atmospheric profiles of pressure, density and temperature between 2 and 35km and humidity profile between 2 and 10km, here we use LS (Lower Stratosphere) to indicate 15-35km, and UT (Upper Troposphere) is denoted to 2-15km.

In Fig.2, average atmospheric temperature reduces from 2km to 17km, but rises even to 236K from 17km to 35km (left), and absolute temperature error reduces versus height (right) in which the maximal of absolute temperature mean bias in UT is 0.03K, the minimal -0.15K, and corresponding standard deviations are 0.65K and 0.008K respectively, but they all come to near zero in LS.

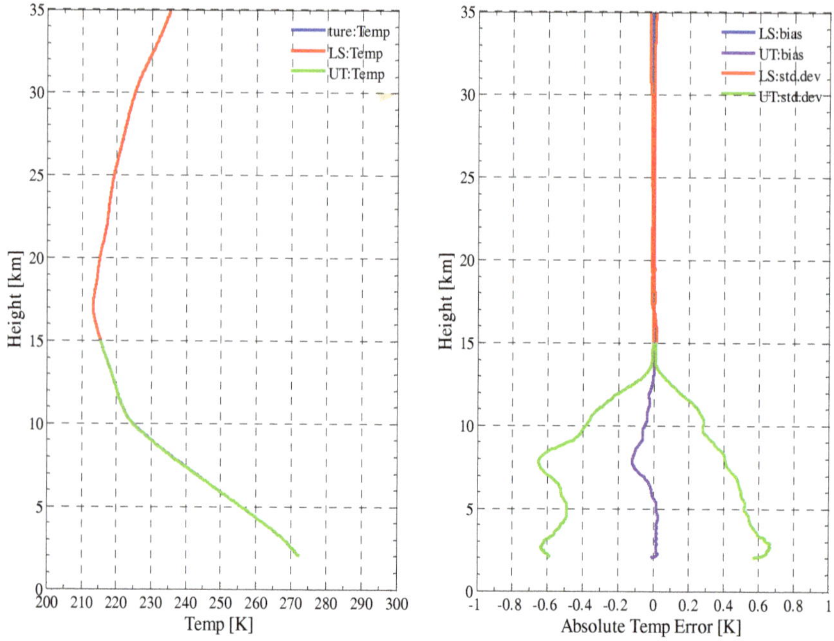

Fig. 2 Atmospheric temperature profile and the error profile.

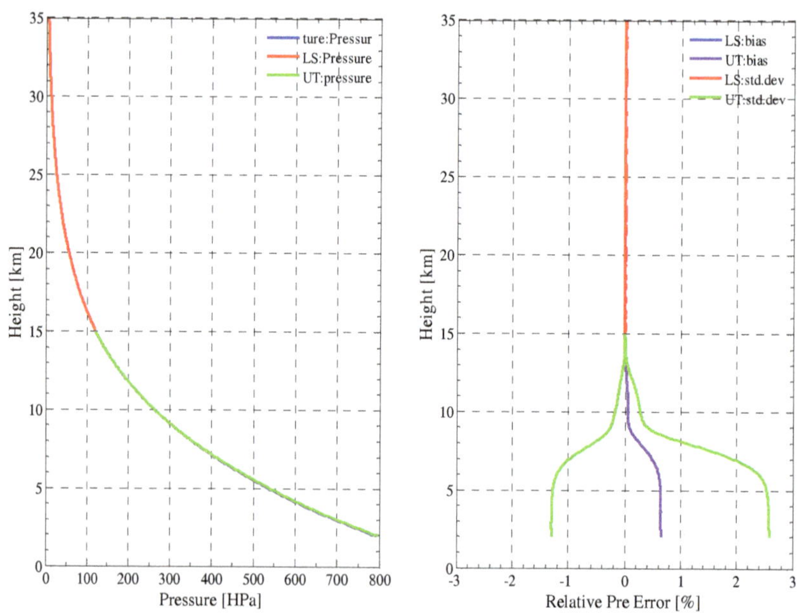

Fig. 3 Atmospheric pressure profile and the error profile.

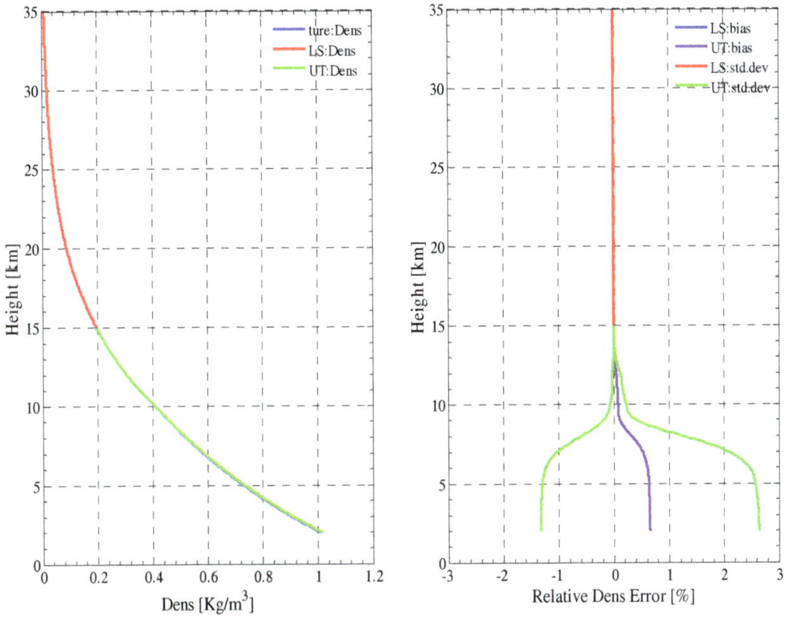

Fig. 4 Atmospheric density profile and the error profile.

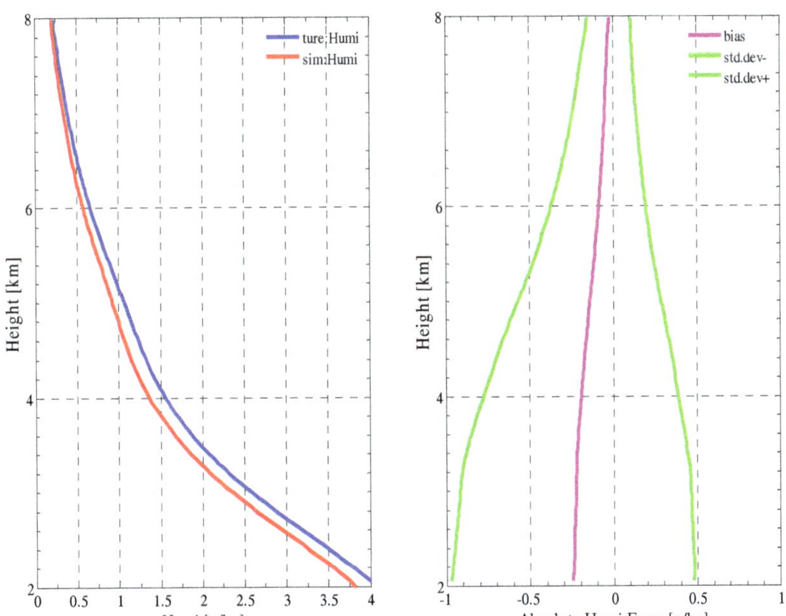

Fig. 5 Atmospheric humidity profile and the error profile.

In Fig.3, atmospheric pressure reduces exponentially from 2km to 35km and eventually verges to zero (left). Relative mean bias of pressure reduces versus height (right), in which the maximal is 0.65%, the minimal -0.0007%, and corresponding standard deviations are 1.95% and 0.0006% respectively in UT, and the maximal of relative mean bias is -0.0005%,the minimal -0.0007%,and corresponding standard deviations are 0.001% and 0.0005% respectively in LS.

In Fig.4, atmospheric density reduces exponentially from 2km to 35km and eventually verges to zero (left).Relative mean bias of density reduces versus height (right), in which the maximal is 0.67%, the minimal -0.0006%, and corresponding standard deviations are 1.98% and 0.0009% respectively in UT, and the maximal of relative mean bias is -0.0005%, the minimal -0.0007%,and corresponding standard deviations are 0.001% and 0.0006% respectively in LS.

In Fig.5, atmospheric humidity reduces gradually from 2km to 8km and eventually verges to 0.22g/kg (left).Relative mean bias of humidity reduces versus height (right), in which the maximal is -0.2g/kg, the minimal -0.2g/kg, and corresponding standard deviations are0.72g/kg and 0.12g/kg respectively between 2km and 8km.

6. Conclusions

This paper introduces LEO-LEO occultation exploration programs and unique advantages over GNSS-LEO occultation techniques. ATOMMS system proposed by NASA is specifically analyzed as well as its detailed algorithm to detect atmospheric parameters such as refractive index, humidity, pressure, temperature and density. Compared with standard atmosphere model, the simulation results show that at the height of 2 to 35 km, the standard deviation of temperature is superior to 0.65K, the standard deviation of atmosphere pressure is superior to 1.95%, and the standard deviation of atmosphere density is superior to 1.98%, while at the height of 2 to 8 km the standard deviation of humidity is superior to 0.72g/kg, which has a greater improvement than the precision of GNSS-LEO occultation observation. The simulation results have a reference value for the development of our country's independent LEO-LEO occultation exploration program.

Acknowledgments

The authors would like to thank Center for Science and Applied Research, Chinese Academy of Science for technical support in data processing.

References

1. Hu Xiong, Zeng Zhen, Zhang Xun Xie. Atmospheric inversion methods of GPS radio occultation. Chinese Journal of Geophysics, 48 (2005).
2. Hajj G A, Kursinski E R, Romans L J, et al. A technical description of atmospheric sounding by GPS occultation. Journal of Atmospheric and Solar-Terrestrial Physics, 64(2002).
3. Jin S, Cardellach E, Xie F. Theory of GNSS Radio Occultation. GNSS Remote Sensing (Springer Netherlands, 2014).
4. Zhao Shijun, Sun Xuejin, Zhu Youcheng, et al. Effect of LEO satellite's orbit parameter on GPS occultation event's number and distribution. Journal of PLA University of Science and Technology, 3(2002).
5. Xu Xiaohua, Li Zhenghang, Luo Jia. Simulation of the impacts of single LEO satellite orbit parameters on the distribution and number of occultation events. Geomatics and Information Science of Wuhan University, 30(2005).
6. Du Xiaoyong, Fu Yang, Xue Zhengang, et al. Simulation of the impacts of satellite orbit parameters on the number and distribution of LEO –LEOoccultation events. Chinese J. Geophys, 50(2007).
7. Martini E, Freni A, Facheris L, et al. Impact of tropospheric scintillation in the Ku/K bands on the communications between two LEO satellites in a radio occultation geometry. Geoscience and Remote Sensing, IEEE Transactions on, 44(2008).
8. Kursinski E R, Ward D, Otarola A, et al.The Active Temperature, Ozone and Moisture Microwave Spectrometer(ATOMMS).(Springer, 2009).
9. Fjeldbo G, Kliore A J, Eshleman V R. The neutral atmosphere of Venus as studied with the Mariner V radio occultation experiments. The Astronomical Journal, 76(1971).
10. Ishimaru A. A new approach to the problem of wave fluctuations in localized smoothly varying turbulence. Antennas and Propagation, IEEE Transactions on, 21 (1973).
11. Liebe H J, Hufford G A, Cotton M G. Propagation modeling of moist air and suspended water/ice particles at frequencies below 1000 GHz. In AGARD, Atmospheric Propagation Effects Through Natural and Man-Made Obscurants for Visible to MM-Wave Radiation 11 p (1993).

A three layer semantic approach for information extraction*

Shu Chen† and Ju-Min Ye
School of Computer, Center China Normal University,
WuHan, HuBei, 430079, P.R. China
†E-mail: Chenshu181@aliyun.com
www.ccnu.edu.cn

This paper presents a novel three layer approach for extracting knowledge from natural language documents. Traditionally, stochastic models are used directly, and they end up causing a low accuracy and recalling rate due to their inability to reflect the true meaning of the words. In our approach, a linguistic data layer is planted on a domain knowledge layer. Domain knowledge is represented by domain ontology while linguistic data is represented by a conceptually-related words structure called HowNet. A stochastic model called ME-2HMM is also proposed, which contributes a mapping from knowledge to linguistics when disambiguating the meanings of words. Experiments show that our approach achieves a better performance in accuracy and recalling rate, especially in domain related scenarios.

Keywords: Information extraction; semantic; ontology; domain.

1. Introduction and Related Works

The process for an information retrieval system that tries to find the right meaning for words in a sentence is called wordsense disambiguation (WSD)[1]. Roughly speaking, there are two categories in dealing with disambiguation in information extraction: pure frequency-based techniques and semantic approaches. The typical usage for the formal approach is NER (named entity recognition) systems[2]. In an NER system, labels are developed in the first place, and a stochastic model is built in order to assign the right labels to the words in run time to deal with ambiguity. For example, Finkel[3]. used CRF to define a local language structure, and it was combined with another model which is used for long-distance language dependencies in a decoding process to achieve the most probable sequence of named-entity labels. The system presented in Guo[4] also use NER system to augment queries. These classical

* This research was financially supported by the Scientific Key project of National Language Committee of China (Project ID: ZDI125-21).

approaches lack of solving the intrinsic problems of language ambiguity. Words with the meanings are always influenced by their syntactic roles are called polysemic while multiple words can have the same meaning are called synonyms. Semantic approaches on the other side focus on extracting the true meanings of words and stuffs, and try to address NERs' shortcomings[5], Dong[6,7] suggests using HowNet, which is a lexical database as vocabulary concept, to simplify the WSD procedure and calculate Chinese and English documents similarities. However, Semantic models such as HowNet are common sense and lack of domain knowledge, which will cause low accuracy.

In this paper, we proposed a three level model called SLMM (Semantic Linguistic Mapping Model) for WSD by applying both stochastic and semantic domain. A domain model represented by ontology is applied to represent the relevant concepts, and a stochastic model is used to deal with linguistic information extracted from NLP tools by mapping words to the right concepts. The main contribution of this paper is defining a novel WSD approach, based on a hybrid and unified environment constructed by ontology, linguistic and stochastic model which will enhance the performance in certain areas.

2. GlobalView

As we can see in Fig.1 which describes the main structure of SLMM model. There are two main actors involved in SLMM through interface, common user in charge of inserting query key words or nature language sentences into the system and obtain the results. Domain expert has the responsibility of defining the domain ontology and linguistic structure, in order of training the stochastic mapping between them. Domain expert is also in charge of discovering and adding new concepts to the domain model.

The main part of SLMM model includes model training, text extractor, information context extractor and semantic mapping extender. Text extractor is a process that accepts multi-format document, convert and extract them into pure plain text. This process can be implemented by some current NLP tools. Information context extractor translates documents into so-called information term sequences from nature texts. Information term sequence carries context of each words that can be extracted from text. Information term sequences are stored in text repository. Information term sequence is defined as following:

Definition2.1. (information term sequence): *Information term sequence which is represented as a sequence* $< X_0, X_1, X_2 .... >$ *is used to represent the context of a word in a document, each team is defined as a three group* $X_j := (W_j, S_j, C_j)$ *in which* W_j *is a lemma that represent a word,* C_j *is its associated concept that* *represent the current meaning of* W_j, S_j *represents a group of synonyms that have the same meanings as* W_j.

Semantic mapping is an implementation of a stochastic model which will solve the linguistic ambiguities that builds a bridge between domain ontology which will help the recognition phase, and linguistic data. Semantic mapping module will also be invoked by information context extractor to establish Information term sequence. Finally, mapping information and data are all stored in mapping index database.

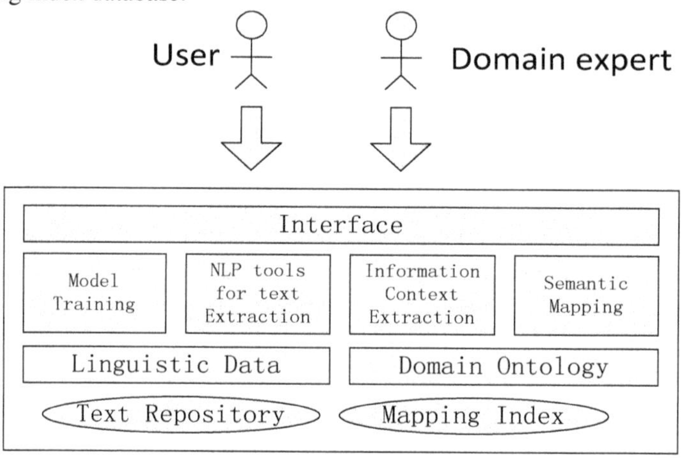

Fig. 1. Main structure of SLMM

3. Domain and Linguistic Model

The SLMM model we have addressed in Section 2 combines domain knowledge with a connected linguistic model. The adding of Domain ontology on linguistic model will help to improve the precision in recognition phase while the stochastic mapping will eliminate ambiguities from linguistic to domain.

An ontology is an explicit specification of a shared conceptualization[8]. The concepts of a given domain is formally represented by a knowledge base. Fortunately, there exist a lot of approaches to develop ontologies such as Uschold and King's method [9], Gruninger and Fox'smethod [10], Noy and McGuiness's method [11], the METHONTOLOGY framework [12], the Cyc methodology, KACTUS, SENSUS, and the On-To-Knowledge Methodology [13]. The common steps for developing a ontology can be simply represented as: identification of purpose and scope, knowledge acquisition, conceptualization, integration, encoding, documentation, and evaluation Stevens et al[14]. Formal items such as relationships and attributes are identified from the natural-language. In our approach we allow four basic items in definition of ontology:

classes, objects, attributes and relationships. Objects are instance of classes and relationships includes the relationship between classes and the relationship between objects. After adding attributes for each class, a lattice is then generated by Grail algorithm, which is implemented in the software Concept Explorer. The lattice will be used to create the ontology. The naming of each class is accomplished by means of the object or attributes labels from the nodes in the lattice. Fig.2 shows a fragment of an example of domain ontology in which specifies a computer. The computer ontology gives the following knowledge: A computer has type, shape and color, Lenovo ThinkPad is an instance of computer with its type as laptop, color as black and shape as rectangle. Lenovo ThinkPad is manufactured by Lenovo Corporation who is located in China. Lenovo ThinkPad also has display as its parts. The display part of a computer contains its knowledge either and has been omitted in this case.

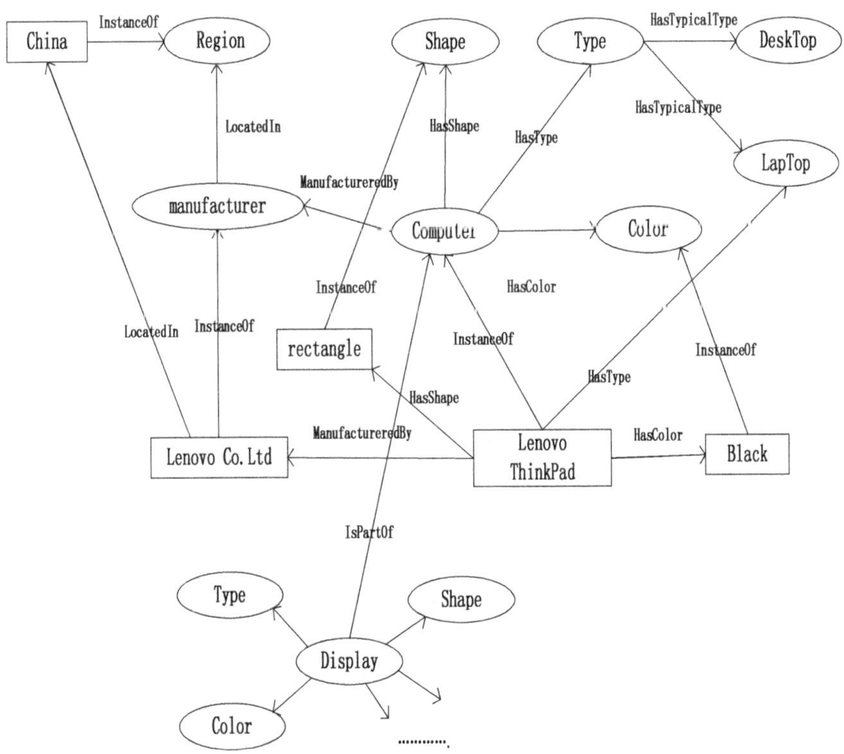

Fig. 2. A fragment of a simple computer ontology

The development of ontology domain is the infrastructure of the system and decided the performance and precision of the system, and it is also a challenge task. In our approach, We use the Protege[15] editor and the OWL DL language.

Ontology represents the conceptual layer, while linguistic data base plays another role that it provides the pure words as semantic dictionary. The reason that we do not use linguistic data directly is that, linguistic data does not contain any domain information. Linguistic data base organize words by means of linguistic relationships such as synonymy, converse, antonym, in a graph structure, somehow, it can be considered as semantic in linguistic level. As we mainly focus on Chinese words, thus, in our approach we use HowNet[6,7] as linguistic data base. HowNet is an on-line common-sense knowledge base unveiling inter-conceptual relations and inter-attribute relations of concepts as connoting in lexicons of the Chinese and their English equivalents[6]. HowNet provides 16 kinds of common relationships of words, these relationships allow us to navigate words and discover their similarities and meanings. As HowNet is Chinese and English equivalents, thus technically, modifying language support from Chinese to English is just a matter of providing the related dictionary databases. The basic building block of HowNet is sememe, a sememe is referred to as the smallest basic semantic term that cannot be divided. In our approach, we take the advantage of relationships and organize the words in HowNet as semset (semantic set), as the author also use concept in HowNet, in order to avoid ambiguity, we uselin_concept instead of concept in linguistic data.

Definition 3.1. (lin_concept):*lin_concept is a meaning of a word, defined by a 8 group (No, W_C, G_C, E_C, W_E, G_E, E_E, DEF). The first item No means the identity, W_C is the Chinese word with that meaning, G_C is the part of speech, E_C is a group of examples. W_E,G_E, E_E have the same meaning but in English. DEF is the definition of that meaning.*

In our approach, we define semset according to lin_concept to organize words in HowNet.

Definition 3.2.(semantic set): *A semset short for semantic set is a set of words with a same meaning and same part of speech.*

Semsets are connected by other 15 kinds of relationships, such as Hypernym, antonym, converse, part-whole, attribute-host, material-product, etc. In Fig.3 we shows a simple example of mapping between semsets and concepts.

Notice that, the noun *Lenovo corporation* is highly specified in domain that represent a computer manufacturer, and can be mapped to the concept *manufacturer*. This shows the fact that a domain knowledge is a must if the IE system is about to achieve a high precision.

4. Stochastic Model

In the above section, we have defined domain model and linguistic model, thus, a mapping procedure between them is needed. Consider the existed polysemic and synonym, one concept may be mapped to one or more semsets, and one semset may be related to one or more concepts. Classes and objects are mapped to semsets that have a same meaning, class and objects are mapped to nouns, adjectives, and adverbs. Relationship of classes and objects are always mapped to adverbs. However, this many-to-many nature will cause ambiguity. In our approach, we apply stochastic technology to solve this problem. Precisely, we call it ME-2HMM: maximum entropy second order hidden markov model. Compared with other stochastic models such as HMM, MEMM, and CRF, ME-2HMM is easier to train and can be efficiently implemented.

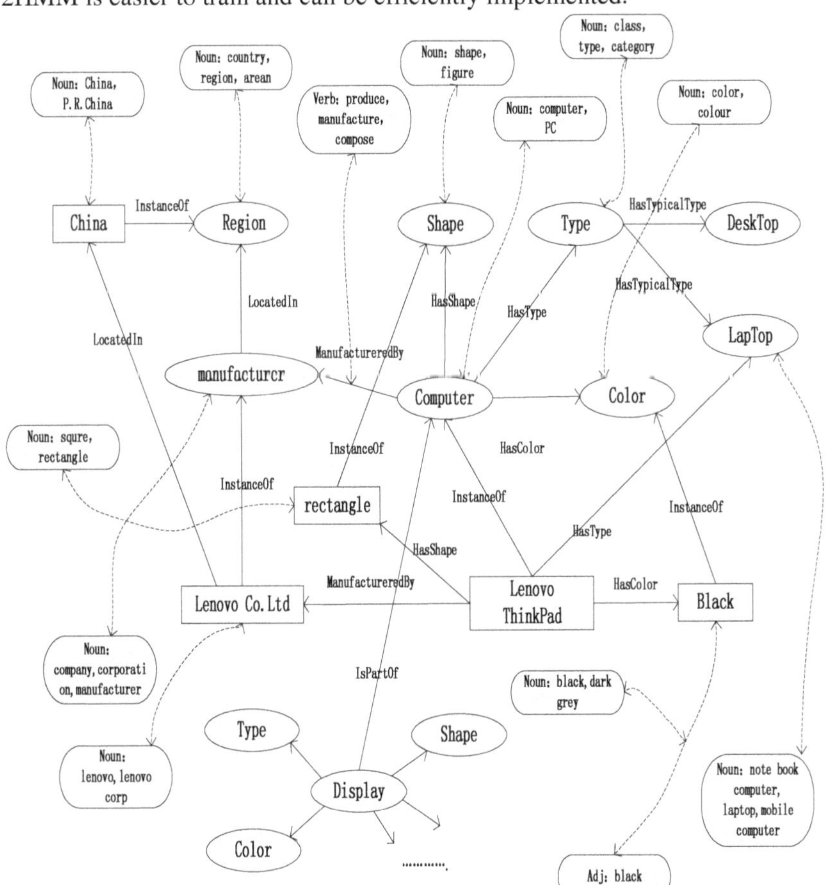

Fig. 3. Mapping between domain and linguistic data

The main purpose of ME-2HMM is to map concepts to semsets, and map semsets to lin_concept, the detail of ME-2HMM is given below. The ME-2HMM model is defined as a 7 group $(C,S,W,Tc,Ts,Tw,\prod)$, in which C is a finite concept set, represents all possible concepts defined as $C \triangleq (C_1, C_2, ..., C_N)$, S is a finite semset set, represents all possible semsets, defined as $S \triangleq (S_1, S_2, ..., S_M)$, W is a finite word set, defined as $W \triangleq (W_1, W_2, ..., W_K)$, Tc is a state transition matrix, represent the transitions of hidden variables, defined as following:

$$Tc \triangleq \begin{cases} Tc_{ij} = P(C_t = C_i \mid C_{t-1} = C_j) & t = 1 \\ Tc_{ijk} = P(C_t = C_i \mid C_{t-1} = C_j, C_{t-2} = C_k) & t > 1 \end{cases} \quad (1)$$

Ts represents the emission probability distribution of semset at time *t,* while *Tw* represents the emission probability distribution of words at time *t:*

$$Ts \triangleq \{Ts_{ij} = P(S_t = S_i \mid C_t = C_j)\}, \quad (2)$$

$$Tw \triangleq \{Tw_{ijk} = P(W_t = W_i \mid C_t = C_j, S_t = S_k)\} \quad (3)$$

As ME-2HMM is based on HMM model, thus it also has some pre-assumptions:

- Stationary assumption: the transition probability is independent of time t, and so does the emission probability of semsets and words.
- Second order assumption: the probability referred to the concept to associate with the word at time t depends solely on the last two concepts.
- Output independence assumption: the probability of semset at time t depends solely on the value of concept at time t. the probability of word at time t depends solely on the value of concept and semset at time t.

The essence of ME-2HMM model is its concept transition matrix built on maximum entropy model. We obtain $P(C_t \mid C_{t-1}) = \dfrac{\exp(\sum_i^n \gamma_i f_i(C_t, C_{t-1}))}{Z(C_{t-1})}$, where normalization function is defined as $Z(C_{t-1}) = \sum \exp(\sum_i^n \gamma_i f_i(C_t, C_{t-1}))$.

A good feature set definition is critical for any maximum entropy model, thus we have designed several feature pattern templates to describe the structure that predicates on concept set. A typical feature pattern template is shown as following:

$$\mathrm{T}(C_t, C_{t-1}, C_{t-2}) = \begin{cases} 1: & C_t = ? \wedge C_{t-1} = ? \wedge C_{t-2} = ? \wedge \\ & W_{t+1} = ? \wedge W_t = ? \wedge W_{t-1} = ? \\ 0: & others \end{cases} \quad (4)$$

GIS algorithm can be applied to train our ME-2HMM model, we have also modified GIS algorithm in order to suit our model, as the limit of space, we omitted the detail. Features defined in the template can predicate the past, current, and future observable which can be extracted from Information term sequence that we have addressed in section2. The training process can be stated briefly as the following steps:

- Document are divided into two sub-sets: the training set and the test set.
- Domain expert associates each word with the right concept and semset.
- Domain expert will evaluate each word to check that if a given word is meaningful and domain related, by its context. If negative, they will be marked as domain non-related.
- Evaluate the accuracy and recalling rate of training set by test set.

By considering the two emission probability distributions of semset and words are always not so huge, because words are usually mapped to a few semset and concepts, thus we can expect a model with good performance that generated from a small training set.

5. Experiment set up

We have implemented our prototype by the theories we have addressed above, and we have evaluated our system in a computer related spare parts product sale scenario. The domain of this scenario composed of 84 classes, 320 objects and 60 relationships. The ontology concepts were mapped to 260 semset in HowNet. We compare it with pure HMM model without concept knowledge, the accuracy and calling rate are showed in Table 1.

Table 1. Comparison between pure HMM model and our model

	HMM		ME-2HMM	
	Accuracy	Recall rate	Accuracy	Recall rate
Product name	0.830385	0.849853	0.930385	0.929853
Manufacturer	0.839483	0.932134	0.924694	0.983418
Affiliation	0.844658	0.912313	0.842341	0.914635
Address	0.833340	0.795757	0.833546	0.795377
Service	0.825731	0.901232	0.820567	0.900465
Features	0.814666	0.762315	0.874730	0.834340
Weight	0.673769	0.909563	0.685465	0.893421
Color	0.755504	0.821285	0.752166	0.827699
material	0.903576	0.931679	0.923416	0.954574
Model	0.833241	0.822314	0.923516	0.911130
Interface	0.734562	0.821230	0.934562	0.924655
Power	0.585437	0.823790	0.688799	0.834265

By analysis the result in table, we can see that our model has a higher performance than traditional HMM model, especially on those terms that highly related to domain, such as features, model and interface. However, on the other side, the time cost for training increase a lot, this is due to the new layer added.

6. Conclusions and Further Works

We have proposed a novel approach for document extraction. We use ontology as domain knowledge, and HowNet as semantic dictionary. We have designed ME-2HMM which combines maximum entropy and second order hidden markov model, in order to establish a bridge between knowledge and words. Our approach is domain specific, and experiment shows that it has a higher performance in domain specific scenario than traditional approach.

Acknowledgments

This research was financially supported by the Scientific Key project of National Language Committee of P.R. China (Project ID: ZDI125-21).

References

1. Navigli, R, Word Sense Disambiguation: A survey. (ACM Computing Surveys, 2009) 41 (2), pp. 1–69.
2. Ratinov, L., Roth, D, Design challenges and misconceptions in named entity recognition. (In: Proceedings of the Thirteenth Conference on Computational Natural Language Learning (CoNLL'09), Association for Computational Linguistics, Stroudsburg, PA, 2009), pp. 147–155.
3. Finkel, J.R., Grenager, T., Manning, C, Incorporating non-local information into information extraction systems by Gibbs sampling. (In: Proceedings of the 43rd Annual Meeting of the Association for Computational Linguistics, 2005), pp. 363–370.
4. Guo, J., Xu, G., Cheng, X., Li, H, Named entity recognition in query. (In: Proceedings of the 32nd International ACM SIGIR Conference on Research and Development in Information Retrieval, SIGIR'09, 2009), ACM, New York, NY, pp.267–274.
5. Wimalasuriya, D.C., Dou, D, Ontology-based information extraction: an introduction and a survey of current approaches, (Journal of Information Science 36, 2010), 306–323.
6. Dong Qiang, Dong Zhen Dong. HowNet, http://www.keenage.com, 2013.

7. Chi-Yung Wang, Knowledge-based Sense Pruning using the HowNet: An Alternative to Word Sense Disambiguation, (Thesis of Master degree, Hong Kong University of Science & Technology, 2002), Hong Kong.
8. Gruber, T.R, A translation approach to portable ontology specifications. (Knowledge Acquisition 5 (June),1993), pp. 199–220.
9. M. Uschold, M. King, Towards a methodology for building ontologies, (Workshop on Basic Ontological Issues in Knowledge Sharing, 1995).
10. M. Grueninger, M.S. Fox, Methodology for the design and evaluation of ontologies, (Workshop on Basic Ontological Issues in Knowledge Sharing, 1995).
11. N.F. Noy, D. McGuinness, Ontology Development 101: A Guide to Creating Your First Ontology, (in: Stanford Knowledge Systems Laboratory Technical Report KSL-01-05. Standford, USA, 2001).
12. M. Fernandez-Lopez, A. Gomez-Perez, N. Juristo, Methontology: From Ontological Art Towards Ontological Engineering, (AAAI Technical Report SS-97-06, 1997).
13. Gomez-Perez, Evaluation of taxanomic knowledge in ontologies and knowledge bases, (12th Workshop on Knowledge Acquisition, Modelling, and Management (KAW '99), Alberta, Canada, 1999).
14. R. Stevens, C.A. Coble, S. Bechhofer, Ontology-based knowledge representation for bioinformatics, (Briefings in Bioinformatics 1 (4), 2000) pp. 398–414.
15. Knublauch, H., Musen, M.A., Rector, A.L, Editing description logic ontologies with the protege owl plugin. (In: Proceedings of International Workshop on Description Logics, Whistler, Canada, 2004).

Saliency optimization via k-means clustering and low rank matrix recovery*

Dong-Jing Shan†, Gui-Bin Zhu, and Bo-Wen Liu
Chongqing Key Laboratory of Emergency Communication,
Chongqing Communication Institute,
Chongqing, China
†*E-mail: shandongjing@pku.edu.cn*

In the field of image saliency detection, previous approaches are mostly built on the low level priors like color and space contrast, boundary or connectivity prior, which are not sufficient to differentiate real salient regions from other independent or high-contrast parts. In this paper, we propose a novel approach which utilizes both high level and low level priors or cues to generate a salient map. Specifically, we obtain a foreground likelihood map via low rank matrix recovery, which incorporates traditional low-level features with higher-level guidance. Then, we compute a background likelihood map via principal component analysis and k-means clustering, which utilize two low level priors: distribution and boundary priors. Finally, the salient values of one image are calculated based on the two likelihood maps via a modified optimization framework. Extensive experiments on numerous publicly available datasets demonstrate that our proposed algorithm outperforms state-of-the-art methods.

Keywords: Salient object detection; Saliency map; Low rank; K-means.

1. Introduction

The human visual system is able to locate objects of interest in a complex scene. In computer vision, numerous computational models have been invented to simulate this system. S. Goferman et al. proposed the local contrast model [3] by comparing a region with its relevant contexts. N. Bruce et al. proposed the information maximization model [4] based on ICA projection and nonparametric density estimation. J. Harel proposed the graph model [5] based on markov chains. The principal component analysis (PCA) and low-rank matrix recovery models are most relevant with our work. The paper [6] designs an algorithm based on PCA, which integrates three factors: patterns, colors and high-level cues. It relies on the inner statistics of the patches in the image for

* This work is supported by science funding of Chongqing communication institute.

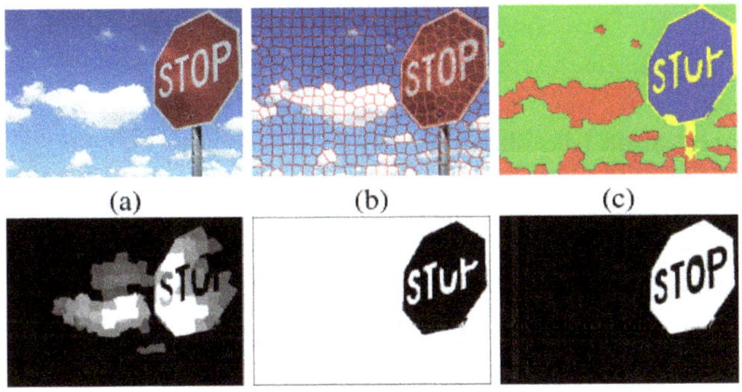

(a) (b) (c)

(d) (e) (f)

Fig. 1. Illustration of saliency detection by our model. (a)the original image, (b) over-segmentation image, (c) k-means clustering map composed of four clustered regions, (d) foreground likelihood map computed by low rank matrix recovery in which only center prior is used, (e) background likelihood map computed by PCA and k-means clustering, (f) the final saliency map.

identifying unique patterns. In [1], an image is represented as a low-rank matrix plus sparse noises in a certain feature space, where the non-salient regions can be explained by the low-rank matrix, and the salient regions are indicated by the sparse noises. In addition, a good feature transformation is learned to make sure most image background lies in a low dimensional sub-space.

In this paper, we propose a novel saliency detection approach via k-means clustering and low-rank matrix recovery. Firstly, feature extraction and segmentation is performed on the image. Then, principal component coefficients of the features are calculated and classified using k-means. The clustered regions are weighted via distribution prior and boundary prior (background likelihood model based on low level priors). Distribution prior measures the divergence of the clustered patches, as we find that the background regions tend to occupy more dispersive area in the image. Boundary prior measures the connectivity of the region to the image boundaries. Next, the foreground likelihood values are computed using low rank matrix recovery, and only the center prior is used here (foreground likelihood model based on high level cue). Finally, the saliency map is computed based on the two kinds of likelihood values via a modified optimization framework. Fig.1 is an illustration of our model.

The main contributions of our work are as follows: 1. We design a novel background likelihood model based on PCA and k-means clustering. For alleviating the instability of k-means, we introduce a criterion that minimizes the coordinate variance of the foreground. 2. We propose a new distribution prior that measures the divergence of the clustered patches, and utilize a new calculating method for boundary prior. 3. We incorporate a foreground likelihood model (low rank matrix recovery) in a modified optimization framework to generate the final saliency map.

2. The Foreground Likelihood Model

We extract multi-scale features around each pixel in a given image as the paper [1] does, including three RGB values, the hue and saturation components, 12 filter responses from the steerable pyramid filters, Gabor filter responses with 12 orientations and 3 scales. The 53 dimensional feature vector characterizes each pixel in the image. Then the image is decomposed into N segments $\{s_i\}_{i=1,\dots,N}$ based on the features by mean-shift clustering. For each s_i, the feature vector of its cluster center f_i is used as its feature representation. Easily we can get $F=[f_i, f_2, \dots, f_N]$ as the feature matrix representation of the entire image. Similar as in [1], we consider the image as a combination of a background residing in a low dimensional space with salient objects as sparse noises.

$$(L^*, S^*) = \arg\min_{L,S}(\|L\|_* + \lambda\|S\|_1) \quad s.t. F = L + S \tag{1}$$

Where $\|.\|_*$ and $\|.\|_1$ indicate nuclear norm and ℓ_1-norm respectively. As the paper [1] declares that most background patches should lie in a low dimensional subspace, a labeled database can be used to train a good feature transformation T (for detail see [1]), and $g_i = Tf_i, G = [g_1, g_2, \dots, g_N] = TF$. Besides, center prior can be used to improve the performance. Generally, objects near the image center are more salient, so we compute a prior map based on the distance of the superpixel (mean-shift cluster center) to the image center, in which $p_i(x) = \exp(-d(x,c)/\delta_1^2), P = diag(p_1, p_2, \dots, p_N)$. The Eqn.1 is advanced to:

$$(L^*, S^*) = \arg\min_{L,S}(\|L\|_* + \lambda\|S\|_1) \quad s.t. TFP = L + S \tag{2}$$

Where L and S can be perfectly recovered in most cases by Robust PCA[7].

$$\omega_i^{fg} = \frac{\sum_j |s_{ij}| - \min_i(\sum_j |s_{ij}|)}{\max_i(\sum_j |s_{ij}|) - \min_i(\sum_j |s_{ij}|)} \tag{3}$$

Where ω_i^{fg} is the foreground likelihood weight of the i-th superpixel, s_{ij} is the element of error matrix S. Fig.1(d) is our generated foreground likelihood map.

3. The Background Likelihood Model

3.1. *K-means clustering on Principal Component Coefficients*

As mentioned above, the feature matrix of the superpixels is extracted and transformed using T, we perform singular value decomposition and principal component analysis on the transformed feature matrix $G = [g_1, g_2, ..., g_N]$. We calculate the centered data matrix $G' = [g_1 - \overline{g}, g_2 - \overline{g}, ..., g_N - \overline{g}]$, where $\overline{g}$ is the mean of N vectors, and we get:

$$Y = \frac{1}{\sqrt{N}} G'^T \tag{4}$$

where each column of Y has zero mean.

$$Y^T Y = \left(\frac{1}{\sqrt{N}} G'^T \right)^T \left(\frac{1}{\sqrt{N}} G'^T \right) = \frac{1}{N} G' G'^T = C_{G'} \tag{5}$$

Where $C_{G'}$ is the covariance matrix of G'. We calculate the *SVD* of Y, $Y = U \Sigma V^T$. The columns of matrix V contain the eigenvectors of $C_{G'}$ and the squared diagonal elements of Σ contain the eigenvalues.

In this paper, we select the eigenvectors associated with the four largest eigenvalues to span a subspace and project the matrix G' onto the subspace, then we get the projection coefficient matrix $A = [a_1, a_2, ..., a_N] \in R^{N \times 4}$, which is classified into a certain number of clusters using k-means [8]. Fig.1(c) illustrates the clustering result of the superpixels (the matrix A corresponds to the superpixels of an image), each cluster is painted in one color and can be non-connected spatially.

3.2. *Distribution and Boundary Priors*

As we know, the k-means is a heuristic algorithm, it is hard to give a initial cluster number and there is no guarantee that it will converge to the global optimum, so it is common to run it multiple times with different starting conditions and select the best one using a specially designed criterion. In this subsection, we compute the background likelihood map using distribution and boundary priors with regard to three different k-means clustering results, in which the cluster number is set to three, four and five respectively.

Distribution prior: obviously, the salient objects always occupy certain areas of the image, while the background usually spreads widely. For each clustered region r_i in a k-means clustering map, we compute its distribution weight based on this prior. The superpixel centers of r_i are $C_{r_i} = [c_{i1}, c_{i2}, ..., c_{in}]$ and the coordinate variance of C_{r_i} is represented as v_{r_i}. The distribution weight of each cluster center is formulated as follows:

Fig. 2. Illustration of k-means criterion. (a) the original image and the final saliency map, (b) the upper image: k-means clustering map with 3 clusters, the lower image: the corresponding background likelihood map, (c) k-means clustering and background likelihood maps generated with 4 clusters, (d) k-means clustering and background likelihood maps generated with 5 clusters. By calculation the result in (d) satisfies the criterion (Eqn.8) and is chosen to be used in Section.4.

$$\omega_{c_{ij}}^{d} = \frac{\| v_{ri} \|_2}{\min(\| v_{ri} \|_2)} \tag{6}$$

Where $\|.\|_2$ means ℓ_2 norm and the superpixels in one clustered region have the same distribution weight.

Boundary prior: it is apparent that salient regions are much less connected to image boundaries than background ones. We adopt a measure proposed in [9] to quantify how heavily the region r_i is connected to the image boundaries and calculate it in our manner with the k-means clustering map.

$$\omega_{c_{ij}}^{b} = \frac{\left|\{s \mid s \in r_i, s \in Bnd\}\right|}{\sqrt{\left|\{s \mid s \in r_i\}\right|}} \tag{7}$$

Where Bnd is the set of image boundary patches and s is an image superpixel segmented by mean-shift. Eqn.7 is the ratio of a region's perimeter on the boundary to the region's square root of area.

The background likelihood weight for each superpixel is formulated as $\omega_{c_{ij}}^{bg} = \omega_{c_{ij}}^{d} * \omega_{c_{ij}}^{b}$. Fig.1(e) depicts the generated background likelihood map. For simplicity, we will use ω_i^{bg} as a replacement of $\omega_{c_{ij}}^{bg}$ in the following text ($i = \{1, 2, ..., N\}$ indicates each superpixel in the image).

3.3. *K-means Criterion*

As we know, we have generated three k-means clustering maps and compute the corresponding background likelihood maps. How to select the most reasonable one? We propose a criterion based on the distribution prior explained above.

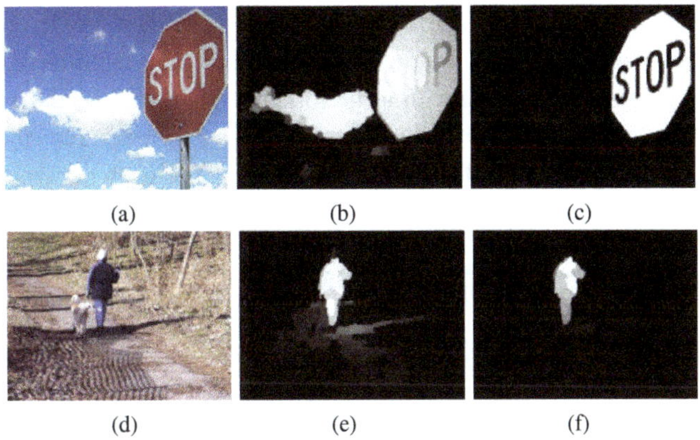

Fig. 3. (a,d) the original images, (b) the result computed by [20], (c) the result computed by [20] via our optimization framework. Smoothness between the adjacent pairs cannot work well when there exists an independent non-salient region, while our framework succeeds owning to the global optimization. (e) the result of our model via the original optimization framework, (f) the result of our model via our framework.

Fig.2 shows the k-means clustering and background likelihood maps, according to the distribution prior, the foreground should occupy a convergence zone spacially. $C=[c_1,c_2,...,c_N]$ represents the center coordinates of image superpixels and $\bar{c}$ is the mean value of c_i. The k-means clustering and the background likelihood maps which minimize Eqn.8 will be chosen as the proper results.

$$\min \| \{ \| c_i - \bar{c} \|_2 * \omega_i \}_{i=1,2,...,N} \|_2$$
$$\omega_i = \exp\left(-\frac{(\omega_i^{bg})^2}{\delta_1^2} \right) \tag{8}$$

Where $\|.\|_2$ indicates ℓ_2-norm, δ_1 is set $\sqrt{2}$ here. In the experiments, we find that every round of k-means clustering and the computation of background likelihood weight costs 0.0088 seconds on average, it is insignificant for the whole running time (see Table 1) of the method, so one can repeat the process more times for better performance according to the requirement.

4. The Optimization Framework

To combine multiple saliency cues, we redesign an optimization framework based on [9]. The objective cost function is minimized by least-square to assign the salient region value 1 and the background region value 0 respectively.

$$\sum_{i=1}^{N} \omega_i^{bg} v_i^2 + \sum_{i=1}^{N} \omega_i^{fg} (v_i - 1)^2 + \sum_{i=1}^{N} \sum_{j=1}^{\mu} \omega_{ij} (v_i - v_j)^2 \tag{9}$$

The background term (the first one) encourages a superpixel with large background weight to take a value close to 1, the foreground term (the second one) encourages a superpixel with large foreground weight to take a value close to 0, the smoothness term (the third one) encourages consistent saliency values between the superpixels with similar appearances. For each superpixel pair, the weight ω_{ij} is defined as

$$\omega_{ij} = \exp\left(-\frac{[d^c(s_i, s_j)]^2}{2\delta_2^2} \right). \tag{10}$$

Where $d^c(s_i, s_j)$ is the Euclidean distance between their average colors in the CIE-Lab color space, and δ_2 is set 10.

5. Experiments

We present experimental results of 7 saliency detection methods including ours on four kinds of datasets. The dataset [10] contains 235 images which are divided into 6 categories (see Table 2); The MSRA [11] contains 5000 images with complex backgrounds and low contrast objects; the CSSD [2] is composed of 200 images which are labeled with pixel-wise ground truth; the ECSSD [2] is more challenging and contains 1000 images with many of them have two salient objects. All the experiments are done with a computer of 2.6GHZ i5 CPU and 4GB memory. Running times of all the methods on images with 400×300 pixels are summarized in Table 1.

We evaluate the proposed method with comparison to OPT [9], SF [12], GS [13], LR [1], BL [14] and PCA [15]. Standard precision-recall curves (PR curves) [9] are used to evaluate the performance quantitatively. A curve is obtained by comparing the binary masks generated from saliency map against the ground truth, and then the curves are averaged on all of the images in the dataset. Fig.4 depicts the curves of CSSD and ECSSD. Also the mean absolute error (MAE) [9] is adopted for evaluation, which is the average per-pixel difference between the binary ground truth and the saliency map. The results are presented in Table 2 (MSRA-subset: with 300 images randomly selected).

Example results of the seven methods on the MSRA or CSSD are depicted in Fig.5. The proposed method works well when there exists an independent background (e.g. the cloud in the image of the first row is a non-salient region), complex background (e.g. the images in the second and third rows), small salient

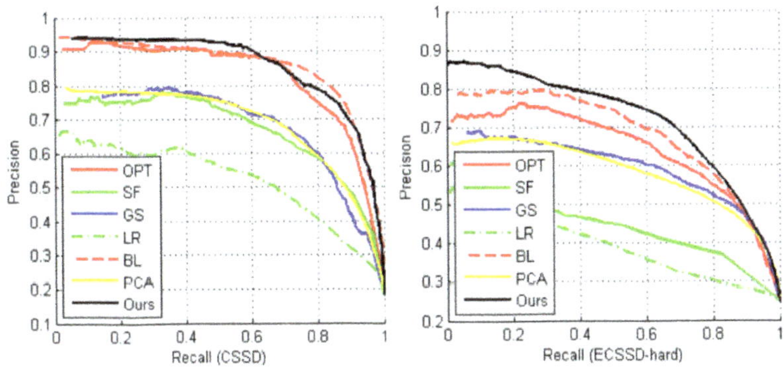

Fig. 4. The precision-recall (PR) curves tested on CSSD and ECSSD-hard.

Table 1. Averaging running time (seconds per image) of different methods.

Method	OPT	SF	GS	LR	BL	PCA	Ours
Time	0.3985	0.2868	0.2740	12.4930	46.2539	5.9352	5.8499
Code	Matlab	Matlab	Matlab	Matlab/ C++	Matlab/ C++	Matlab /C++	Matlab /C++

Table 2. Comparison of the mean absolute error (MAE) on four datasets.

	OPT	SF	GS	LR	BL	PCA	Ours
C1*	0.092559	0.134622	0.146312	0.164597	0.150775	0.140371	0.128047
C2*	0.122919	0.149448	0.169291	0.116124	0.247393	0.110553	0.097828
C3*	0.133060	0.063498	0.178899	0.103199	0.334363	0.110165	0.107339
C4*	0.112887	0.061646	0.226444	0.085179	0.212740	0.111964	0.082880
C5*	0.090514	0.090712	0.194553	0.101134	0.189823	0.101498	0.083030
C6*	0.091294	0.106337	0.121042	0.157168	0.217056	0.116030	0.090406
CSSD	0.112136	0.157563	0.168048	0.262365	0.170864	0.203622	0.125993
ECSSD	0.206833	0.247864	0.235506	0.332521	0.247843	0.267854	0.184789
MSRA	0.142927	0.214219	0.182581	0.258883	0.203180	0.172513	0.160468

* Dataset in [21]

object (e.g. the person in the images of the fifth row) or multiple salient objects (e.g. the flowers in the images of the seventh and eighth rows).

6. Conclusion

In this paper, we propose a saliency detection method based on k-means clustering and low rank matrix recovery. In this model, the foreground likely hood map is generated using low rank matrix recovery; the background likelihood map is computed via principal component analysis and k-means clustering, in which we have proposed a novel prior model. In the end, a modified optimization framework is utilized to generate a saliency map.

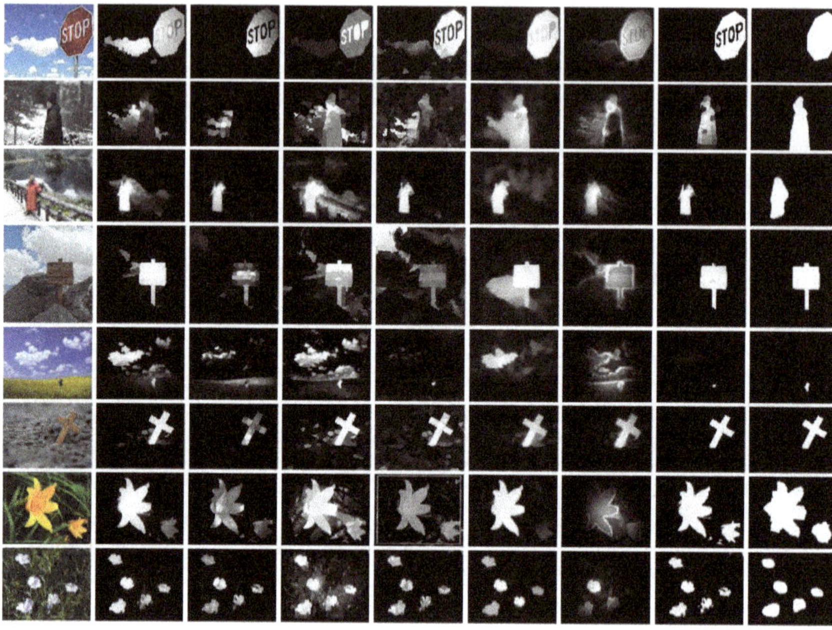

(Image) (OPT) (SF) (GS) (LR) (BL) (PCA) (Ours) (GT)

Fig. 5. Examples of extracted saliency from six images by seven models, where GT means the ground truth.

The model behaves well when it encounters the situation of complex background, multiple or small objects. In addition, the background likelihood model actually provides a saliency map and it works really fast (8.8 milliseconds), it is reasonable to be combined with another foreground likelihood model to achieve better performance.

References

1. Xiaohui Shen and Ying Wu, "A unified approach to salient object detection via low rank matrix recovery," in Proceedings of the 2012 IEEE Conference on Computer Vision and Pattern Recognition (CVPR), Washington, DC, USA, 2012, CVPR '12, pp. 853–860, IEEE Computer Society.
2. Qiong Yan, Li Xu, Jianping Shi, and Jiaya Jia, "Hierarchical saliency detection," in Proceedings of the 2013 IEEE Conference on Computer Vision and Pattern Recognition, Washington, DC, USA, 2013, CVPR'13, pp. 1155-1162, IEEE Computer Society.
3. Stas Goferman, Lihi Zelnik-Manor, and A. Tal, "Context-aware saliency detection," IEEE Trans. Pattern Anal. Mach. Intell., Vol. 34, No. 10, Oct. 2012 pp. 1915-1926.

4. Neil Bruce and John Tsotsos, "Saliency based on information maximization," in Advances in Neural Information Processing Systems 18, Y. Weiss, B. Schölkopf and J.C. Platt, Eds., pp. 155–162. MIT Press, 2006.
5. Jonathan Harel, Christof Koch, and Pietro Perona, "Graph-based visual saliency," in Advances in Neural Information Processing Systems 19. 2007, pp. 545–552, MIT Press.
6. R. Margolin, A. Tal, and L. Zelnik-Manor, "What makes a patch distinct?," in CVPR, 2013.
7. John Wright, "Robust principal component analysis: Exact recovery of corrupted low-rank matrices via convex optimization," in Advances in Neural Information Processing Systems 22, 2009.
8. Tapas Kanungo, David M. Mount, Nathan S. Netanyahu, Christine D. Piatko, Ruth Silverman, and Angela Y. Wu, "An efficient k-means clustering algorithm: Analysis and implementation," IEEE Trans. Pattern Anal. Mach. Intell., Vol. 24, No. 7, pp. 881–892, July 2002.
9. Wangjiang Zhu, Shuang Liang, Yichen Wei, and Jian Sun, "Saliency optimization from robust background detection," in Proceedings of the 2014 IEEE Conference on Computer Vision and Pattern Recognition, Washington, DC, USA, 2014, CVPR'14, pp. 2814–2821, IEEE Computer Society.
10. Xiangjing An Jian Li, Martin Levine and Hangen He, "Saliency detection based on frequency and spatial domain analyses," in Proceedings of the British Machine Vision Conference. 2011, pp. 86.1–86.11, BMVA Press, http://dx.doi.org/10.5244/C.25.86.
11. Tie Liu, Jian Sun, Nan ning Zheng, Xiaoou Tang, and Heung Yeung Shum, "Learning to detect a salient object," in In CVPR, 2007.
12. Federico Perazzi, Yael Pritch, and Alexander Hornung, "Saliency filters: Contrast based filtering for salient region detection," in In CVPR, 2012.
13. Yichen Wei, Fang Wen, Wangjiang Zhu, and Jian Sun, "Geodesic saliency using background priors," in ECCV (3)'12, 2012, pp. 29–42.
14. Xiang Ruan Na Tong, Huchuan Lu and Ming-Hsuan Yang, "Salient object detection via bootstrap learning," in In IEEE Conference on Computer Vision and Pattern Recognition (CVPR15). IEEE Computer Society, 2015, pp. 1–8.
15. Ran Margolin, Ayellet Tal, and Lihi Zelnik-Manor, "What makes a patch distinct?," in Proceedings of the 2013 IEEE Conference on Computer Vision and Pattern Recognition, Washington, DC, USA, 2013, CVPR'13, pp. 1139–1146, IEEE Computer Society.

A robust image watermarking scheme using multi-objective particle swarm optimization

Yong-Gang Fu[†]
College of Computer Engineering,
Jimei University, Xiamen, 360121, China
[†]*E-mail: yonggangfu@jmu.edu.cn*

Yun-Fei Jiang
College of Computer Engineering,
Jimei University, Xiamen, 360121, China

This paper proposes a novel image watermarking scheme based on the positive and negative coefficients numbers of the blocked DCT coefficients. The watermark is embedded into the middle frequency of DCT coefficients. In order to achieve better robustness and imperceptibility, multi-objective particle swarm optimization (MPSO) has been introduced to the watermark embedding and extracting procedure. The particle swarm optimization is applied to obtain optimum multiple scaling factors and the embedding strength. The experimental results show that the proposed scheme has significant improvement in term of imperceptibility and robustness under various attacks.

Keywords: Watermarking; MPSO; Robust.

1. Introductions

Computer and networks have been a significant way to exchange information in this era. Various digital multimedia products are spread on the Internet. Nowadays, people can copy, alter, and transmit multimedia properties easily, including texts, images, audios, videos, and so on. Modern information technology brings not only convenient and cheap digital transmission services, but also various challenges. One of the most important is how to protect the properties of the legal owner [1]. In this case, digital watermarking is proposed, which means hiding information or anti-counterfeit mark into the host image to protect copyright. It will be flourishing in the next decades with the urgent market demand. It is the digital watermarking technology that may be utilized as a supplement for cryptography encryption and scrambling technology [2].

During the last decade, many robust image watermarking algorithms have been proposed, which can be divided into three categories according to the domain what the watermark can be embedded: spatial domain watermarking, transform domain and hybrid domain watermarking. The spatial domain watermarking algorithms [3, 4] mainly embed the watermark by modulating the image pixels or altering its gray values directly, such as the least significant bit (LSB) and so on. These algorithms are fragile to manipulations, but they have little computation work and can be easily implemented. The transform domain watermarking schemes [5] are mainly concentrated on some signal transforms such as DCT(Discrete Cosine Transform) [6], DFT(Discrete Fourier Transform) [7] and DWT(Discrete Wavelet Transform) [8], etc. The watermark is embedded into the host by modifying some or all frequency domain coefficients, especially middle frequency coefficients. These methods have better robustness and can resist common attacks [9–11]. The hybrid domain can employ the advantages in two or more domain. The typical scheme is based on two transform domains. The first transform may have some good characteristics to some common attacks, and the second transform can be robust to geometric attacks. The hybrid domain algorithms can achieve good performance, but they are complex in implementation. It may be a future development direction because computer performance is strengthening on calculation. Now, most of the proposed watermarking schemes are based on the transformation domain. In this paper, the proposed algorithm can effectively resist some geometrical attacks.

The paper is organized as follows. In Section II, the particle swarm optimization is reviewed. Section III describes the proposed watermarking scheme in detail. The experimental results are shown in Section VI. And Section VII conclude the paper.

2. The Particle Swarm Optimization

The particle swarm optimization is an evolutionary technique widely used in parameter optimization, which is proposed by Kennedy and Eberhart [12]. The particle swarm optimization mimics the simplified social models such as fish schooling and bird flocking. A swarm is defined as a set of mobile agents that collectively carry out problem solving in a distributed manner. Each agent is called one particle. In a swarm, each particle keeps track of their own attributes. Initially, the position $x_{(i,j)}(k)$ and the velocity $v_{(i,j)}(k)$ of each particle p are randomly created. Each particle represents a candidate solution to optimize problem by calculate its fitness. The fitness of each particle is evaluated by an objective function. During iteration, the best location visited by each particle is kept as the local best position $P_{best(i,j)}$, while the best location visited by all particles is kept as the global best position $G_{best(i,j)}$.

Therefore, a new population is created based on a preceding one and the particles are updated by the following equations:

$$\begin{cases} v_{(i,j)}(k+1) = \omega v_{(i,j)}(k) + c_1 r_1 [P_{best(i,j)}(k) - L_{(i,j)}(k)] + c_2 r_2 [G_{best(i,j)}(k) - L_{(i,j)}(k)] \\ L_{(i,j)}(k+1) = L_{(i,j)}(k) - v_{(i,j)}(k+1) \end{cases} \quad (1)$$

Where parameters can be depicted as follows:

i : particle index;

j :index of the position in the particle;

k : iteration number;

$v_{(i,j)}(k)$: velocity of the ith particle in the swarm on the jth index of the position in the particle;

$L_{(i,j)}(k)$: position of the particle;

r_1 and r_2 : random numbers uniformly distributed between 0 and 1;

c_1 and c_2 : acceleration number, which control the influence of $P_{best(i,j)}$ and $G_{best(i,j)}$;

ω : inertia weight, usually with in $0.8 \le \omega \le 1.4$.

3. Proposed Watermark Embedding and Extracting Scheme

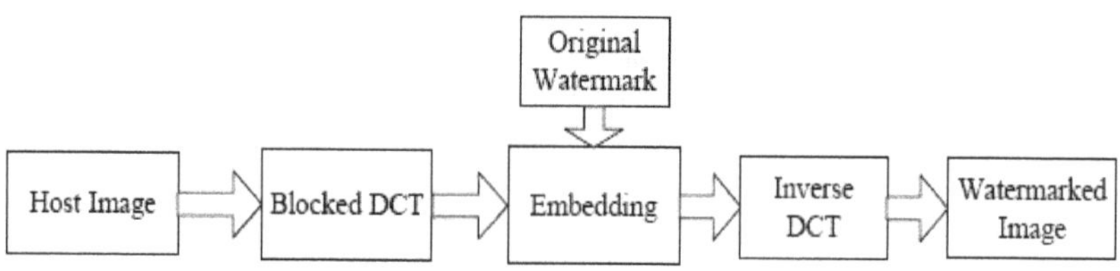

Fig. 1. Watermark Embedding Diagram

The discrete cosine transform coefficients can be divided into DC coefficient and AC coefficients. The DC coefficients of mth block can be expressed as $F_m(0,0)$ and the rest can be divided into low, middle and high frequency coefficients. Since tiny modifying the DC or low frequency coefficients may degrade image subjective quality, embedding watermark into the DC or low frequency coefficients should be avoided. The watermark information embedded into high frequency coefficients can be easily removed during the general image processing. So the middle frequency coefficients should be a better choice according to the necessity of compromise between robustness and invisibility. In this paper, the watermark is embedded into the middle frequency coefficients. Watermark embedding and extracting block diagrams are shown in Fig. 1 and Fig. 2.

Firstly, the host image is divided into disjoint blocks and then transformed to frequency domain by block-DCT. And mid-frequency DCT coefficients are selected for watermark embedding. The Zigzag scanning of 8×8 coefficients and embedding positions are shown in Fig. 3.

Since slight modification of the 1st, 2nd and 3rd coefficients may affect the

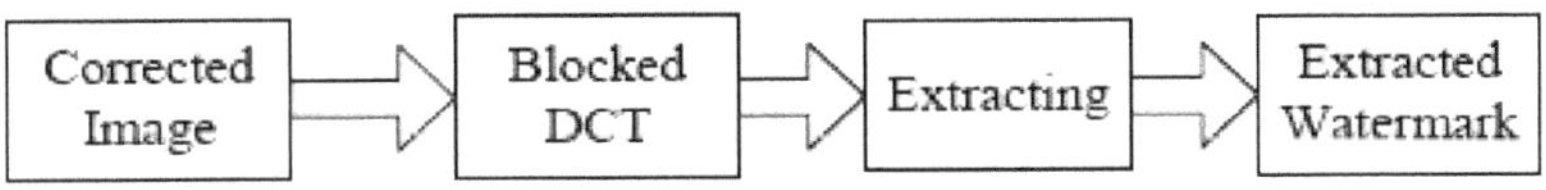

Fig. 2. Watermark Extracting Diagram

image quality serious, middle frequency coefficients are selected for watermark embedding. The proposed watermark embedding and extraction procedure can be depicted as following steps:

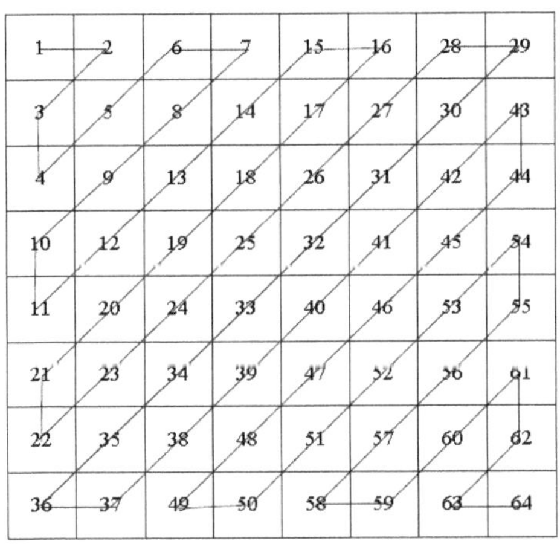

Fig. 3. The zigzag scanning of the 8×8 DCT

Embedding algorithm:

(1) Divide the original image into 8*8 blocks denoted as $f_m(i, j)$, and transformed to DCT domain denoted as $F_m(u, v)$. And the logo watermark denoted as *watermark* (i, j);

(2) Scan $F_m(u, v)$ into $F_m(s)$ with zigzag order, and select coefficients between k_1^{th} and k_2^{th}. The positive and negative numbers among these coefficients can be denoted as $Tp(i, j)$ and $Tn(i, j)$, respectively.

(3) Modify the coefficients and getting modified coefficients $F_m'(u, v)$ as following method:

(i) If $watermark(i,j)=1, Tp(i,j)=\alpha(p,q)Tn(i,j)$ (2)

(ii) If $watermark(i,j)=0, Tn(i,j)=\alpha(p,q)Tp(i,j)$ (3)

(4) Do the inverse DCT on $F_m'(u,v)$ to merge into the watermarked image I'.

Where, $0 \le i, j, u, v \le N/M$, $0 \le k_1 \le k_2 \le 63$, modifying the selected coefficients by changing the sign of the least coefficients in ascending order of their absolute values until the numbers of positive or negative coefficients to be Tp or Tn, respectively.

In a word, the watermark embeds one bit in each block. The total capacity to be embedded is 1024 bits.

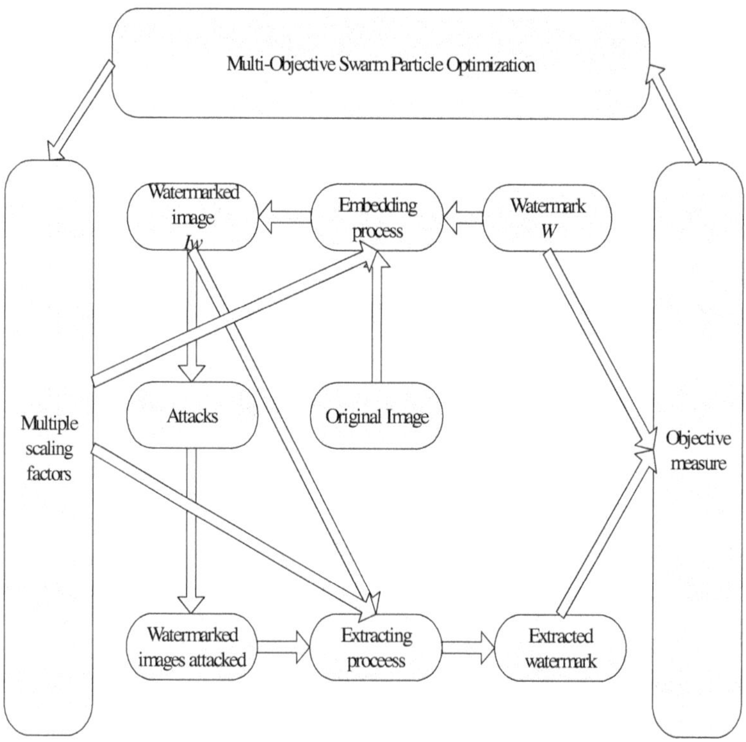

Fig. 4. The block diagram of multi-objective optimization

After analyzing the characteristics of middle frequency coefficients, the absolute value of them are very close, so we introduce an embedding strength factor, which can not only adjust the contradiction between robustness and invisibility but also eliminate rounding error from the function uint8. This error may affect the watermark extraction.

1) Determining multiple scaling factors:

A parameter $\alpha(p,q)$ which is used to control the trade-off between the imperceptibility and robustness, generally used as scaling factor, should be decided firstly. If the scaling factor is different in each watermark bit to be embedded, α can be a matrix, i.e. multiple scaling factors. To determine the optimal values of these multiple scaling factors can be viewed as an optimization problem. Multi-objective particle swarm optimization may be applied.

The block diagram of multi-objective optimization is a closed-loop control system. The system input is multiple scaling factors and objective measure as system output. This measure is calculated from original image *I*, watermarked image I_w, watermark *W* and the (T+1) extracted watermarks.

The steps for applying a multi-objective particle swarm optimization into the proposed watermarking scheme are enumerated.

2) Extracting algorithm:

(i) Divide the suspected image into 8×8 blocks and perform the DCT transform on each block. The m-th block coefficients denoted as $F_m'(u,v)$;

(ii) Scan $F_m(u,v)$ into $F_m(s)$ with zigzag, and select coefficients between the $k_1 th$ and $k_2 th$. The numbers of them are $Tp'(i,j)$ and $Tn'(i,j)$, respectively.

(a) If $Tp'(i,j) > \alpha(p,q)Tn'(i,j)$, then $ewatermark(i,j) = 1$ (4)

(b) If $\alpha(p,q)Tp'(i,j) < Tn'(i,j)$, then $ewatermark(i,j) = 0$ (5)

(c) Merge $ewatermark(i,j)$ into logo image.

Where $0 \le i, j, u, v \le N/M$, $0 \le k_1 < k_2 < 63$.

4. Experiments and Results

In order to evaluate the performance of the proposed method, we carried out experiments on MATLAB Ra2008. The experiments are carried out on several images including "Lena", "Pepper", "Plane", "Baboon", which represent different texture complexity. Due to the limitation of space, here only demonstrate the experimental results on gray Lena image with size 256*256. Fig. 5 shows the original image, watermarked image, logo watermark and the extracted watermarked.

4.1. *Objective image visual quality evaluation*

Image visual quality evaluation methods can mainly be divided into two categories which are subjective ones and objective ones. The subjective ones are the methods of grading subjective quality or video expert group, etc. The objective one are the methods of the mean square error (MSE), the signal-to-noise ratio (SNR), the peak signal-to-noise ratio (PSNR), the structure similarity (SSIM), and other methods based on human visual characteristics. Here mainly uses the mean square error (MSE) and the peak signal-to-noise ratio (PSNR) to measure the image quality.

Hypothesis of $f(i,j)$ being the gray value of L-level image whose size is $N \times N$, $f'(i,j)$ is the gray value of the watermarked image, and $0 \le i, j \le N-1$, MSE and PSNR are defined as follows:

(a) Original Lena

(b) Watermarked Lena

(c) Original Watermark

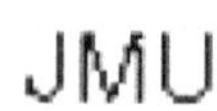

(d) Extracted Watermark

Fig. 5. Experimental Results

$$MSE = \frac{1}{N \times N} \sum_{i=0}^{N} \sum_{j=0}^{N} | f(i,j) - f'(i,j) |^2 \tag{6}$$

$$PSNR = 10 \log_{10} \frac{(L-1)^2}{MSE} \tag{7}$$

The smaller MSE is, the distortion is less and the PSNR is larger; the larger MSE is, the distortion is more serious and the PSNR is larger.

4.2. *Extracted watermark quality evaluation*

The original watermark and the extracted watermark can be denoted as $watermark(i,j)$ and $ewatermark(i,j)$ respectively, whose size is $N \times N$. The bit error ratio (BER) is used to evaluate the quality of extracted watermark. The BER can be defined as follows:

$$BER = \frac{1}{N \times N} \sum_{i,j} ewatermark(i, j) \neq watermark(i, j) \tag{8}$$

Where $\sum .$ is the sum on all the bit errors, and $0 \le i, j \le N-1$.

4.3 *Experimental results*

We choose five 256*256 gray images (Lena, Cameraman, Couple, Peppers, and Barbara) to do the simulation experiments. The block size is 8*8. The results are shown in Table 2. The watermark embedded is 1024 bits. The strengths are set to $d = k_1 - k_2 + 1 = 28 - 7 + 1 = 22$. The PSNR and the BER result of watermark extraction non-using MPSO and using MPSO method are shown in Table 1.

Table 1. Experimental result of Proposed Scheme

Attacks	PSNR(dB)		BER(Bit Error Rate)	
	Non-using PSO	Using PSO	Non-using PSO	Using PSO
JPEG(75%)	29.472	33.4573	0.772	0.8573
JPEG(85%)	32.5443	34.6574	0.8443	0.8874
Gauss noise (2%)	31.5785	32.6342	0.8785	0.9342
Scaling (reduced 25%)	31.3235	33.5343	0.8235	0.9343
Salt&pepper (2%)	30.5241	32.7645	0.8241	0.9545
Adding Noise (2%)	31.5363	32.6465	0.8563	0.9465
Cropping (10%)	29.4232	31.5433	0.8632	0.9533
Blurring (4×4)	30.3423	33.4554	0.8423	0.9154
Sharpening (4×4)	32.2136	34.5765	0.9136	0.9365
Rotating (3°)	29.5712	31.6357	0.8712	0.9457

5. Conclusions

A novel DCT based image watermarking scheme is proposed in this paper. MPSO is used to optimize the embedding scaling factors to achieve a good performance. The experimental results show that the proposed method based on PSO is robust to many attacks and resilience to some geometric attacks.

Acknowledgments

This work was partly supported by the Natural Science Foundation of Fujian Province of China (No. 2014J01244), the Science Project of Xiamen city (No. 3502Z20075054) and the Social Science Foundation of Fujian Province (No. 2014B131).

References

1. I.J. Cox, Joe Kilian, et al., Secure Spread Spectrum Watermarking for Multimedia. IEEE transactions on image processing, 1997, Vol.6(12): pp. 1673–1687.
2. H. Berghal and L.O. Gorman, Protecting ownership rights through digital watermarking. IEEE Computer, 1996, Vol.29(7): pp. 67–72.
3. Chi-Kwong Chan and L.M. Cheng, Hiding data in images by simple LSB substitution. Pattern Recognition, 2004, Vol.37: pp. 469–474.
4. Dipti Prasad Mukherjee, Subhamoy Maitra, and Scott T. Acton, spatial domain digital watermarking of multimedia objects for buyer authentication. IEEE transactions on multimedia, 2004, Vol. 6(1): pp. 1–15.
5. Chinh-Chin Lai and Cheng-Chih Tsai, Digital image watermarking using discrete wavelet transform and singular value decomposition. IEEE transactions on instrumentation and measurement, 2010, Vol. 59(11): pp. 3060–3063.
6. Chunhua Li and Zhiying Qin. A blind digital image watermarking algorithm based on DCT, IET International conference on smart and sustainable city. 2013. pp. 372–374.
7. Vassilios Solachidis and Loannis Pitas, Circularly symmetric watermark embedding in 2-D DFT domain. IEEE transactions on image processing, 2001, Vol.10(11): pp. 1741–1753.
8. Chen Mingzhi, Yan Li, et al., A combined DWT and DCT watermarking scheme optimized using genetic algorithm. Journal of Multimedia, 2013, Vol.8(3): pp. 299–305.
9. E.Esen and A.A.Alatan, Robust video data hiding using forbidden zone data hiding and selective embedding. IEEE transactions on circuits and systems for video technology, 2011, Vol.21 (8): pp. 1130–1137.
10. Yonggang Fu, Robust oblivious image watermarking scheme based on coefficient relation. Optik – International Journal for Light and Electron Optics, 2013, Vol.124 (6): pp. 517–521.
11. Lingling An, Xinbo Gao, et al., Robust lossless data hiding using clustering and statistical quantity histogram. Neurocomputing, 2012, Vol.77: pp. 1–11.
12. Kennedy J. and Eberhart R., Particle swarm optimization, in Proceedings of IEEE International conference on neural networks 1995. pp.1942–1948.

Application research of neural networks based on map-reduce on cloud computing clusters*

Hai-Jun Zhang† and Qi-Min Zhu, Nan-Feng Xiao
School of Computer Science and Engineering, South China University of Technology, Guangzhou 510006, China
†E-mail: nihaoba_456@163.com
www.scut.edu.cn

With the advent of the "big data" era, the traditional stand-alone serial-based training machine learning has had difficulties in meeting the needs of "big data" applications. To this end, this paper discusses and applies a neural network learning algorithm based on cloud computing. First, several experiments are designed for verifying the feasibility of neural network machine learning based on cloud computing. Next, the algorithm is applied to specific face recognition, speech recognition, and affective computing issues, among others. Finally, through experimental verification, it was found that the neural network learning algorithm based on cloud computing has a faster training speed, higher recognition accuracy, and greater data processing capabilities than that of the traditional learning algorithm.

Keywords: Map-Reduce; Cloud computing; Hadoop; Neural network; Face recognition; Speech recognition; Affective computing.

1. Introduction

With the advent of the era of "big data", data which many applications need to deal with is growing at the speed of geometric progression. By Moore's law, computing capacity which a single processor can provide is now close to the limit. For this, cloud computing technology arises at the historic moment, it combines the characteristics of distributed computing, parallel computing and grid computing, uses computing clusters set up by huge amounts of ordinary servers and storage clusters set up by a large amount of cheap equipment to provide scalable computing resources and storage space for uses, so as to realize the massively parallel computing of "big data". In addition, the neural network as an important tool of machine learning needs to train samples by a lot of iterations therefore needs a lot of storage space and computing time. Because the

* This work is supported by the National Natural Science Foundation (Grand 61171141, 61573145), the Public Research and Capacity Building of Guangdong Province (Grand 2014B010104001) and the Basic and Applied Basic Research of Guangdong Province (Grand 2015A 030308018)

neural network has the characteristics of distributed storage and parallel processing, this paper discusses and implements the neural network machine learning method based on cloud computing: 1) the neural network machine learning method based on cloud computing validated by the experiment is feasible; 2) applied to the specific problems–face recognition, speech recognition, emotion calculating. Through experimental verification the neural network machine learning method based on cloud computing has faster training speed, higher recognition accuracy and larger data processing ability than the traditional neural network machine learning method based on a single machine.

2. The Existing Research of Domestic and Foreign

Due to the characteristics of inherent in the neural network for the cloud platforms to implement machine learning attracts the attention of many researchers. Taochu Cheng et al. proposed an kind of BP (Back-Propagation) network algorithm based on Map-Reduce on multi-core computer in 2006 [1]; Zhiqiang Liu et al. handled BP network algorithm based on Map-Reduce to deal with huge amounts of data of mobile communications, which obtained higher training speed , better ability to scale and resist noise than the traditional serial training implementation in 2010 [2]; Chenjie Zhu et al. Combining with the HDFS (Hadoop Distributed File System) storage characteristics, presented a kind of BP neural network with the characteristics of local iteration and basing on genetic optimization in 2012, verified by the experiment the algorithm increases markedly in speed of convergence, accuracy and efficiency of parallel than the algorithm in [1, 3]. The study above concentrated to realize the application of BP network on the cloud platform, and for other neural networks with wide range uses, such as on the cloud platform to realize the neural network of RBF (Radix-based Function), lack of adequate research.

3. The Cloud Platform of Hadoop and Map-Reduce

Apache built an open source project based on HDFS and Map-Reduce, Hadoop [4]. Hadoop clusters use the architecture of master-slave (Master-Slave Pattern), provide functions of storing large files and reading and writing data with high throughput rate in clusters consisting of general hardware, as shown in (a) of Fig.1, it consists of NameNode, DataNode, Secondary NameNode, Job Tracker and Task Tracker. NameNode and Job Tracker are running on the Master, and a DataNode and Task Tracker are deployed on each Worker.

Map-Reduce was introduced by Google, it is a good way to offer programming model for the program development of solving the problem of large-scale data [5], its core idea is to make task divide into two processes of Map and Reduce, its process is shown in (b) of Fig.1.

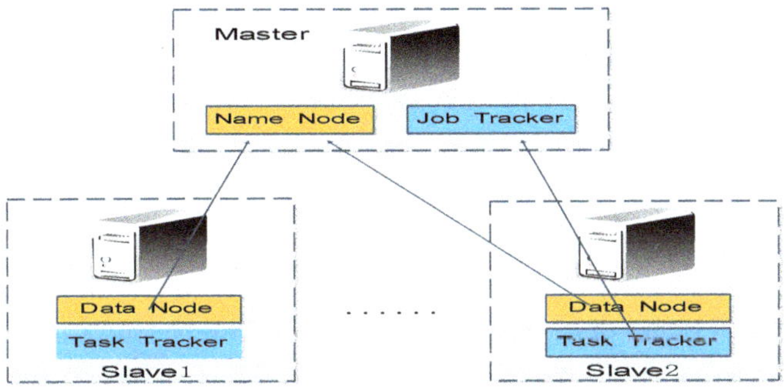

(a) Master-Slave architecture of Hadoop

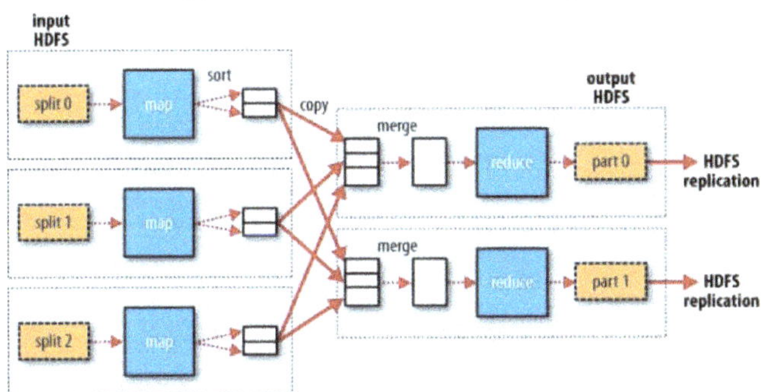

(b) Map-Reduce processing flow

Fig. 1 Master-Slave architecture of Hadoop and Map-Reduce processing flow

4. The Model of Neural

Aleksander and Morton defined the neural network as a kind of adaptive machine [6]: the neural network is a large scale parallel distributed processor consisting of a number of simple processing units, having features of storing and applying empirical knowledge. The neuron is made of three basic parts of connection weights, adder and activation function, the model is shown in Fig.2.

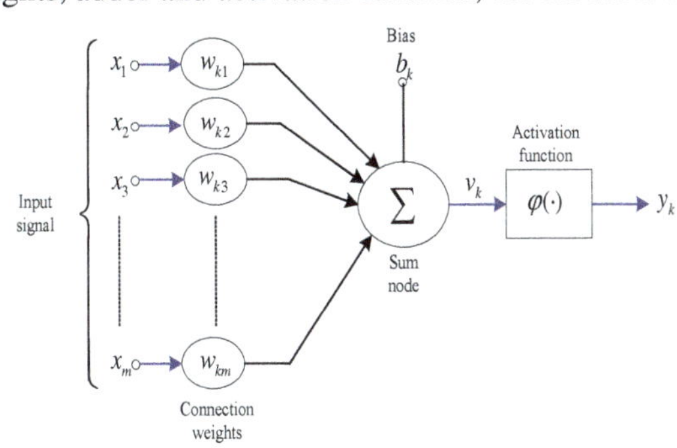

Fig. 2 Diagram of neuron model

Mathematical relationship of all variables in the neurons as shown in Eq.(1),

$$u_k = \sum_{j=1}^{m} w_{kj} x_j ,\ y_k = \varphi(u_k + b_k) \tag{1}$$

where, x_1 to x_m is input signal, w_{k1} to w_{km} is network weights of neuron k, u_k is output of input signal processed by linear combiner, b_k is bias, $\varphi(\cdot)$ is activation function, y_k is output of neuron.

5. Implementation of the Neural Network Based on Cloud Computing

This section introduces implementation and experiment results of the BP neural network based on cloud computing.

5.1. *Map-Reduce parallel implementation of BP neural network*

By the following Eq.(2) and Eq.(3) it can know BP network will update weights after completion of a round iteration of sample [7], so which can make the training algorithm of BP parallel.

$$\xi_{av} = \frac{1}{2N} \sum_{n=j}^{N} \sum_{i \in c} e_i^2(n) \tag{2}$$

$$\Delta w_{ij} = -\eta \frac{\partial \xi_{av}}{\partial w_{ij}} = -\frac{\eta}{N} \sum_{n=1}^{N} e_i(n) \frac{\partial e_j(n)}{\partial w_{ij}} \tag{3}$$

where, $e_i(n)$ is the output error, ξ_{av} is the mean square error, Δw_{ij} is correction weight after batch training. Making BP Map- Reduce parallel can be divided into the following two steps: First, in the process of Map using batch training for a certain number of iterations then outputs weights of the network; Second using Reduce to merger all outputs of each Map obtains total weights the network, and according to the adjustment quantity of network weights decides whether to iterate again or not [8].

5.2. *Experimental results and analysis*

The experimental configuration is described below; data sets are from UCI [9]. Experimental samples are divided to each Mapper to train; each experiment repeats for 10 times and achieves average. Experimental results are shown in Table 1.

Table 1. Experimental Results Of Bp And Bp Based On Map-Reduce

Algorithm	The number of rounds	Error	Classification accuracy	The number of Mapper	Mapper training iterations
Serial BP	2	0.171	0.7639	--	--
Serial BP	10	0.136	0.7639	--	--
Serial BP	50	0.122	0.9399	--	--
Map-Reduce-BP	2	0.04	0.9914	3	(1,2,2), (1,1,1)
Map-Reduce-BP	2	0.13	0.9828	9	(69, 50, 6, 5, 3,3, 35, 31, 6), (10, 18, 1,2,1,1,17, 10, 1)

From data of the second, third and fourth columns of the Table it can be seen that Map-Reduce-BP achieves better training effect after two rounds of training, better than serial BP. Also, from data of the second, third and fourth columns of the table it can be seen that Map-Reduce-BP improves training speed of BP neural network, increase the testing accuracy. The experiment confirmed the feasibility of the neural network based on the cloud platform.

5.3. *Map-Reduce parallel implementation of RBF neural network*

According to the determination method of the center of RBF, RBF has a variety of ways to learn. In the bellow experiment, using the method based on self-organizing learning center and the method based on supervision learning center respectively makes Map-Reduce parallel implementation of RBF neural network and compares learning efficiency and accuracy of training of the two implementations.

5.3.1. *Learning method based on self-organizing learning center*

Using this method to parallelize can be divided into two parts: the parallelization of clustering algorithm and the learning of network weights. First, using the parallelization of k-means^{++} algorithm implements clustering of the training samples [10], consisting of selecting the initial clustering center and clustering according to the initial center. The difficulty of selecting initial center is how to select a sample from the training samples obtained from Map as candidates center. In this paper, according to the following method to select center: according to the characteristics of Map-Reduce assigning training set s_k for each compute node, then the total sample set S as follows,

$$S = s_1 + s_2 + \cdots, + s_M \qquad (4)$$

where, M is the total number of Map. For sample points $\{x_i \in s_m\}$, assuming the distance near recent clustering center c is d_i, the sum of the distance of each ample d_i in subsets s_m is ds_m, the ds_m as follow,

$$d_{s_m} = \sum_{i=1}^{N} d_i \tag{5}$$

And the probability of point x_i selected as a candidate is as follow,

$$P_i = \frac{d_i}{d_{s_m}} \tag{6}$$

In Eq.(6), when point x_i participates in the election of the next round the distance is replaced with the ds_m, so the probability chosen in the next round campaign is as follow,

$$P = P_i * P_m = \frac{d_i}{d_{s_m}} * \frac{d_{s_m}}{\sum_{k=1}^{M} d_{s_m}} = \frac{d_i}{\sum_{i-1}^{N} d_i} \tag{7}$$

Eq.(7) shows the greater of the distance, the greater of the probability of selected as next initial center, is the same of the selection principle of k-means^{++} algorithm.

In the Map function, the input is the sum of subset of training samples, and the output is the classe SampleWriTable of self-definition realizing WriTable interface, recording the distance of next initial center selected with the recent center. The Map function process is as follows: ①Read the initial center from HDFS, initialize variables of DistanceSum = 0 and Distance[] = [0]; ②For each sample point N calculate the distance of sample points with the initial center, choose the smallest number as distance; Update variable, DistanceSum+ =distance, Distance[*i*]=distance; ③In the range of [0, Distance Sum] get a random number of random, when random>0, random-=Distance[*i*++]; ④Use point *i* obtained and DistanceSum to structure output of SampleWriTable sample.

Input of Reduce function was the candidate initial center and its distance obtained from each Map function, choose an initial center according to the algorithm adopted to choose an initial center above and write in HDFS, there is not to make specific requirements for output. Reduce function process is as follows: ①For each output of Map function: read sample points of candidate to the array of D[n], cumulate the distance of sample to gain SumDistance; ②Get a random number random1 in the range of [0, SumDistance] and a random number random2 in the range of [0, DistanceSum]; ③ When random>0, calculate random-= Distance [i++]; i obtained just is the selected sample point, write it into HDFS.

After obtaining the initial clustering center, apply it to k- means clustering algorithm based on Map-Reduce for clustering, the input of Map function is the sample point of training subset, the output is the clustering center classified and data of sample points. To do this, define a class SampleSumWriTable implementing the WriTable interface to store the subset, the within sample vector and the number of samples as output. The Map function process is as follows: ①Read clustering center to $c[k]$ from HDFS, initialize a HashMap <int, SampleSum>; ②For each sample point being read construct the sample data and compute shortest distance of it with the existing center, classify it to the cluster with the shortest distance; ③Output each cluster.

The input of Reduce function is the index of clustering center and the data SampleSumWriTable in cluster, the output is the clustering center adjusted. Reduce function process is as follows: ①Initialize the counter count, array sum[center-Length], the centerLength is dimensions of the samples; ②When read data, make count^{++} and sum^{++}, obtain the mean of clustering according to the sum and the count, as the clustering center newCenter after updating; ③Output updated clustering center.

Use the algorithm above to obtain hidden layer center, then learn weights of the network using the method of gradient descent. The input of Map function is the sample subsets, the output is the adjustment value of network weights in the form of WeightWriTable. The Map function process is as follows: ①Read the network information from HDFS, instantiate RBF network according to the information; ②Analysis the data and extract input and target output of the training sample, train the network; ③Access the network data after training, instantiate WeightWriTable and output.

Reduce function receives the <LongWriTable, Weight Wri- Table> of Map output as the input and output is <Long Wri- Table, IntWriTable>, IntWriTable decide whether or not to the next round of training, IntWriTable for 1 train, for 0 finish training. Reduce function process is as follows: ①Read network weight values of WeightWriTable from HDFS, accumulate each output data from the Map function; ②Apply the cumulative record to the network information, obtain new network weight of weight; ③Written the new network information in HDFS, cover original information and calculate the change quantity of the new network information; ④According to the amount of change output end of identity of training.

5.3.2. *Learning method based on supervised learning center*

In the Map function, define an IntermediateWriTable class implementing WriTable interface as the Value type of Map. The Map function process is as follows: ①Read the network information from HDFS, according to the information instantiate RBF network; ②Analysis reading data, extract the sample input and target output and train the network; ③Use result of network learning, instantiate IntermediateWriTable and output.

Reduce function receives the <LongWriTable Intermediate WriTable> output by Map as input; the output is <Long- WriTable, IntWriTable>. Reduce function process is as follows: ①Read the network information Intermediate-WriTable from HDFS, accumulate each record read from the Map function; ②Apply the cumulative record to the network information, get a new network information intermediate, writte the new network information to HDFS and cover the original information; ③Calculate the change quantity of new network information, according to the change amount set the symbol of training end.

5.4. *The experimental results and analysis*

Experiment uses 4 PCs into a machine groups, the configuration of every machine is as follows. The master node NameNode: 1, CPU 2.3 GHz, memory 2G, hard disk 300G; the data node DataNode: 3, CPU 2.67 GHz, memory 1G, hard disk 80G. Software environment: Ubuntu10.10, JDK 1.6, Hadoop 0.20.1, HBase 0.90.3. Experimental tests respectively approximation ability of function and ratio of speed up of the RBF network based on Map-Reduce. Experimental data is respectively Hermit function and data from KDD cup 1999 [11]. Hermit function is a kind of linear function, and its expression is,

$$h(x) = 1.1 * (1 - x + 2 * x * x) * \exp(-1 * x * x / 2) \tag{8}$$

For function approximation, 100 data points are took from [-4, 4] as the training samples, 9 data points are randomly selected from [4, 4] as the test data. The following data are average for 10 times test, use learning method based on self-organization choose center and the experimental results are shown in Table 2. Use the method based on supervised learning selecting center, the experimental results are shown in Table 3. KDD cup 1999 is data set about anomaly detection of network intrusion. Experiments uses the learning method based on supervised learning selecting center, is mainly used for testing training speed of Map-Reduce, the experimental results are shown in Table 4.

Table 2 Function approximation based on the self-organizing selection center

Rounds of training	Average error
5	125%
50	56.2%
250	13.82%
400	12.65%

Table 3 Function approximation based on supervised learning selection center

Rounds of training	Average error
5	161%
50	161%
250	53.4%
400	7.5%

Table 4 Training time based on supervised learning selection center

DataNode number	Operation time of cluster	Stand-alone operation time	Ratio of speed up
1	9336s	7842s	0.84
2	5724s	7842s	1.37
3	3647s	7842s	2.15

The data of second column of Table 2 and Table 3 shows the method based on self-organization selecting learning center has faster training speed than that based on supervised learning selecting center, but generalization ability is worse. From the data of the first, second, fourth column of Table 4 it can be seen that the learning speed of the network based on Map-Reduce increases with the increase of the number of DataNode, there is no condition to test whether there have the optimal number of cluster to now.

6. Application Research of the Neural network Based on Cloud Computing

6.1. *Face recognition research*

This section applies RBF network based on Map-Reduce above, combines feature extraction method based on PCA (Principal Component Analysis) and LDA (Linear Discriminant Analysis) to design a set of parallel face recognition algorithm, and compare with the traditional neural network recognition algorithm. The detail process for feature extraction based on PCA and LDA of Map-Reduce, and the Hybrid RBF network learning algorithm and classifier design are omitted. The following experiment used the ORL face database created by AT&T laboratory of university of Cambridge [12], it contains facial expression image data of 40 peoples, each people 10, a total of 400. Experimental configuration described above trains the network respectively for 5, 10, 30, 50 rounds, and then use the test set to test the network, the results are shown in Table 5.

Table 5. The effect of face recognition

Rounds of training	Identification number	Error	Recognition rate
5	15	0.15	85%
10	3	0.03	97%
30	2	0.02	98%
50	2	0.02	98%

Recognition rate in the Table is basically same with the recognition rate of [13], considering both algorithms principle adopted is basically consistent, so the result is predictable. At the same time, from data of the first, second and third column of Table 5 it can be seen that RBF network classifier has higher training efficiency and better classification performance.

6.2. *Speech emotion recognition research*

Speech emotion recognition is a kind of technology based on the analysis of voice information to identify the speaker emotional state at that time. This section applies the network based on Map-Reduce above for speech emotion recognition research. The detail process for the emotion classification, the choice of characteristic parameters, the speech database, as well as the speech signal preprocessing based on cloud computing are omitted. The experimental configuration as stated earlier, the experimental results as shown in Table 6.

The experimental results show that speech emotion recognition based on Map-Reduece have average recognition rate of 67.57%; Recognition rate of the state of anger, sorrow and calm is slightly higher than the rate of other emotions, it has a relationship with the characteristic parameters of speech signal used in the experiment.

Table 6. The effect of speech emotion recognition

Desired state \ Output state	Anger	Fear	Happiness	Sadness	Surprise	Calm	Recognition rate
Anger	186	12	12	8	20	4	77.5%
Fear	22	126	11	45	13	23	52.5%
Happiness	19	7	165	6	32	11	68.75%
Sadness	3	23	7	177	6	24	73.75%
Surprise	25	6	33	21	145	10	60.42%
Calm	6	8	12	34	6	174	72.5%

6.3. *Affective computing research*

For smart computers and robots, if need to provide a harmonious human-computer interaction environment, only having the emotion recognition ability is not enough, still need them to be respond to human emotion state, so as having emotion ability. Therefore this section will make affective computing research. The detail process for the emotional calculation model based on hidden Markov, the three layers model of emotion process, the emotion computing based on cloud computing, as well as the D-S evidence theory are omitted. It built a smart agent with vitriolic character in the Experiment, the emotion transfer probability matrix A is as follows, the experimental results are as shown in Table 7.

$$A = \begin{bmatrix} 0.675 & 0.074 & 0.004 & 0.001 & 0.157 & 0.092 \\ 0.071 & 0.507 & 0.214 & 0.003 & 0.201 & 0.004 \\ 0.16 & 0.114 & 0.7 & 0.001 & 0.007 & 0.018 \\ 0.012 & 0.003 & 0.001 & 0.653 & 0.001 & 0.66 \\ 0.013 & 0.143 & 0.026 & 0.008 & 0.803 & 0.007 \\ 0.161 & 0.055 & 0.019 & 0.247 & 0.006 & 0.512 \end{bmatrix}$$

Table 7. Experimental results of emotion computing

Expression image	Voice emotion	Voice content	Expression identify	Voice emotion identify	Voice text identify	Results of emotion identify	Stimulus intensity	Mood reaction	The end state
Big smile	happy	"I am very happy"	happy	happy	"I'm very happy"	happy	49	happy	calm
Calm	calm	"ok!"	sad	calm	"ok!"	calm	311	a little bit fear	calm
Stare、grind teeth	anger	"You dare to do so"	anger	anger	"You dare to do so"	anger	453	fear	calm

From data of the ninth column of Table 7 it can been see that emotion computing simulation system can react according to the mood for different external stimuli, so as to whether the emotion response is proper or not, there is not yet a recognized evaluation standard. At the same time, regardless of the initial emotion is what, the state eventually tends to such a stable state of "calm" after state transition of Markov model, which is consistent with human emotion change.

7. Conclusions and Future Work

This paper discusses and implements machine learning method of the neural network based on cloud computing, machine learning method of the neural network based on cloud computing verified by the experiment is feasible; And apply it to specific problems–face recognition, speech recognition, emotion calculation. Experiments prove that machine learning method of the neural network based on cloud computing than the traditional neural network machine learning method based on a single machine has faster training speed, higher recognition accuracy and larger data processing ability. It can be seen from the above study, machine learning based on cloud computing still has a lot of content to be worth being studied, such as application research of deep learning based on cloud computing, combining resource scheduling to increase the training speed of machine learning, etc., these all are the directions of study in the future.

References

1. Chu, Kim, Y.A. Lin, Y Yu, G Bradski, A Y. Ng, K Olukotun.2006. Map-Reduce for Machine Learning on Multicore. [M] In Proceedings of Neural Information Processing Systems Conference (NIPS). Vancouver, Canada.

2. Zhiqiang Liu, Hongyan Li, Gaoshan Miao. MapReduce-based Back propatagion Neural Network over Large Scale Mobile Data[J]. ICNC 2010.
3. Zhu Chenjie, Rao Ruonan. The Improved BP Algorithm Based on MapReduce and Genetic Algorithm[J]. CSSS 2012.
4. White T. Hadoop: The definitive guide [M]. O'Reilly Media, Inc., 2012.
5. Dean J., Ghemawat S. MapReduce: simplified data processing on large clusters[J]. Communications of the ACM, 2008, 51(1): 107–113.
6. Aleksander and Morton, An introduction to neural computing [M]. 2nd Edition. 1990.
7. R.Hecht-Nielsen, Theory of the back propagation neural network [M]. Proceeding of IJCNN, 1989.
8. Zhu Chenjie, Rao Ruonan. The Improved BP Algorithm Based on Map-Reduce and Genetic Algorithm[R].International Conference on Computer Science and Service System, 2012, 1568:1570.
9. UCIdataset[EB/OL]https://archive.ics.uci.edu/ml/datasets/Breast+Cancer+Wisconsin+(riginal).
10. WeizhongZhao, H Ma, Q He – Springer. Parallel K-Means Clustering Base on MapReduce[J]. CloudComputing, 2009.
11. KDD Cup 1999 dataset: [EB/OL] https://kdd.ics.uci.edu/databases/kddcup99/kddcup99.html.
12. ORL Faces database [EB/OL] https://www.cl.cam.ac.uk/research/dtg/attarchive/facedatabase.html.
13. Nan-Feng Xiao, Yong-Gang Yao. Face recognition algorithm based on RBF neural network [J]. Journal of Chongqing University of Science and Technology (natural science), 2010, 24(10):56–67.
14. R. Fernandez, R. Picard. Recognition affect from speech prosody using hierarchical graphical models [J]. Speech Communication, 2011, 53 (9–10): 1088–1103.
15. Sahidullah. Md, Saha, Goutam. Design, analysis and experimental evaluation of block based transformation in MFCC computation for speaker recognition [R]. Speech Communication. 2012-05, 54 (4): 543–565.
16. CASIA Chinese emotional corpus [EB/OL]http://www.chinip.csdb.cn/resource_info.php?rid=76.
17. Picard, R.W. Affective computing [M]. Sen-Lin Luo Beijing institute of technology press, 2005:12–13.

A novel approach of time series trend extraction based on global constrained multi-segment polynomial fitting

Chang-Feng Yan[1,†], Jian-Fang Fang[2], Li-Long Zhou[1] and Li-Xiao Wu[1]

[1]*School of Mechanical and Electrical Engineering,*
Lanzhou University of Technology, Lanzhou, Gansu, China
[†]*E-mail: changf_yan@163.com*

[2]*Shanxi Tai Gang Stainless Steel Co. Ltd*
Taiyuan, Shanxi, China

Trend variations are an important feature of time series, allowing more information to be mined from the time series directly. The process of qualitative trend analysis for time series includes trend extraction and identification. These help to extract qualitative trend features from quantitative data before representing them by using episodes. Integrating the polynomial fitting and piecewise linear representation based on local maxima, minima and extrema, a new approach of time series trend extraction based on global constrained multi-segment polynomial fitting is proposed in this paper. The experimental results from two different sets of simulated data indicate that this method is effective for extracting the trend in time series with high signal-noise ratio and low complexity.

Keywords: Time Series; Qualitative Trend Extraction; Global Constrained Multi-Segment Polynomial Fitting; Scaled Average Global Error; Scaled L_∞ Error.

1. Introduction

Time series contain valuable information about the state, operation, and behavior of the process plant. Trend variation is an important feature of time series. It is helpful for us to understand the evolution condition of time series directly[1]. By using qualitative trend analysis (QTA) technique, the useful feature information is extracted from signals, and trend feature is represented based on sign language[2]. Qualitative trend analysis includes trend extraction and trend identification. It is necessary to define a map set to describe the trend by language or symbol.

Williams[3] introduced an episode-based concise trend representation in qualitative reasoning. Cheung[4] proposed formal trend representation schemes consisting of seven triangular episodes and two trapezoidal episodes. Janusz[5] presented a framework for the automatic generation of qualitative process trend descriptions directly from sensor data. The qualitative process behavior is constructed by identifying four levels of the behavior, namely, primitives, episodes, trends and profiles. Yan *et al.*[6] proposed an approach of time series piecewise linear representation based on local maximum minimum and extremum (LMME). Joan[7] addressed a method of process trend extraction based on the qualitative representation. Charbonnier *et al.*[8] proposed an effective method of trend description. Three primitives are defined: increasing, decreasing and steady. Those primitives can combine into seven trend shapes.

Konstantinov[9] presented an approach of trend extraction based on fixed-width window polynomial fitting. Bakshi[10] proposed an approach multi-scale trend extraction by using wavelet. Rengaswamyn[11] addressed syntactic pattern-recognition approach for fault diagnosis. The syntactic approach provides a capability to describe complex patterns using a set of simple primitive pattern. Vedam and Venkatasubramanian[12] proposed a nonlinear adaptive algorithm based on wavelet theory to identify the trend. The 9 primitives were used to represent the trend.

This paper presents a novel trend extraction method based on global constrained multistage polynomial fitting (GCMPF). The reasonable segment points are extracted by using local maximum, minimum and extremum (LMME) method in the current time series. Every end-point of the piecewise is chosen as a constraint point in constrained polynomial match. Then, the process of constrained polynomial fitting is carried out in the current point. And the quadratic fitting function of each subsequence is obtained. Finally the first and second derivatives of signs are used to identify trend, and current time series is represented as a set of primitives.

2. Trend Extraction Based On Constrained Polynomial Fitting

2.1. *Trend language description*

Trend extraction is primary for the qualitative trend analysis. The qualitative information is extracted from the process data and represented by symbolic language. The 9 primitives were adopted to extract and identify the process data in this paper. Each primitive is determined by the value and sign of the first and second derivatives of fitting function, such as (**A**(0, 0), **B**(+, +), **C**(+, 0), **D**(+, -), **E**(-, +), **F**(-, 0), **G**(-, -)), H((+,-),-), I((-,+),+). It is obvious that the primitives H and I are determined by the sign of first derivatives of segment endpoints.

2.2. *Trend extraction*

In order to extract trends from the current time series data, the original signal should be transformed into a form by symbolic language. The signal is represented to symbolic language by specified function. The least-square polynomial fitting is used to obtain the fitting function of the data in this paper.

Given a time series $X = \langle(y_1, x_1), (y_2, x_2), \cdots, (y_n, x_n)\rangle$, then X can be approximated by a polynomial$P(y_i)$, where x_i is the value of time series, y_i is the time of sampling data. The function of $P(y_i)$ is shown in Eq. 1.

$$P(y_i) = \sum_{k=0}^{n} a_k {y_i}^k. \quad (1)$$

Where a_k is k-th power coefficient of y_i, n is the highest power of $P(y_i)$.

The purpose of least-square polynomial fit is to obtain a set of coefficientsa_k which minimize the mean square error between the fitting data and the original data as shown in Eq.2.

$$\min_{a_k} \sigma = \sum_{i=1}^{m} \left(\sum_{k=0}^{n} a_k y_i^k - x_i\right)^2. \quad (2)$$

The mean square error can be considered as a function $f(a_0, a_1, \ldots, a_k)$. If f is minimum, the optimal coefficients $a_k$will be obtained.

Time series is divided into several sub-segment sequences by least squares global constraint multi-segment polynomial. Piecewise point between each sub-segment point can be regarded as constraint point. The time series is fit by polynomial fitting method combined with Lagrange operator λ_I. The global optimum polynomial fitting function of each sub-segment can be obtained.

Time-series $X = \langle(y_1, x_1), (y_2, x_2), \cdots, (y_n, x_n)\rangle$ can be noted as $X = \langle x_1, x_2, \cdots, x_n \rangle$, in order to simplify the expression. Piecewise point set is$X_i = \langle x_{i_1}, x_{i_2}, \cdots, x_{i_m} \rangle$, where$x_{i_1} = x_1$, $x_{i_m} = x_n$. Both the first and the last point of time sequence are also regarded as piecewise point. The current time series is divided into J sub-sequences, where $J = m - 1$.The j-th sub-segment in the time series set is represented as $\left(x_{i_j}, \ldots, x_{i_{j+1}} - 1\right)$ in which the data is between the j-th piecewise point and former data point of $j+1$-th piecewise. The number of data points is m_j, and the last segment is $\left(x_{i_{m-1}}, \ldots, x_{i_m}\right)$. The j-th sub-segment can be approximated by polynomial $P_j(y)$as shown in Eq.3.

$$P_j\left(y_j\right) = \sum_{k=0}^{n_j} a_{j,k} {y_j}^k. \quad (3)$$

Where $a_{j,k}$ is the k-th power coefficient of the j-th sub-segment, n_j is the highest power of the polynomial fitting of the j-th sub-sequence.

If the constrain conditions $g_I(y) = 0\ (I = 1,2, \ldots, l)$ is considered, the optimum coefficient of polynomial fitting should also satisfy the Eq.4.

$$\begin{cases} \min_{a_{j,k}} \sigma = \sum_{j=1}^{J} \sum_{i=1}^{m_j} \left(\sum_{k=0}^{n_j} a_{j,k} y_{j,i}^k - x_{j,i}\right)^2 \\ g_I(y) = 0\ (I = 1,2, \ldots, l) \end{cases}. \quad (4)$$

For the constraint condition processing, the method of Lagrange multipliers is applied in this paper as shown in Eq.5.

$$\min_{a_{j,k}\lambda_I} \sigma = \sum_{j=1}^{J} \sum_{i=1}^{m_j} \left(\sum_{k=0}^{n_j} a_{j,k} y_{j,i}^k - x_{j,i}\right)^2 - \sum_{j=1}^{J} \lambda_I g_I(y) . \tag{5}$$

Where λ_I is the *I*-th Lagrange multipliers and *y* is the constraint points.

Constraint points are extracted from the time series by using LMME method[6]. The set of piecewise points is represented as $X_i = \langle x_{i_1}, x_{i_2}, \cdots, x_{i_m} \rangle$. The set of constrained points$\bar{X} = \langle x_{i_2}, \cdots, x_{i_{m-1}} \rangle$ has J-1 points. The set of constrained point can also be represented as$\bar{X} = \langle t_1, \cdots, t_j \rangle$, where $t_1 \equiv x_{i_2}$, $t_j \equiv x_{i_{m-1}}$, and the maximum of j is J-1.

The smooth connection of constraint point is that the values of neighbor sub-sequences polynomials (X_{i-1}, X_{i+1}) and their first derivative are equal. Constraint equation $g_j(x)$ of the *j-th* constraint point consists of two parts: equation $g_{1j}(t_j)$ ensures the fitting curve at the constraint point continuous; equation $g_{2j}(t_j)$ ensures the first derivatives of fitting curve beside constraint point equal. They are given by Eq.6 and Eq.7.

$$g_{1j}(t_j) = \sum_{k=0}^{n_j} a_{j,k} t_j^k - \sum_{k=0}^{n_{j+1}} a_{j+1,k} t_j^k = 0. \tag{6}$$

$$g_{2j}(t_j) = \sum_{k=1}^{n_j} k a_{j,k} t_j^{k-1} - \sum_{k=1}^{n_{j+1}} k a_{j+1,k} t_j^{k-1} = 0. \tag{7}$$

In order to ensure each sub-sequence fitting function continuous and smooth at piecewise point, the constrained point must be connected smoothly. In this paper, the practical application is also taken into consideration. Equipment or some chemical process monitoring is usually continuous and the collected time series date is updated on-line. Because the current and former the time series trend are continuous, the former constraints trend should be added in when the current time series are under constrain polynomial matching. Constrain equation is given by Eq.8.

$$g_3(t_1) = \sum_{k=0}^{n_1} a_{1,k} t_0^k - x_0 = 0. \tag{8}$$

Where t_0 is the time of the last data of former time series, x_0 is the fitting value.

Under the constraint conditions mentioned above, global polynomial fitting of the current time series is used to obtain the optimum coefficients $(a_{1,0}, a_{1,1}, \ldots, a_{1,k}, \ldots, a_{j,0}, a_{j,1}, \ldots, a_{j,k})$ of all sub-sequences. The optimum coefficients make the fitting data and the original time series data conform to the least squares as given by Eq.9.

$$\min_{a_{j,k},\lambda_j,\alpha_j,\beta} \sigma = \sum_{j=1}^{J} \sum_{i=1}^{m_j} \left(\sum_{k=0}^{n_j} a_{j,k} y_{j,i}^k - x_{j,i}\right)^2 - \sum_{j=1}^{J-1} \lambda_j g_{1j}(t_j) - \sum_{j=1}^{J-1} \alpha_j * g_{2j}(t_j) - \beta * g_3(t_1). \tag{9}$$

where λ_j , α_j and β are multipliers of each constraint equation respectively.

The coefficient $a_{j,k}$ of each sub-sequence fitting function is obtained by solving linear equations. These coefficients not only satisfy the smooth transition between each sub-sequence, but also make the current time series trend and former time series trend continuous. At the same time, the fitting error between fitting curve and current time series is also minimum.

2.3. *Trend extraction process*

In this paper, trend extraction method based on global constrained quadratic polynomial fitting includes three parts. The process of trend extraction is shown in figure 1.

(1) Data preprocessing

After load the data of time series, the standard deviation of noise σ_{noise} is calculated, and then the contaminated signal is de-noised. In this paper, morphological filter is used to de-noise. The shape of structure element of the filter *SE* choose disk. The control parameters *R* and *N* of *SE* choose 1 and 4 respectively.

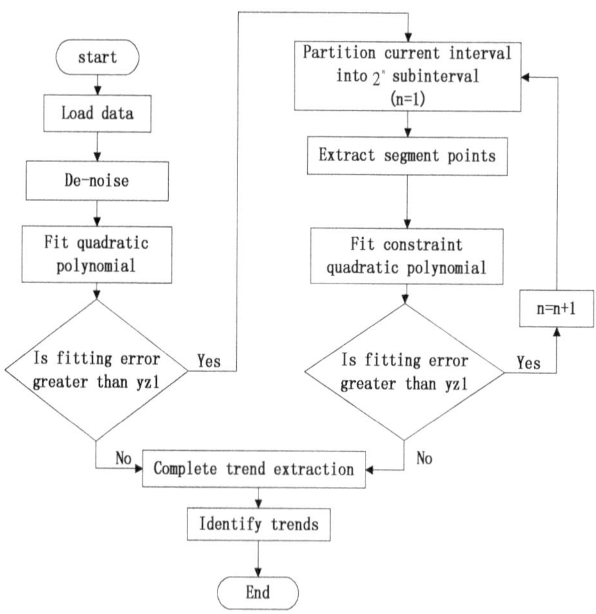

Fig. 1. Flow chart of trend identification

(2) Trend extraction

The trend extraction algorithm proposed in this paper is used to extract trend. Quadratic polynomial fitting is employed for entire data. If the fitting error is greater than the setting threshold $yz1$, the entire data will be split into two parts and the reasonable piecewise point is extracted. After the piecewise point is obtained, the global constrained multi-segment polynomial fitting is conducted for the data and the fitting error is calculated. If the fitting error is greater than the setting threshold $yz1$, each sub-sequence will continue to halve and the fitting error will calculated again until the fitting error is less than the threshold $yz1$.

After the trend extraction, the current time series is divided into several sub-sequences and the fitting function of each sub-sequence is obtained.

(3) Trend identification

The unimodal interval means that the interval data has only one extreme or monotonically decreasing/increasing. Unimodal interval set is obtained by using trend extraction algorithm mentioned above. Every unimodal interval matches with primitive set according to the sign of the first and second derivatives of the unimodal interval. The current time series would be represented by primitives.

2.4. *Reasonable piecewise point extraction*

The reasonable selection of piecewise point and the number of piecewise will affect the trend extraction directly. Therefore, LMMES is used to extract the key-point in this paper. On different time scales, the key-point which has the greatest influence on the morphology of time series is regarded as the reasonable piecewise points. In order to get the desired threshold of fitting error, the length of time window of piecewise point is adjusted continuously. The length of time window of piecewise point Δt is given by Eq.10.

$$\Delta t=\left(\frac{1}{2}\right)^{n} T. \tag{10}$$

where n is the number of fitting, T is length of the sequence of the current time series.

Generally, under the low signal-noise ratio condition, the shortest length of sub-sequence l_{th} is10. In this paper, although the experimental data added with white noise, the noise is very low after filtering. Therefore, the minimum length of the sequence l_{th} is 10 in this paper.

2.5. *Trend identification*

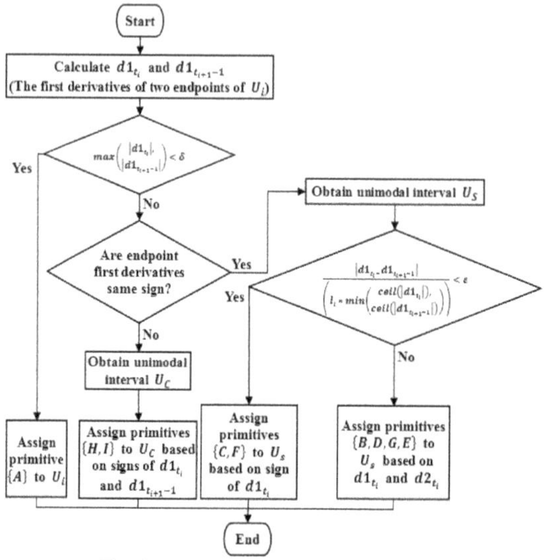

Fig. 2. Flowchart of trend identification

$$\left\{\left(x_{i_1},\ldots,x_{i_2}-1\right),\ldots,\left(x_{i_j},\ldots,x_{i_{j+1}}-1\right),\ldots,\left(x_{i_{m-1}},\ldots,x_{i_m}\right)\right\}\xrightarrow{\text{Trend extraction}}\{P_1,\ldots,P_j,\ldots,P_J\}$$

$$\{P_1,\ldots,P_i,\ldots,P_J\}\xrightarrow{\text{Trend identification}}Q\ \left(Q\subseteq\{A,B,C,D,E,F,G,H,I\}\right)$$

Fig. 3. Diagram of qualitative trend analysis

The process of trend identification is shown in Fig.2. The global optimum fitting function of each sub-sequence is obtained by trend extraction. Trend identification is that the fitting data of each sub-sequence match with the 9 primitives. The current time series is represented by the primitives as shown in Fig. 3.

The shape of the fitting function for each unimodal interval is determined by the magnitude and sign of the first derivative ($d1_t$) and the second derivative ($d2_t$). In this article, each unimodal interval is fitted by quadratic polynomial and the second power coefficient of polynomial coefficients is not zero.

3. Experimental Verification and Discussion

In order to evaluate the effectiveness of this method, two different complexity simulation data are applied to extract simulated data trends. Data 1 is Gaussian signal given by Eq.11. Data 2 is the sum of several slope, step and sinusoidal signal given by Eq.12. The complexity of simulated data is from low to high. The length of data is 1024.

$$f_1(t)=800e^{-((t-u)^2)/2\delta^2} \quad \mu=512, \delta=120. \tag{11}$$

$$f_2(t)=90\sin\frac{t}{60}+\begin{cases} 240+1.2t & 1\leq t\leq 170 \\ & 340<t<512 \\ 1200 & 512<t<680 \\ & 850<t<1024 \\ 120 & 170\leq t\leq 340 \\ 1800 & 680\leq t\leq 850 \end{cases} \tag{12}$$

3.1. *Performance evaluation of the algorithm*

It is also very important to evaluate the performance for trend extraction algorithm. Root mean square error (RMSE) and L_∞ error often are applied on the least square fitting and data compression and regarded as evaluation indicators of performance. Root mean square error and L_∞ error are defined as Eq.13 and Eq.14 respectively.

$$RMSE=\sqrt{\frac{\Sigma(f_{fit}-f)^2}{N}}. \tag{13}$$

$$L_\infty\text{-error}=\max\left|f_{fit}-f\right|. \tag{14}$$

Where f_{fit} is fitting data, f is the original data, N is the data length.

In order to quantitatively evaluate the performance of trend extraction method proposed in this paper, *RMSE* and L_∞ error are normalized by using estimated standard deviation of added noise (ESTDN)σ_{noise}. The result was used as performance indictor to evaluate the method proposed in this paper as shown in Eq.15 and Eq.16.

$$SAGE=\frac{1}{\sigma_{noise}}\sqrt{\frac{\Sigma(f_{fit}-\hat{f})^2}{N}}. \tag{15}$$

It is shown that SAGE is closely related to RMSE. If the value of SAGE is close to 1, it means that the most of noise is suppressed and the extracted trend is approach to the original time series.

$$SLE=\frac{1}{\sigma_{noise}}\max\left|f_{fit}-f\right|. \tag{16}$$

It is shown that *SLE* is closely related to L_∞error. If the value of *SLE* is large, it means that the deviation between the fitting data f_{fit} and filtered data f is large, which indicate that the performance of trend extraction is poor.

3.2. *Experiment verification*

In order to verify the effect of the method presented in this paper. The simulation data are mixed with different standard deviation of Gaussian noise. Simulated data contaminated by noise is filtered by using the morphological filter.

3.3. *Experiment results and discussions*

(1) Trend extraction on simulated data 1

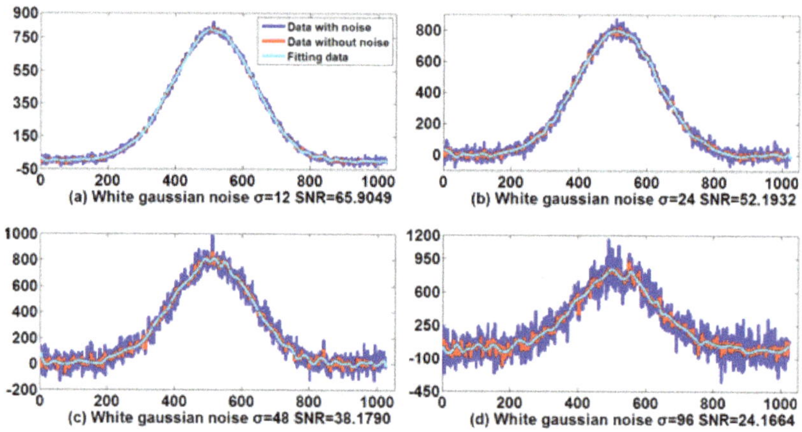

Fig. 4. Trend extraction of simulated data 1

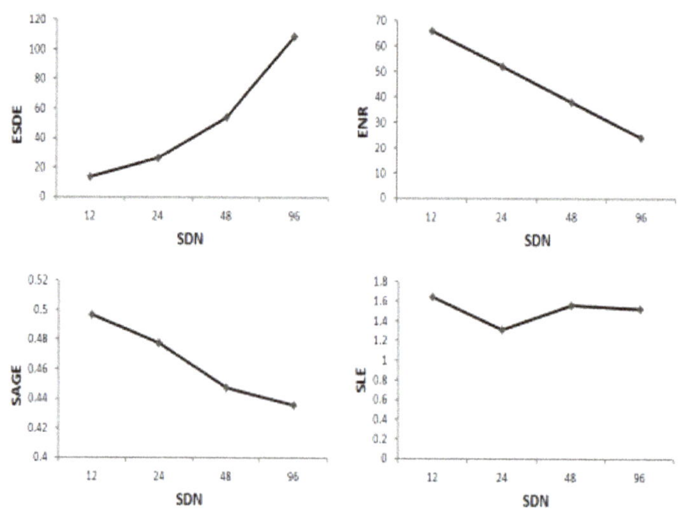

Fig. 5. Experimental results of simulation data 1

The simulated data 1 is Gaussian signal. SNR, ESDN, SAGE and SLE are calculated under the different standard deviation of noise (SDN). The results are shown as Fig.4 and Fig.5.

In Fig.4(a), the simulated data is mixed with small SDN (σ=12, *SNR*=65.9049) and the results show that the fitting data coincide with de-noising data fitting. It is shown in Fig.4(b) and (c) that the fluctuation of de-nosing data and fitting data are increased with the decreasing of SNR, but the whole effect of trend extraction is still good. In Fig.4(d), because the original simulated date suffers from the strong noise interference, the fluctuations between fitting data and de-noising data are very large. The deviation between the trend of original data and the extracted trend based on GCMPF is large.

It is seen from Fig.5 that, with the SDN increasing, the SNR is decreasing and the SDEN getting larger. The performance index of SAGE and SLE are relatively small.

(2) Trend extraction on simulated data 2

The simulated data 2 is complex simulated signal. The experimental results are shown in Fig. 6 and 7. It can be intuitively seen from Fig.6 (a)-(d) that, with the decreasing of SNR, the fluctuations of fitting data and de-nosing date are increased. The deviation between fitting data and de-noising data in step area is larger than other area. With the fluctuation of de-noising data increasing, the piecewise points are increasing and the deviation between fitting data and de-noising data is decreased. If the standard deviation of noise is 96, the value of SLE is very small. In Fig.6(d), due to the influence of strong noise, the step trend in data points (500-550) is almost disappeared. Although the trend of fitting data is consistent with de-noising data, great distortion occurs between the fitting data trend and the original data.

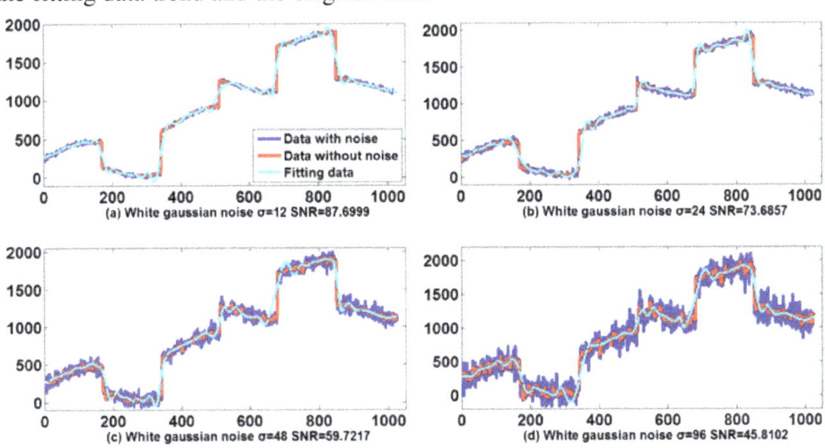

Fig. 6. Trend extraction of simulated data 2

The value of SAGE and SLE are increased compared with data 1. The fitting effect of algorithm is weakened with increasing complexity of simulated signal. As the SNR decreases, the values of SAGE and SLE are decreased rapidly. Finally, the values will reach a reasonable range (SAGE<1, SLE≈3), which indicates that the effect of trend extraction is improved.

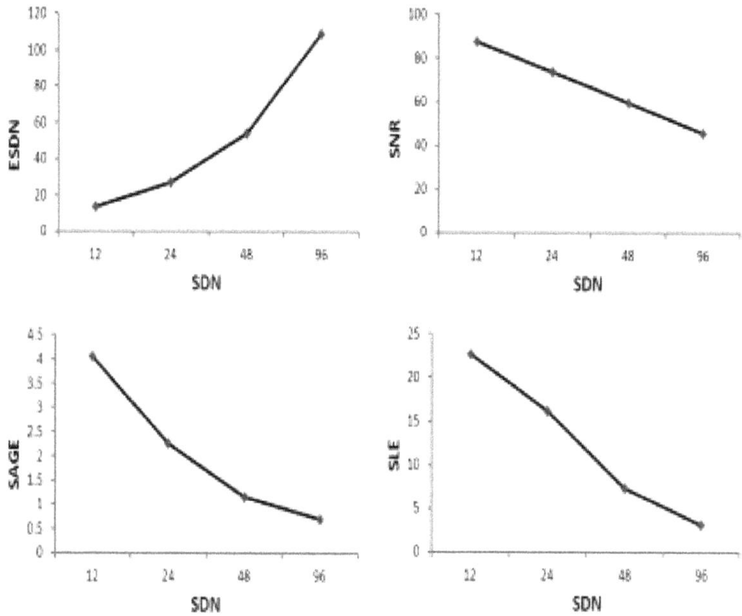

Fig. 7. Experimental results of simulation data 2

The experiment result of data 1 show that if the complexity of fitting data is lower and the trendy of data point varies slowly, the effect of trend extraction based on GCMPF is better. With the SNR decreasing, the effect of trend extraction declines. The experiment result of data 2 shows that the effect of trend extraction declines and the performance indicators of SAGE and SLE are relatively large if the complexity of fitting data is higher. Compared figure 4 with figure 6, it is easy to conclude that the effect of trend extraction based on GCMPF is relatively improved with the SNR decreasing. Although the value of SAGE and SLE are decreasing, some trend contained in original data is lost in local area.

4. Conclusion

A trend extraction method of time series based on GCMPF is proposed in this paper. The effectiveness of trend extraction is verified by using two simulated data with different complexity. Both the complexity of the data and the SNR have directly influence on the performance of trend extraction. Experimental results show that GCMPF method has a good performance in trend extraction, especially the data with low complexity and high SNR.

Acknowledgment

This work supported by the National Nature Science Foundation of China (No. 51165018).

References

1. Q. Zhou and T. J. Wu, Trend feature extraction method based on important points in time series, Journal of Zhejiang University, 41 (2006) 1782–1787.
2. H. Y. Zhang, X. Ma, Applied research of qualitative trend analysis in process fault diagnosis, Journal of System Simulation, 20 (2008) 2750–2753.
3. B. C. Williams, Doing Time: Putting Qualitative Reasoning on Firmer Ground, AAAI, (1986) 105–112.
4. J. T. Cheung and Stephanopoulos, A formal representation framework. Computers and Chemical Engineering, Computers & Chemical Engineering, 14 (1990) 495–510.
5. J. Venkatasubramanian, Automatic generation of qualitative description of process trends for fault detection and diagnosis, Engineering Applications of Artificial Intelligence, 4 (1991) 329–339.
6. C. F. Yan, J. F. Fang, L. X. Wu and S. M. Ma, An approach of time series piecewise linear representation based on local maximum, minimum and extremum, Journal of Information & Computational Science,10 (2013) 2747–2756.
7. J. Meléndez and J. Colomer, Episodes representation for supervision: application to diagnosis of a level control system, In workshop on principles of diagnosis DX, 1 (2001) 119–126.
8. C. Sylvie, C. B. Carlos, C. Cadet and S. Gentil, Trends extraction and analysis for complex system monitoring and decision support, Engineering Applications of Artificial Intelligence, 18(2005) 21–36.
9. Konstantinov, B. Konstantin and Y. Toshiomi, Real-time quantitative analysis of the temporal shapes of (Bio) process variables, AIChE Journal, 38 (1992) 1703–1715.
10. Bakshi and Stephanopoulos, Multi-scale extraction of trends from process data, Computers & chemical engineering, 18 (1994) 267–302.

11. R. Rengaswamy and V. Venkatasubramanian, A syntactic Pattern recognition Approach for process Monitoring and Fault Diagnosis, Engineering Applications of Artificial Intelligence, 8 (1995) 35–51.
12. Vedam and Venkatasubramanian. A wavelet theory-based adaptive trend analysis system for process monitoring and diagnosis, in Proc. of the American Control Conference, 8 (1997) 309–313.
13. H. Vedam, An intelligent operator decision support system for diagnosis and abnormal situation in process plants, Ph.D. dissertation West Lafayette: Purdue University, 1999.

Resource allocation model based on Dijkstra's algorithm

Wen-Zhi Chen[†]

International School, Beijing University of Posts and Telecommunications, Beijing, China
[†]*E-mail: 15501121368@163.com*
www.bupt.edu.cn

The increasingly frequent occurrence of natural, man-made and environmental disasters in recent years has resulted in both a large number of casualties and widespread property damage. To mitigate the damage caused, a lot of emergency supplies are required. However, due to the limited quantity of emergency supplies, the allocation of rescue resources is extremely important. Using the March 2014 Ebola outbreak in western Africa as an example, this paper uses Dijkstra's algorithm to build a system which determines the optimal allocation of emergency resource to cities in Sierra Leone. Each city's weightage is calculated using data provided before Dijkstra's algorithm is applied.

Keywords: Dijkstra's algorithm; emergency resource allocation; Ebola; Sierra Leone.

1. Introduction

Dijkstra's algorithm is an algorithm for finding the shortest paths between nodes in a graph, which may represent, for example, road networks. It was conceived by computer scientist Edsger W. Dijkstra in 1956 and published three years later [1][2].

The Ebola virus causes an acute, serious illness which is often fatal if untreated. Ebola virus disease (EVD) first appeared in 1976 in 2 simultaneous outbreaks, one in Nzara, Sudan, and the other in Yambuku, Democratic Republic of Congo. The latter occurred in a village near the Ebola River, from which the disease takes its name [3].

The system uses the number of the people who infected with Ebola as the weight to run the algorithm. Adopt the method of operating system analysis [4] which is combined with graph theory model and the distribution of logistics engineering [5] and production distribution content, we set up a system. The scheme included drug production, the locations that we accept the drug, the method of drug distribution, give drugs to patients.

First divide all cities into three trees, and then choose the city Freetown to negotiate in a further step. Then we get the best way to allocate resource.

2. Assumption

2.1. *Assumption one*

The selection site of drugs should also consider traffic condition and disease condition.

2.2. *Assumption two*

B. Drug distribution and delivery is based on the gradation. A country deliver the drugs from the accepted city to large cities in other area, and then from the large cities to the small cities.

3. Select Sites

By assumption, we should consider more accept sites for an infected country. When there is a high degree of infection, we give priority to choose this area as the principle point, We use graph theory [6], take the example of West African country of Sierra Leone, in Figure 1, the correlated important cities is labeled in the map and have corresponding distance (here only consider valid up to the line).

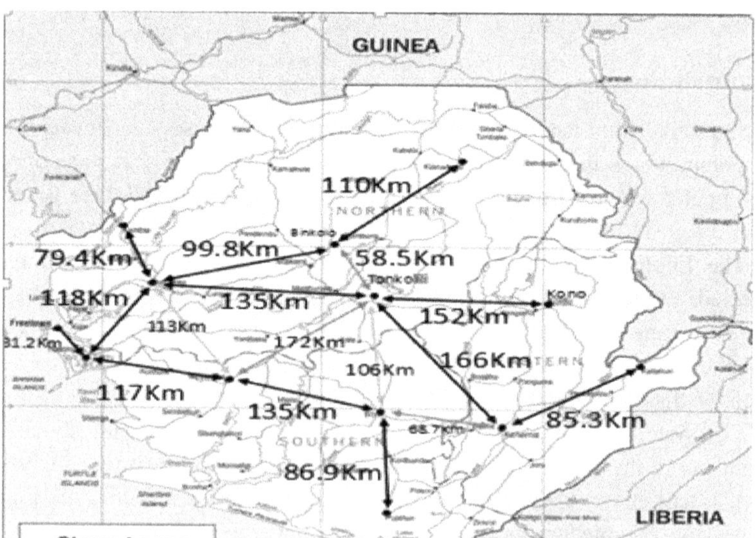

Fig. 1. Distance between important cities in Sierra Leone with background.

Set up a weighted undirected connected graph G (N, W), N is the vertex set, represents the city set, W is a weighted edge set, represents the distance between cities. We give Fig.1 above the city from left to right and from top to bottom.so they are numbered 0, 1, 2..... 12, which represent the vertex, W can be expressed as <n1, n2>, <n2, n3>.... (n1, n2 is the number of vertex) which represents the distance.

4. Algorithms and System

4.1. *Algorithm1*

(1) The first step, enter the weighted Figure G (N, W), N is the set of vertices, W is the set of edges.

(2) The second step, the initialization: V is a new set of vertices, V = {v0}, v0 is any vertices in set N, E is a new set of edges, E = { }, is an empty set.

(3) The third step, the iteration: Repeat the operations a and b, until V = N.

(a) Select the smallest weights edges <v, u> in the set W where v is an element in V, and V, and u is not in V, but in N. If there is more than one edge with the same weight, you can choose any one of them.

(b) Add v to V, add the minimum edge <v, u> to E

(4) The fourth step, with a new set of output V, and set E describes the minimum spanning tree. The results obtained are the edge set E{ }. Figure expressed as Figure 2:

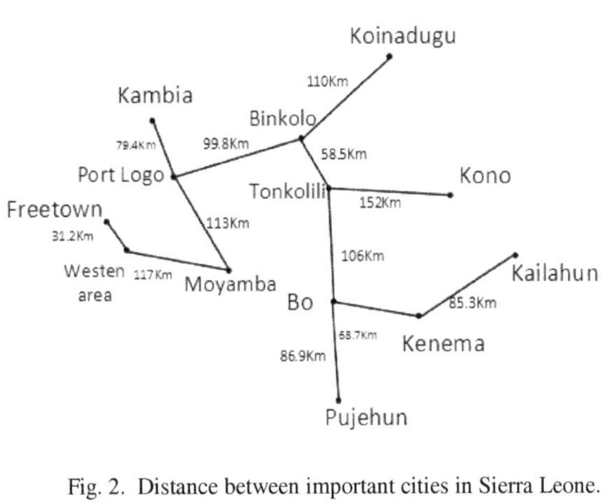

Fig. 2. Distance between important cities in Sierra Leone.

We have put the size of the cumulative number of infections in all regions as a measure of the level of severity of the infection, we preferentially select the city which infections is high to be as an acceptance place, but taking into account the backwardness of African countries' air transportation, such as Sierra Leone, it only has three airports, in order to make drug distribution more uniform, more coverage, the three airports in the country as an acceptance place, and then find the point at the center to receive from its nearest or most convenient to reach a wider distribution of the drug or the principles of urban sub-root points above the minimum spanning tree is divided into three sub-tree, choose Bo, Freetown, Port Logo and other three cities as root, and the Selection principle:

- Freetown is the capital and there is more than 2000 people infected.
- Port Logo is the closest city from the airport.

We choose Bo, taking into account coverage and distribution, while the overall resistance, regardless of the distance of the airport in order to achieve minimize tree layers role. As shown in Figure 3:

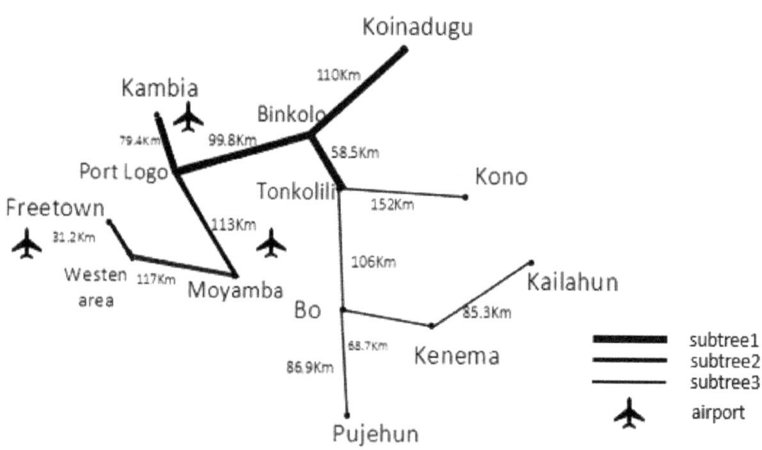

Fig. 3. Distance between cities in Sierra Leone with three hierarchies.

According to the tree hierarch, for each of the sub-trees, we traverse from near and far to visit each node [7], which means the goods to be distributed according to level, every city is different according to the distribution of the various regions and the number of infections fixed, the principle is the preferred the areas whose drug demand is large.

4.2. *Algorithm2*

(1) Steps:

(a) The first step, select the root node of each sub-tree, according to the number of infections in the region and the next location of a sub-goods to the next area.

(b) The second step, when it arrives at a new location (if the location is not the end of the sub-tree), based on the number of infections in the region and the next place, according to the proportion of sub-goods again.

(c) The third step, the process is repeated until each sub-tree of all nodes have been visited completely.

(2) Completed process by hands as follows:

(a) Chose Freetown city as the root node, distribute drugs to Westen area, Chose Port Logo city as the root node, distribute drugs to Kambia and Binkolo, Chose Bo city as the root node, distribute drugs to Tonkolili, Kenema and Prujehun.

(b) Chose Westen area as center, distribute drugs to Moyamba.

Chose Binkolo area as center, distribute drugs to Koinadugu.

Chose Tonkolili area as center, distribute drugs to Kono.

(c) The specific distribution ratio by the number of infections as Table 1.

Table 1. Confirmed cases and its rate in each places.

Place	Total confirmed cases	Rate
PUJEHUN	31	0.003825271
KOINADUGU	104	0.012833169
KAMBIA	151	0.018632774
MOYAMBA	205	0.02529615
KONO	247	0.030478776
BO	314	0.038746298
TONKOLILI	499	0.061574531
KENEMA	502	0.061944719
KAILAHUN	565	0.069718657
BOMBALI	990	0.122161895
WESTERN RURAL	1121	0.138326752
PORT LOKO	1334	0.164610069
FREETOWN	2041	0.251850938
SUM	8104	

There rate means place confirmed cases or Country confirmed cases. We distribute drugs in the each level according to the Table1, and we can get the sub-tree' proportion which have not been allocated in Figure 4:

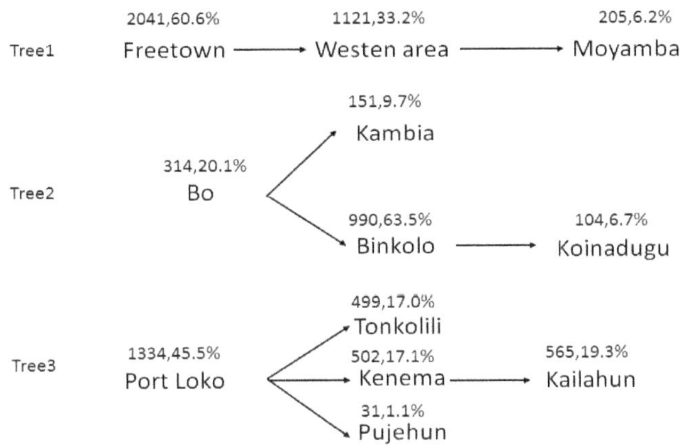

Fig. 4. The proportion of three trees.

(d) The process is the result of January, 2015, to be more specifically, consider changes in the number of infections and predict trends [8] [9].

(3) Allocation model for each disaster area:

Model above only gives each disaster area's transportation between cities, for big cities, our drug should distribute to reach the city's all institutions, such as hospitals, clinics, etc… drugs should be designed to better transportation solutions, if we consider which is nearest to the central hospital, to make the transport time shortest, we have designed an algorithm based on the idea of the Dijkstra algorithm, calculate the distance between the relevant areas of the Sierra Leone's capital -Freetown.

Specific process is: similar to the model above, we construct a graph diagram about Freetown, remark the transport distance between the relevant areas in the figure, choose the region as the initial point. According to the actual situation, the nearest distance of the airport from the Freetown is Wellington (as shown in Figure 5). We want to calculate the shortest path from the various regions of Freetown.

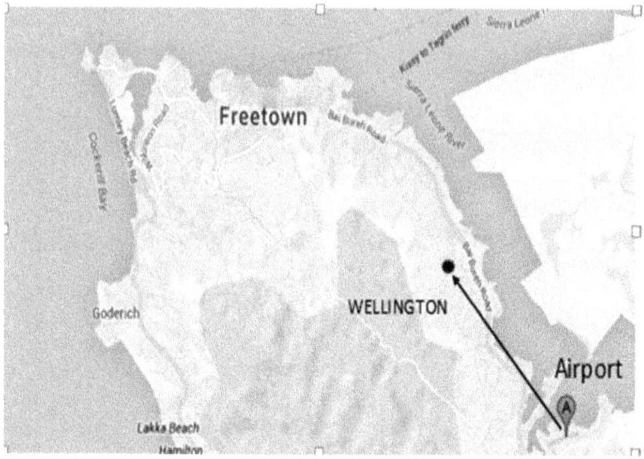

Fig. 5. Distance between Airport to Wellington.

Assuming a weighted undirected connected graph G (N, W), N is the vertex which represents a collection of various regions. W is a weighted line which represents the distance between the various regions. In the same way, set the number 0, 1, 2..... 12 from left to right and from top to bottom to mark each region, we can represent as <n1, n2>, <n2, n3>.... (n1, n2 is the number of vertices), which value represents the distance.

In this case, we actually abstracted it into a problem of area network connective, you can use graph theory to solve it. The results obtained are shown in Figure 6:

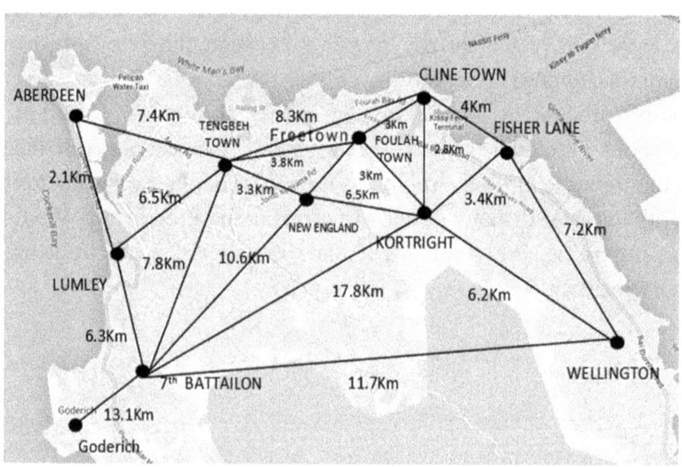

Fig. 6. Distance between small regions in Freetown.

(4) Explanation of schematic:

The straight line distance in the picture does not represent the displacement of each individual region. The value of the two points is not the geometry displacement, we only consider the weight.

4.3. *Algorithm3*

Based on the Dykstra algorithm, we find the shortest path from a point to each point. Using the greed idea to calculate and compare the distance between each point.

(1) The first step: Initialization process, divided the set of vertices into two groups N, S: We already determined set of vertices (initially containing only source point), T = NS: Not determined set (other than when the initial source of point), record the distance from the source point to the various points in D.

(2) The second step:

(a) We first visit the point that we do not visit in S (initial source point) all neighbor point s (s belong to T), to find the right neighbor minimum weight line and access. (You can set the mark to record access status).

(e) To find the smallest adjacent point, and sequentially calculated the sum of the distance through this adjacent point to other point, and then compared with the previously recorded distance, if less than the original record distance, then, record the sum of the distances recorded in D.

(3) The third step: Repeat a, b process until all the nodes have been visited, then the distance D is recorded as the shortest distance.

Execution of the program, we get the following results:

D={9,7.2,13.0,6.2,11.0,12.7,18.4,17.5,0,11.7,24.8}

P={3 0 0 1 3 4 4 0 0 9}

4.4. *Algorithm4*

(1) The first step: record this number, find the number which is corresponding to the second location, if it is not 0, continuous to find the number's location and record.

(2) The second step: repeat the first step until we found the number which is correspond to the number 0 and record. Then write the number we recorded in a reverse sequence. The system gets the corresponding path [10].

Example: 5->2->4->9

5. Summary

This paper establishes a system according to the situation of western Africa of the disaster. Other similar disaster can also use the same way to solve, and the system can be used to describe the problems that need to be studied, we can research and improve the system for more complicated situation.

References

1. Thomas J. Misa, *An Interview with Edsger W. Dijkstra*, Communications of the ACM (2011), pp. 41–47.
2. E. W. Dijkstra, *A note on two problems in connexion with graphs*, NumerischeMathematik (1959), pp. 269–271.
3. World Health Organization, Ebola virus disease, http://www.who.int/mediacentre/factsheets/fs103/en/.
4. L. D. Bentley, Systems Analysis and Design for the Global Enterprise 7th Edition, pp. 160.
5. S. Dowlatshahi, *A modeling approach to logistics in concurrent engineering*, European Journal of Operational Research (1999), pp. 59–76.
6. A. R. Mashaghi, A. Ramezanpour, V. Karimipour, *Investigation of a protein complex network*, The European Physical Journal B — Condensed Matter and Complex Systems (2004), pp. 113–121.
7. F. Baader, G. Brewka, T. Eiter, *KI 2001: Advances in Artificial Intelligence*, Joint German/Austrian Conference on AI, Vienna, Austria (2001).
8. K. H. Rosen, Discrete Mathematics and Its Applications (1999).

9. M. A. Weiss, Data Structures and Algorithm Analysis in C (2nd Edition, 1996).
10. X. R. Chai, R. J. Wang, Research of emergent material dispatching algorithm based on multi-depot and multi-material, Computer Engineering and Applications (2010), pp. 224–226.

An approach to modeling software architecture based on SOA*

Jun Ding
Yunnan Normal University, Kunming, 650500, China

Ye Qian†
School of Basis and Information Engineering,
Yunnan Agricultural University, Kunming 650091, China
†E-mail:qy198403@163.com

Ji-Hong Sun
Academy of Sciences in Yunnan Province,
Kunming 650091, China

Na Li
Kunming Medical University Haiyuan College,
Kunming,650106, China

Software architecture model play an important role in improving the quality of software products. A software architecture model is used for defining software. In this paper, an approach to modeling multi-level software architecture based on SOA is put forward. Firstly, four important properties of SOA are discussed. Secondly, based on Workflow-net, the formal definition of service is proposed. Finally, three rules to model composite service are introduced. The software model based on SOA lays the foundation for analyzing and reasoning it mathematically.

Keywords: Software Model; SOA; Service; Workflow-Net.

1. Introduction

Under the influence of the globalized economy, the requirements and environment of computing system has been undergoing tremendous changes. Because the elements of the system change frequently, the software is facing a

---

* This work is supported by the Applied Basic Research Programs of Yunnan Province (No. 2012FD020);
This work is supported by the Science Foundation of Education Department of Yunnan Province (No. 2012C086);
† Corresponding author.

challenge to regulate itself promptly for the changing demands, SOA and web service has come into being under the circumstances[1]. According to the first and second paper, SOA (service-oriented architecture) is now capturing more and more attention in the software industry and academia. SOA is used to completely solve the problem of information isolation, for application and resources will be transformed into standard services which can be shared in the Internet environment[2].

Nowadays, software systems change continuously with the changes in techniques and requirements[3]. We used to analyze evolution of traditional software system based on modeling software architecture. Because there are many different between traditional software and software based on SOA, the model of service-oriented architecture (SOA) is indispensable. In this paper, an approach to modeling software architecture based on SOA is put forward. Firstly, four important properties of SOA are discussed. Secondly, based on Workflow-net, the formal definition of service is proposed. Finally, three rules to model composite service are introduced. The software model based on SOA lays the foundation for analyzing and reasoning it mathematically.

2. Related Work

2.1. *SOA*

So far there is no agreement on a universally accepted definition for service-oriented architecture (SOA), and this paper identifies with the definition of SOA by W3C.

Definition1: (SOA) In W3C, SOA is an application architecture in which all services with Well-defined callable interface defined as independent functional, and these services are called in a particular order to form a business process.

There are four important features of service-oriented software architecture following:

a) Multi-level architecture[5]. With the increasingly large-scale software systems, software systems become increasingly complex, and SOA, as a form of distributed systems architecture, is also increasingly complex. Therefore, systems architecture is usually stratified to reduce the complexity of it.

b) Loosely coupled[4]. SOA is loosely coupled system architecture, which makes quarantine service users and service providers from the service implementations and how to use services by customers.

c) Different granularity service in scale. Granularity of service is the scope of a service public function, and services are divided into fine granularity services which can provide a small usability service in business processes and coarse granularity services which can provide usability service in high-level business logic[4]. Because of different granularity service in scale, SOA should be hierarchical with the same size granularity service in one layer, and it is easier to analyze the service evolution in each layer.

d) Interaction and cooperation[6]. All services in SOA interact and cooperate with each other to accomplish the mission of users, so each part of the software system is complementary to a whole, and not isolated.

There are other characteristics in SOA, the above features which this paper argues is most important for modeling SOA[7].

2.2. *Petri nets*

Petri Nets are graphical formalisms which have gained popularity as tools for the representation of complex logical interactions (like synchronization, sequentiality, concurrency and conflict) [3] among services in software based on SOA. In this paper, Workflow-net, a kind of Petri Nets is used as formal tool for formalizing software architecture based on SOA[8].

Definition 2 [3, 10]: A triple N = (P,T,F) is called a net iff

(1) P,T are disjoint sets.

(2)$F \subseteq (P \times T) \cup (T \times P)$is a binary relations.

Definition 3[3, 10]: A triple N = (P,T,F)is called a net

(1)For $x \in N$

$\bullet x = \{ y \mid y F x \}$is called t8he preset of x;

$x\bullet = \{ y \mid x F y \}$is called the postset of x.

(2) For $x \subseteq N$

$\bullet X = \cup_{x \in x} \bullet x$ and

$X\bullet = \cup_{x \in x} x \bullet$.

(3)A pair (c, a) $\in P \times T$ is called a self-loop iff $c F a \wedge a F c$. N is called pure iff F does not contain any self-loop.

(4) $x \in N$ is called isolated iff $\bullet x \cup x\bullet = \emptyset$.

(5) N is called simple iff$\forall$ x, $y \in N$:($\bullet x = \bullet y \wedge x\bullet = y\bullet) \Rightarrow x = y$.

Definition 4 (Workflow-net) [8, 9]: A Petri net WFN = (P,T,F)is called workflow-net iff:

There is an initial place $i \in P$, $\bullet i = \emptyset$;

There is an end place $o \in P, o\bullet = \emptyset$;

Each node$x \in P \cup T$ are located in the a path from i too.

In Workflow-net, atomic-service or sub-composite service is formalized as transition; the condition which can trigger transition is formalized as place; flow relationship regulates the logical structure of interaction relations among atomic-services or sub-composite services. The state of Workflow-net is defined as token distribution in all places, and | P |-dimensional vector M indicates the real-time state of a running instance in the Workflow-net, every element in M is the number of tokens in the corresponding place, for instance, M P indicates the number of tokens in place p.

3. An Approach to Modeling Multi-level Software Architecture Based on SOA

According to the above characteristics in SOA, SOA is Multi-level architecture, a formal, multi-level software architecture model based on SOA is proposed, which is formalized by Workflow-net, the hierarchical structure can be divided into three layers from the bottom to the up: Atomic services layer, sub-composite service layer and composite service layer composition. According to the above characteristics in SOA, SOA is Multi-level architecture, a formal, multi-level software architecture model based on SOA is proposed, which is formalized by Workflow-net, the hierarchical structure can be divided into three layers from the bottom to the up: Atomic services layer, sub-composite service layer and composite service layer composition[9].

Definition 5: (Atomic service) atomic service can't be split, an atomic service is mapped to an transition.

3.1. *Sub-composite service*

Atomic service is the smallest granularity service, and it can provide extremely limited function. In order to supply more function, many atomic services can constitute a sub-composite service.

Definition 6: (Sub-composite service) Sub-composite service is mapped to a transition or another workflow-net, and it is composed of two or more atomic services through certain composition rules. There are four rules of composition services following.

(Sequential composition) t and t' are two atomic services or sub-composite services, t•=•t'= {v}, transition t and t' execute in sequence. After the occurrence oft, t' can trigger, and M v=1, which is shown in Figure1.

(Concurrent composition) t1, t2, t and t' are atomic services or sub-composite services, t1• = {u, k}, •t={u}, t•={g}, • t' ={k}, t' •= {u'}, •t2 = {u', g}, transition t and t' can trigger at the same time. When transition t and t' can be triggered, Mu= Mk=1, which is shown in Figure 2.

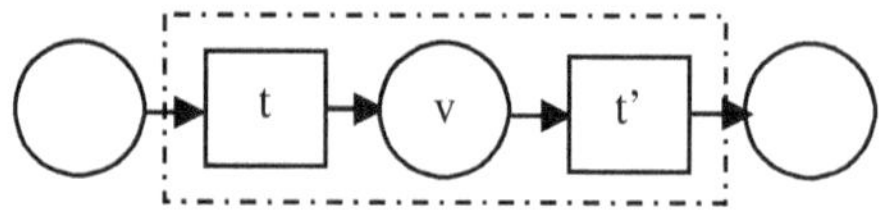

Fig. 1. Sequential composition

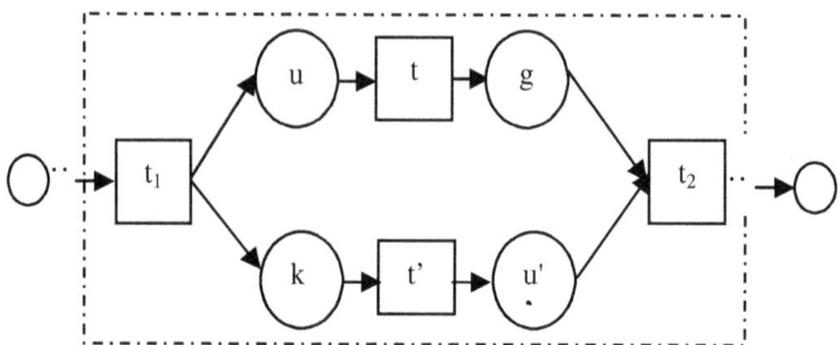

Fig. 2. Concurrent composition

If there is a match between two concurrent atomic services in their port, it is necessary to synchronized interact between them, so concurrent composition is one of the most important composition methods.

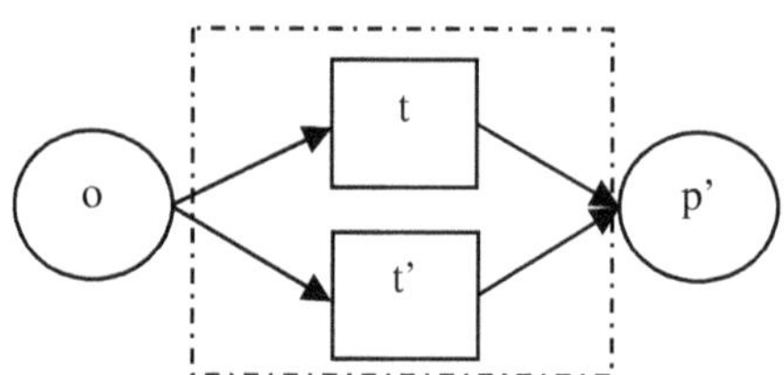

Fig. 3. Choice composition

(Choice composition) t and t' are two atomic services or sub-composite services, •t=•t'= {o}, t•=t'•= {p'}, transition t and t' are chosen to execute. When transition t or t' can be triggered, Mu= Mk=1, which is shown in Figure 3.

When only one atomic service or sub-composite service can be chosen, they would be generally assembled into choice composition service. According to the running environment or other services with which the choice composition service interacts, it selects one atomic service or sub-composite service to execute[10].

(Iteration composition) t1, t2, t and t' are atomic services or sub-composite services, •u= {t1, t'}, u•={t}, •t={g}, g•={ t2,t' }, transition t and t' can be triggered repeatedly. After the occurrence oft, t' or t2 can trigger, and M_g=1. If t_2 is chosen to execute, iteration is terminated, which is shown in Figure 4.

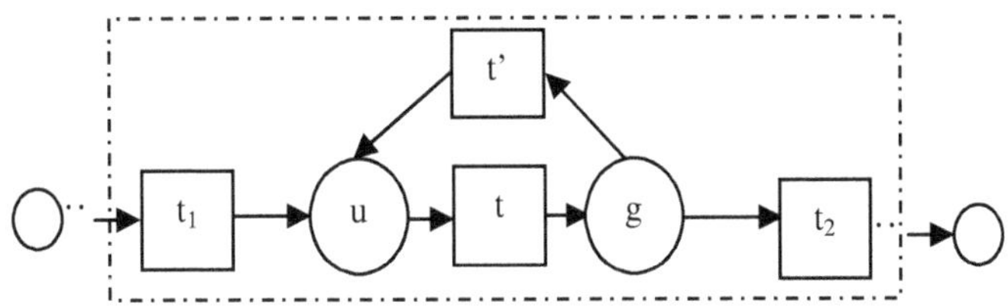

Fig. 4. Iteration composition

The seventh paper points out that sequential composition, concurrent composition, choice composition, and iteration composition do not introduce non-deterministic.

Definition 7: (composite service) By the above rules, atomic services can constitute sub-composite service, and composite service that satisfies the modeler is formed through utilizing the above rules finitely, which makes service itself to be hierarchic.

The service is shown in definition 8.

Definition 8: (service)

An atomic service is a service;

it is also a service that is composed of two or more atomic service through sequential composition rule, concurrent composition rule, choice composition rule, and iteration composition rule;

atomic services which constitute a service are increasing in quantity until any atomic service is no longer added into the service.

3.2. *The process to modeling multi-level software architecture based on SOA*

Based on the only one import and export in service, the process to modeling Multi-level software architecture is proposed. Atomic services which constitute composition service are collected, then by utilizing finitely sequential composition rule, concurrent composition rule, choice composition rule, and iteration composition rule, the hierarchic services is formed from the bottom to up.

The Process to Modeling Multi-level Software Architecture Based on SOA as follows:

Input: the collection N of all Atomic services

Output: the hierarchic service model

```
BEGIN
While(!M)   // M is hierarchic services model which   satisfies the modeler
    BEGIN
      For services in N which is needed to be sequentially composed
      Call sequential composition rule;
      The sequential composition services are put into N and M';
      For services in N which is needed to be concurrently composed
      Call concurrent composition rule;
      The concurrent composition services are put into N and M';
      For services in N which is needed to be choice composed
      Call choice composition rule;
      The choice composition services are put into N and M';
      For services in N which is needed to be iteratively composed
      Call iteration composition rule;
      The iteration composition services are put into N and M';
     The most complex services M' is put into M;
   END WHILE
  Output M;
 END
```

4. Conclusion

In this paper, an approach to modeling multi-level software architecture based on SOA is put forward. Firstly, four important properties of SOA are discussed. Secondly, based on Workflow-net, the formal definition of service is proposed. Finally, three rules to model composite service are introduced. The software model based on SOA lays the foundation for analyzing and reasoning it mathematically.

References

1. Xu Xiao Fei, Wang Zhong Jie, *The development and impact of the services Calculation and service project*. 2008 China Computer Science and Technology Development Report
2. Han Ding, Shen Jian Jing, Research on encapsulation of service component based on service—oriented architecture. 2009, 30(7) Computer Engineering and Design
3. Tong Li. An approach to modeling software evolution process. Tsinghua University Press.

4. Zhang, Yousheng. *Software architecture*. Tsinghua University Press 2006. PP.183–184.
5. Xie Zhongwen, Li Tong, Dai Fei, Qin Jiang long, Yu Yong. *A paraconsistent meta-model of requirements for software evolution*. Journal of Jiangsu University (Natural Science Edition), 2011, 32(5): 562–568 (EI Accession number: 20114214432313)
6. Zhong Wen Xie, Tong Li, Fei Dai, Qian Yu, Yong Yu, Na Zhao and Ping Lu. *A Formal Meta-Model of Software Process*. Proceedings of the 2nd International Conference on Information Science and Engineering, December 4-6, 2010, Hangzhou, China, 4245–4248. (EI Accession number: 20110913705666)
7. Girish Chafle, Koustuv Dasgupta, Anm Kumar et a1. *Adaptation in WebService Composition and Execution*[C]. Proceedings of the IEEE International Conference on Web Services, ICWS 2006. PP.549–557.
8. Zeng Jin, Sun Hai-Long, Liu Xu-Dong, Deng Ting, Huai Jin-Peng. *Dynamic Evolution Mechanism for Trustworthy Software Based on Service Composition* [J]. Journal of Software, Vol.21, No.2, February 2010, pp.261–276.
9. von de Aalst WMP, von Hee K. *Workflow Management Models, Methods and Systems*. Cambridge: The MIT Press, 2002.
10. Reisig W (1985) *Petri Nets: an introduction*, Springer Verlag, Berlin.

Fractional-order chaotic system: Analysis and application*

Jun Ding
Yunnan Normal University,
Kunming, 650500, China

Na Li†
Kunming Medical University Haiyuan College,
Kunming,650106,China
†*E-mail: linakm@qq.com*

An integral-order hyper-chaotic system with four-dimension is expanded into a fractional-order system whose chaotic behaviors are analyzed. Based on the stability theory of fractional-order linear system and the idea of tracking control, a synchronization method for two fractional-order systems with different structures is proposed, and an analytic expression for a synchronization controller is given. The proposed synchronization method is applied to encryption and decryption of digital images. The simulation results show that this method has many advantages for encrypting and decrypting digital images, such as sensitive secret keys, random uniform distribution of pixels and low correlation between adjacent pixels.

Keywords: Fractional-order chaotic system; chaotic synchronization; image encryption.

1. Introduction

Chaos is a random, uncertain and never repeated movement happened in particular systems and it expresses the properties of a system, such as complexity, disorder and randomness [1]. At present, the research on the control of chaos synchronization, hyper-chaos, chaotic neural networks, chaotic secure communication, and chaotic economics etc. shows potential value of chaos.

In recent years, some unique properties that only exist in fractional-order rather than integer-order dynamics system have been found via incorporating fractional-order differential operator with any real constants into nonlinear dynamic system. For example, an integer-order nonlinear system whose order is

---

* This work is supported by the Applied Basic Research Programs of Yunnan Province (No. 2012FD020);
This work is supported by the Science Foundation of Education Department of Yunnan Province (No. 2012C086);
† Corresponding author.

less than three does not induce any chaotic phenomenon, but it appears in Chuas circuits with 2.7 orders [2], it also appears in the heteronomy Duffing system whose order is less than two [3] and Rossler equation with 2.4 order [4]. It has been found that a chaotic system still presents chaotic state after fractional-order differential operator is incorporated into it. Furthermore, a fractional-order chaotic system has more complex dynamic characteristics and shows more physical phenomena than integer-order chaotic system. The research on fractional-order chaotic systems has received a great deal of attention.

In this paper, a four-dimensional integral-order hyper-chaotic system proposed in [5] is expanded to be a fractional-order system. Firstly, some chaos characteristics of the fractional-order system are theoretically analyzed and quantitatively simulated. Secondly, based on the stability theory of fractional-order linear system and the thinking of tracking control[6], a method to synchronize two fractional-order chaos systems with different structures is put forward. Thirdly, the analysis formula of synchronous controller is given, and the synchronization between the extended fractional-order system and Rőssler hyper-chaotic system is quantitatively simulated. Finally the proposed synchronization method is applied to encrypt and decrypt digital images[7][8]. The results show that this method has many advantages for encrypting and decrypting digital images.

2. The Description and Analysis of Fractional-Order Chaotic System

The integral-order differential is replaced with fractional-order differential and a state variable$x_3$is added, by this way, a integral-order system is expanded to be a fractional-order system that is shown in formula (1).

$$\begin{cases} \frac{d^\alpha x_1}{dt^\alpha} = a(x_2 - x_1) + x_3, \\ \frac{d^\alpha x_2}{dt^\alpha} = bx_1 + x_1x_3 - x_4, \\ \frac{d^\alpha x_3}{dt^\alpha} = -x_1x_2 - cx_3 + x_4, \\ \frac{dx_4}{dt^\alpha} = dx_1 + x_2. \end{cases} \quad (1) \qquad \begin{cases} a(x_2 - x_1) + x_3 = 0, \\ bx_1 + x_1x_3 - x_4 = 0, \\ -x_1x_2 - cx_3 + x_4 = 0, \\ dx_1 + x_2 = 0. \end{cases} \quad (2)$$

Where α is the fraction, i.e. $0 < \alpha \text{and} n - 1 \leq \alpha < n$, n is an integer. Let the parameters of system (1) be $\alpha = 8, b = 20, c = 1.1, d = 3$. In order to find equilibrium points of the system (1), set each equation in (1) equals to zero, which is shown in formula (2).

Two equilibrium points, $O_1(0,0,0,0)$ and $O_2(14.7210,0.4343,$ $-1.3029,13.8971)$, can be found by solving equations (2). According to Routh-Hurwitz condition, O_1 is an unstable saddle point, because the eigenvalues of Jacobian matrix at O_1 , $\lambda_1 = -17.2942, \lambda_2 = 9.1148,\ \lambda_3 = 0.0950$, and $\lambda_4 = -1.0155$, are all real, but not all are negative. The eigenvalues of Jacobian matrix at O_2 are $\lambda_1 = -13.7606, \lambda_2 = -0.2580,\ \lambda_3 = 2.2013 + 11.6412$, and $\lambda_4 = 2.2013 - 11.6412j$, and $\lambda_4 = 2.2013 - 11.6412j$, where λ_1 and λ_2 are negative real, λ_3 and λ_4 are conjugate complex roots with positive real part, and their argument is $|arg(\lambda_3)| = |arg(\lambda_4)| = arcgtg\left(\frac{11.6412}{2.2013}\right) = \frac{0.8810\pi}{2}$. According to the stability theory of fractional-order chaos [9], O_2 is an unstable saddle point when $\alpha > 0.8810$. So when $\alpha > 0.8810$, both O_1 and O_2 are all unstable saddle points. By using Wolf method [10], we can calculate the maximum Lyapunov exponent of this fractional-order system and the result is 0.3059.

Based on the method defined by Grünwald-Letnikov (G-L) for fractional-order differential coefficient numerical value [11], numerical simulation of equations (1) is carried out in Matlab. We set the initial state to (0.1,0.1,0.1,0.1) andα =0.95. The trajectories of system (1) in phase portrait of$x_1 - x_2 - x_3$ and $x_2 - x_3 - x_4$ are shown in figure 1(a) and Figure 1(b) respectively, and the Poincare section of system (1) is shown in Figure 2.

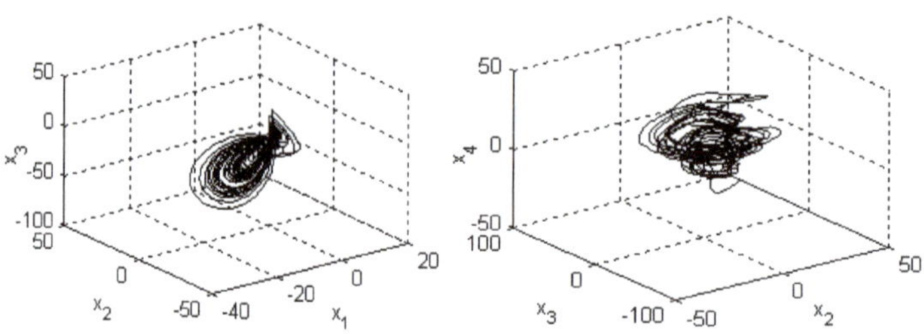

(a)x1-x2-x3 (b)x2-x3-x4

Fig. 1. The trajectory of system (3), $\alpha = 0.95$

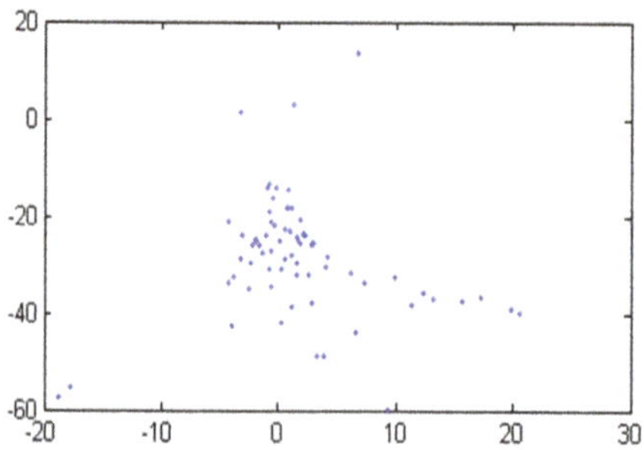

Fig. 2. Poincare section of system (2)

In Figure 1(a) and Figure 1(b), the movement in phase space corresponds to random-separation trajectory with no loop. In Figure 2, there are dense points in the Poincare section. Both Figure 1 and Figure 2 illustrate that the system (1) is in chaotic state.

3. The Method to Synchronize Two Fractional-Orders System with Different Structures

In this section, system (1) is appointed as driving system whose state variables are denoted by $\mathrm{x}(t) = (x_1, x_2, x_3, x_4)^T$, Rőssler hyperchaotic system is appointed as responsive system whose state variables are denoted by $\mathrm{y}(t) = (y_1, y_2, y_3, y_4)^T$. A controller is designed to regulate$\mathrm{y}(t)$ to trace$\mathrm{x}(t)$.

In order to realize the chaotic synchronization of the two systems, that is, to realize $\lim_{t\to\infty}\|e(t)\| = \lim_{t\to\infty}\|\mathrm{y}(t) - \mathrm{x}(t)\| = 0$, the fractional-order Rőssler hyperchaotic system is constructed as[12]:

$$\begin{pmatrix} \frac{d^\alpha y_1}{dt^\alpha} \\ \frac{d^\alpha y_2}{dt^\alpha} \\ \frac{d^\alpha y_3}{dt^\alpha} \\ \frac{dy_4}{dt^\alpha} \end{pmatrix} = \left(\begin{cases} -(y_2 + y_3), \\ y_1 + \tilde{a}y_2 + y_4, \\ \tilde{b} + y_1 y_3, \\ -\tilde{c}y_3 + \tilde{d}y_4. \end{cases} \right) + \mathrm{u}(\mathrm{x}(t)) + \mathrm{u}(\mathrm{y}(t), \mathrm{x}(t)) \tag{3}$$

$$\mathrm{f}(\mathrm{x}(t)) = \begin{pmatrix} -(x_2 + x_3) \\ x_1 + \tilde{a}x_2 + x_4 \\ \tilde{b} + x_1 x_3 \\ -\tilde{c}x_3 + \tilde{d}x_4 \end{pmatrix} \tag{4}$$

In formula (3), $\mathrm{u}(\mathrm{x}(t))$+ $\mathrm{u}(\mathrm{y}(t), \mathrm{x}(t))$ is a tracking controller, in which $\mathrm{u}(\mathrm{x}(t))$ is compensator and $\mathrm{u}(\mathrm{y}(t), \mathrm{x}(t))$ is feedback controller. According to the output of driving system (3), the compensator of responsive system (3) is defined as $\mathrm{u}(\mathrm{x}(t)) = \frac{d^\alpha \mathrm{x}(t)}{dt^\alpha} - \mathrm{f}(\mathrm{x}(t))$, where formula (4), so controlled fractional-order Rőssler hyperchaotic system (3) can be written as follows:

$$\begin{pmatrix} \frac{d^\alpha y_1}{dt^\alpha} \\ \frac{d^\alpha y_2}{dt^\alpha} \\ \frac{d^\alpha y_3}{dt^\alpha} \\ \frac{dy_4}{dt^\alpha} \end{pmatrix} = \left(\begin{cases} -(y_2+y_3), \\ y_1+\tilde{\alpha}y_2+y_4, \\ \tilde{b}+y_1y_3, \\ -\tilde{c}y_3+\tilde{d}y_4. \end{cases} \right) + \begin{pmatrix} \frac{d^\alpha x_1}{dt^\alpha} \\ \frac{d^\alpha x_2}{dt^\alpha} \\ \frac{d^\alpha x_3}{dt^\alpha} \\ \frac{dx_4}{dt^\alpha} \end{pmatrix} - \begin{pmatrix} -(x_2+x_3) \\ x_1+\tilde{\alpha}x_2+x_4 \\ \tilde{b}+x_1x_3 \\ -\tilde{c}x_3+\tilde{d}x_4 \end{pmatrix} + \mathrm{u}(\mathrm{y}(t),\mathrm{x}(t)) \quad (5)$$

Let the synchronized error between responsive system and driving system is $e_{i=}y_i - x_i (i = 1,2,3,4)$, we can obtain the synchronization error system by moving an item in formula (5):

$$\begin{pmatrix} \frac{d^\alpha e_1(t)}{dt^\alpha} \\ \frac{d^\alpha e_2(t)}{dt^\alpha} \end{pmatrix} = \begin{pmatrix} B_1e_1(t)+F_1(x(t),e_2(t)) \\ B_2e_2(t)+F_{21}x(t),e_1(t),e_2(t)+F_{22}x(t),e_2(t) \end{pmatrix} + \mathrm{u}(\mathrm{y}(t),\mathrm{x}(t)) \quad (6)$$

where

$F_1\mathrm{x}(t),\mathrm{e}_2(t) = \begin{pmatrix} -\mathrm{e}_3 \\ \mathrm{e}_4 \end{pmatrix}$, $F_{21}\mathrm{x}(t),\mathrm{e}_1(t),\mathrm{e}_2(t) = \begin{pmatrix} -\mathrm{x}_3\mathrm{e}_1+\mathrm{e}_1\mathrm{e}_3 \\ 0 \end{pmatrix}$, $F_{22}\mathrm{x}(t),\mathrm{e}_2(t) = \begin{pmatrix} \mathrm{x}_1\mathrm{e}_3 \\ 0 \end{pmatrix}$, $B_1 = \begin{pmatrix} 0 & -1 \\ 1 & \tilde{\alpha} \end{pmatrix}$, $B_2 = \begin{pmatrix} 0 & 0 \\ -\tilde{c} & \tilde{d} \end{pmatrix}$, $\mathrm{u}(\mathrm{y}(t),\mathrm{x}(t)) = \begin{pmatrix} A_1\mathrm{e}_1(t)-F_1(\mathrm{x}(t),\mathrm{e}_2(t)) \\ A_2\mathrm{e}_2(t)-F_{22}(\mathrm{x}(t),\mathrm{e}_2(t)) \end{pmatrix}$, the system error $\mathrm{e}(t) = (\mathrm{e}_1(t),\mathrm{e}_2(t))^T$, where $\mathrm{e}_1(t) = (\mathrm{e}_1,\mathrm{e}_2)^T, \mathrm{e}_2(t) = (\mathrm{e}_3,\mathrm{e}_4)^T$.

We get formula (7) by substituting the feedback controller $\mathrm{u}(\mathrm{y}(t),\mathrm{x}(t))$ in formula (6):

$$\begin{pmatrix} \frac{d^\alpha e_1(t)}{dt^\alpha} \\ \frac{d^\alpha e_2(t)}{dt^\alpha} \end{pmatrix} = \begin{pmatrix} (B_1+A_1)e_1(t) \\ (B_2+A_2)e_2(t) + F_{21}x(t),e_1(t),e_2(t) \end{pmatrix} \quad (7)$$

According to the stability theory of fractional-order linear system [14], if there is a matrix $A_1 \in R^{2\times 2}$ which in formula (7) meet the need that arguments of all eigenvalues of $B_1 + A_1$ are more than $0.5\pi\alpha$, then $\lim_{t\to\infty}\mathrm{e}_1(t) = \lim_{t\to\infty}(\mathrm{y}_i - \mathrm{x}_i) = 0 \ (i = 1,2)$, and when $t \to \infty$, $\lim_{\mathrm{e}_1(t)\to 0} F_{21}\mathrm{x}(t),\mathrm{e}_1(t),\mathrm{e}_2(t) = 0$; if there is a matrix $A_2 \in R^{2\times 2}$ that meet the need that all eigenvalues of $B_2 + A_2$ are more than $0.5\pi\alpha$, then $\lim_{t\to\infty}\mathrm{e}_2(t) = \lim_{t\to\infty}(\mathrm{y}_i - \mathrm{x}_i) = 0 \ (i = 3,4)$. From the above analysis, the aim of $\lim_{t\to\infty}\|\mathrm{e}(t)\| = 0$ can be achieved, so the Rőssler hyperchaotic system can synchronize with driving system (1)[13].

We have implemented synchronization simulation in Matlab. In the experiment, we set $A_1 = \begin{pmatrix} -5 & 2 \\ -1 & -3.32 \end{pmatrix}, A_2 = \begin{pmatrix} -3 & 0 \\ 0.5 & -4 \end{pmatrix}$. The eigenvalues of matrix $B_1 + A_1, B_2 + A_2$ are -5， -3 and -3, -3.95 respectively with the arguments of them satisfying the condition $|arg\lambda_\pm| = \pi > 0.5\pi\alpha$. Initial state of driving system (1) and responsive system (3) are set to $(8,4,2,1)$ and $(-50,-10,160,40)$ respectively, and the synchronous courses of two systems variables are shown in Figure 3, and the synchronization error curve of these two systems is shown in Figure 4. We can see that there is difference between each pair of variables of two systems at the beginning, but they can achieve synchronization step by step by tracking.

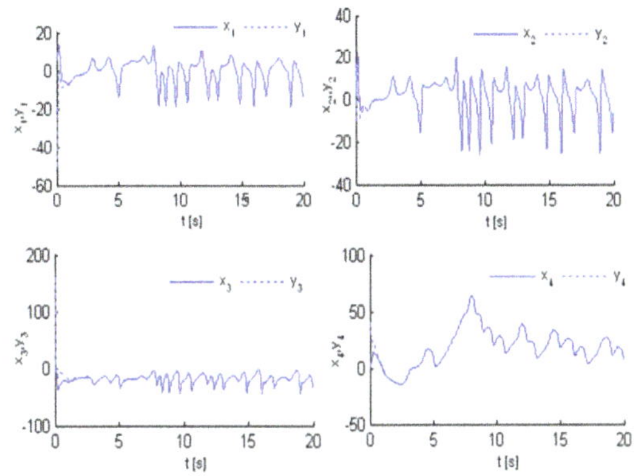

Fig. 3. synchronous process of variables

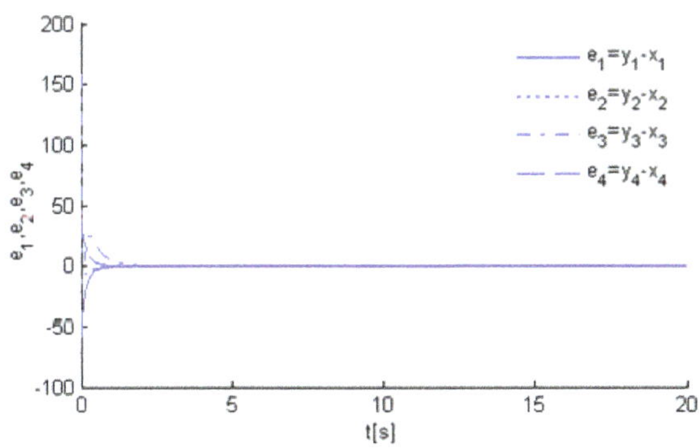

Fig. 4. the error curve of these two systems

4. The Application of Chaotic Synchrony in Image Encryption

The process of the applying chaotic synchrony to encrypt and decrypt digital image is shown in Figure 5.

The original image is denoted by matrix $G = (g_{ij})_{M\times N}$, where M and N represent row and column of the image respectively, and g_{ij} is the gray value of the pixel which is at row i and column j of this image, with gray value depth set to A (typically A=256). The encryption process is completed according to the following steps:

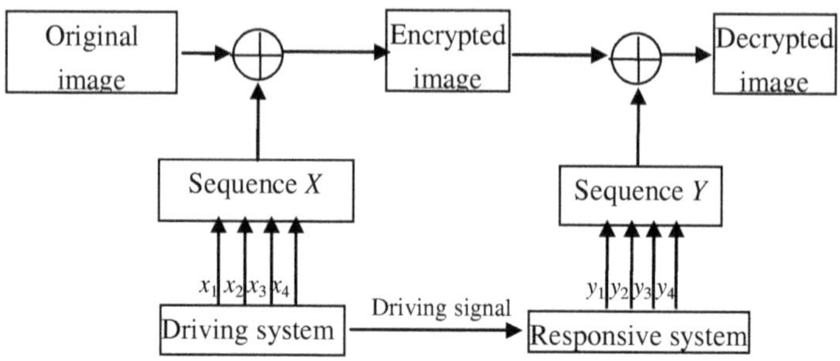

Fig. 5. The process of applying chaotic synchrony to encrypt and decrypt Image.

(1) Let the time steph = 0.005, and make driving system into chaotic state via iteration under given initial state $[x_1^{(0)}, x_2^{(0)}, x_3^{(0)}, x_4^{(0)}]$, then the driving system will produce four chaotic sequences, denoted by $[x_1, x_2, x_3, x_4]$.

(2) Four chaotic sub-sequences can be get via keeping $(M \times N)$ / 4 points in each chaotic sequences fromt = 5000h. Join the four chaotic sub-sequences end to end to be a sequence $B_{1\times(M\times N)}$.

(3) In order to make operations between the sequence and image pixel matrix value, after amplifying sequence $B_{1\times(M\times N)}$, it is taken modular arithmetic with A, and$B'_{1\times(M\times N)}$would be converted into the unsigned eight bit integer datafor making the sequence value in the range from 0 to A , that is, $B'_{1\times(M\times N)} = mod(1000 * B_{1\times(M\times N)}, A)$, $B''_{1\times(M\times N)} = uint8(B'_{1\times(M\times N)})$;

(4) Carry out XOR operation between each element of $B''_{1\times(M\times N)}$ and g_{ij}, the value of g_{ij} is arranged in regular rows, then we get a new matrix$G' = g'_{ij\times(M\times N)}$ which denotes the image after being encrypted.

The process of decryption process is the inverse process of encryption process. At receiving terminal, use the secret key (the initial value of driving system) to make responsive system into chaotic state via iteration, and we can get four chaotic sequences. Then, according to the above step (2) - (4) of encrypting, assemble these sequences into a new sequence, finally, perform bitwise XOR between each element of the sequence and corresponding encrypted image G' in turn, then we can get the decrypted image.

4.1. *The simulated results of image encryption and decryption*

In this section, we will apply tracking control synchronization in to encrypting Lena whose size is 256×256, and we will do simulation with matlab. Let $x_1^{(0)} = 8, x_2^{(0)} = 4, x_3^{(0)} = 2, x_4^{(0)} = 1, \alpha = 0.95$, and the simulation results are shown in Figure 6.

(a) the original image (b) the encrypted image (c) the decrypted image

Fig. 6. The simulation results of encryption and decryption

From Figure 6(b), we can see that the encrypted images are very different from the original images. From Figure 6(c), we can see that the encrypted images can be decrypted at the receiving terminal.

In order to furtherly check the validation of the method presented, Lena original image, its encrypted image and its decrypting image are compared in histogram and in correlation between adjacent pixels. We also analysis the sensitivity of the encrypted image to the secrete keys.

4.2. *The Comparition of Histogram*

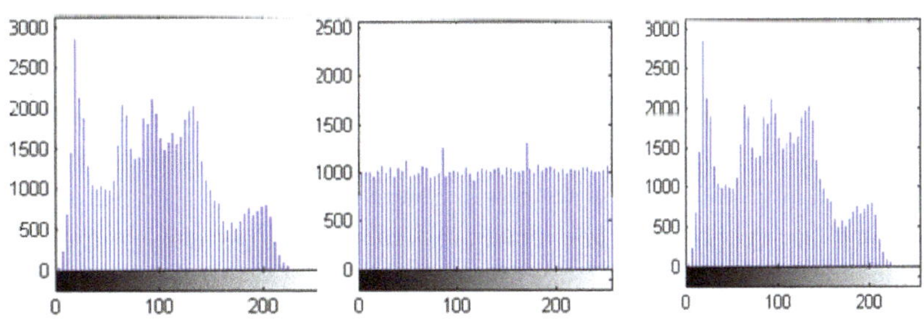

(a) original image (b) encryption image (c) decrypted image

Fig. 7. The gray histogram

The histograms of Lena original image, its encrypted image and its decrypted image are shown in Figure 7 (a), (b) and (c) respectively. It is clear that the histogram of the Lena original image is extremely similar to the one of the Lena decrypted image, but the histogram of Lena encrypted image is nearly even-distributed in the entire range of gray. It shows that statistical property of the original image has been broken, so the encrypted image has a good ability to resist the attack that is based on statistical characteristics.

4.3. *The Sensitivity of the Encrypted Image to the Secrete Keys*

A chaotic system is very sensitive to initial state. If there are not completely correct secrete keys, the encrypted information cannot be restored correctly. During the simulation experiment, if figure 8(a) the encrypted Lena image is decrypted using the correct secrete key $x_1^{(0)} = 8, x_2^{(0)} = 4, x_3^{(0)} = 2, x_4^{(0)} = 1$, we can get the right image shown in figure 8(b); if it is decrypted using another secrete key $x_1'^{(0)} = 8.0000000001, x_2'^{(0)} = 4, x_3'^{(0)} = 2, x_4'^{(0)} = 1$, we can get another image shown in figure 8(c). Because of a tiny difference 10^{-10} between the two keys, we can get two completely different results. This shows that the encrypted image is sensitive to secrete keys, and the encrypted image can be decrypted correctly only by the right keys [15].

(a) the encrypted (b) the decrypted under precision keys (c) the decrypted under changed keys

Fig. 8. Encrypt Lena image and its decrypted images under different keys

5. Conclusion

In this paper, we expanded a four-dimensional integral-order hyperchaotic system to be a fractional-order system, then we theoretically analyzed and quantitatively simulated some chaos characteristics. We also put forward a method to synchronize two fractional-order chaos systems with different structures on the basis of the stability theory of fractional-order linear system and the tracking control ideas, and we gave the analysis formula of synchronous controller and carry out quantitative simulations on the synchronization between the extended fractional-order system and Rőssler hyperchaotic system. The experimental results show that this method has many advantages for encrypting and decrypting digital images.

References

1. Puyun Gao. *Nonlinear Dynamics*. Changsha: National Defense University Press. 2005.
2. Hartly T, Lorenzo C F, *Qammer H K. Chaos in a fractional order Chua's system*. IEEE Trans CAS-I. 1995; 42(8): 485–490.

3. Arena P, Caponetto R, Fortuna L, Porto D. *Chaos in a fractional order duffing system*. In: Proceedings of ECCTD. Budapest. 1997; 1259–1262.
4. Li C G, Chen G R. *Chaos and hyperchaos in the fractional-order Rössler equations*. Physica A. Volume 341, 55–61.
5. Liu L, Liu C X, Zhang Y B. *Analysis of a Novel Four-Dimensional* Hyperchaotic System. Chinese Journal of Physics. 2008; 46(4): 386–393.
6. Zhou Ping, Kuang Fei. *Synchronization between fractional-order chaotic* system and chaotic system of integer orders. ActaPhysicaSinica. 2010; 59(10): 6851–6858.
7. Rőssler, O.E. *An Equation for continuous chaos*. Physics Letters A. 1976; 57: 397–398.
8. Rőssler, O.E. *An Equation for Hyperchaos*. Physics Letters A. 1979; 71 (2-3): 155–157.
9. Jianbing Hu. *Stability theory of fractional-order chaos and synchronization method*. China National Knowledge Infrastructure. 2008.
10. A. Wolf, J.B. Swinney, H.L. Swinney, et al. *Determining Lyapunov exponents from a time series*. Physica D. 1985;16: 285–317.
11. Ivo Petrαš. Fractional-Order Nonlinear Systems: Modeling, Analysis and Simulation. Beijing, China: Higher Education Press. 2011,3.
12. Christos K. Volos. Image Encryption Scheme Based on Coupled Chaotic Systems. Journal of Applied Mathematics & Bioinformatics, 2013; 3(1); 123–149.
13. X. Wang, L. Teng and X. Qin. A Novel Colour Image Encryption Algorithm Based on Chaos. Signal Process, 2012; 92; 1101–1108.
14. G. Sansone and R. Conti. Non-linear Differential Equations. Pergamon Press, 1964. (International Series of Monographs in Pure and Applied Mathematics, Volume 67).
15. Chen Guanrong, Mao Yaobin, Charles K. A Symmetric Image Encryption Scheme Based on 3D Chaotic Cat Maps. Chaos, Solitons and Fractals. 2004; 3(21): 749–761.

Application of an AES encryption and decryption system to an embedded processor*

Min Feng, Jun Ding†, Yun-Xia Ge,
Liang-Yu Chen, Yuan He and Yan-Jiao Wei

Yunnan Normal University, Kunming, 650500, China
†E-mail: dinxj@qq.com

Using the advanced encryption algorithm AES [1][2], an encryption and decryption IP core is designed for embedded processors. Using the FPGA on-chip storage module and the reconfigurable S box, the key and the original data are managed with a dedicated storage method, to be locked in the kernel. In order to make the IP core able to be flexibly used in the Nios II embedded system, the corresponding interface and its address mappings are designed in accordance with the Avalon bus interface specifications. The IP core can thus be easily customized in the Nios system. The Nios II processor, customized encryption and decryption components, network controller, memory and other corresponding peripheral equipment are composed to a hardware platform by using SOPC technology. For the purpose of automation control, corresponding software is written to make it a real-time network encryption and decryption system.

Keywords: Embedded Structure; Encryption and Decryption; FPGA; AES algorithm.

1. Introduction

The rapid development of network communication has provided people with convenient and flexible communication, at the same time also brought a lot of security problems, which are mainly concentrated on the network communication channel. The encryption technology can prevent illegal listening in the network communication, largely avoiding the important data giving away [1]. The encryption and decryption system designed in this paper is mainly for the bit stream data, so it belongs to the security of the physical layer of the network. At present, the research of network security mainly aimed at the network layer and above level. And in the physical layer and data link layer the

* This work is supported by the National college students innovation training program (No. 201310681012).
This work is supported by the Applied Basic Research Programs of Yunnan Province (No. 2012FD020);
This work is supported by the Science Foundation of Education Department of Yunnan Province (No. 2012C086);
† Corresponding author.

plaintext data is still used to transmit, so the security is still not guaranteed. The network real-time encryption and decryption system can ensure the data security in the physical layer transmission. Therefore, the encryption and decryption function is expected to be used in the network card or a similar device, to guarantee the data safety in the underlying network. So this system can is widely used in the field of data security.

2. Algorithm Introduction

According to the different encryption methods, the symmetric cipher algorithm can be divided into two categories: block cipher and stream cipher. The AES algorithm belongs to the block cipher algorithm. The block cipher divides the message into the fixed length, and the output of the cipher block is the same as the length of the input. The length of the key "K" is 128, 192 or 256 bits. Using "Nk"=4,6,8 to represent the word size of the key string(1 word =32 bits), chosen before the hardware description language synthesis in the IP core. Using "Nr" to represent the round number of block encryption, and each round requires an expansion key "Ke" whose length is the same as the input block[2][3]. Because in the AES the external input encryption key "K" has limited size, use a key expansion program to expand the K into longer bit strings, to generate the encryption key. AES-128/192/256 algorithms are basically the same, only have two differences: key length and encryption round are different. To the three different types of AES, encryption key block size and corresponding encryption round are different[4], as shown in Table 1.

Table 1. AES-128/192/256 parameter comparison table

AES Type	Key length	Block length	Encryption round Nr
AES-128	4 Byte	4 Byte	10
AES-192	6 Byte	4 Byte	12
AES-256	8 Byte	4 Byte	14

The AES algorithm contains the Key expansion process and the encryption and decryption processes. Encryption process includes round key XOR operation, row shift operation, byte substitution operation and column mixing operation. The algorithm carries on the Nr iterations, and the final round iteration does not do the column mixing operation[5]. The decryption process uses inverse operation of all aspects of the encryption process. AES encryption and decryption processes are shown in Figure 1.

3. Encryption/Decryption Module

The data block of AES algorithm is the research object, and each sub module of the whole encryption and decryption IP core are designed. First, we need to know the IP core interface. Taking into account portability of the IP core, we use the Avalon bus interface specification. The control bus contains clock signal, reset signal, control enable signal, function selection signal, status signal and interrupt signal. Second, in accordance with encryption and decryption process of the data block the IP core can be divided into the finite state machine module, the key expansion module, the round key XOR operation module, the byte substitution module, row shift transformation module, the column mixing operation module and the multiplexer module seven parts. They are shown in Figure 2. The finite state machine module and the multiplexer module both control the whole iterative process of the IP core, making the data block correctly run to complete the encryption and decryption operation in each module.

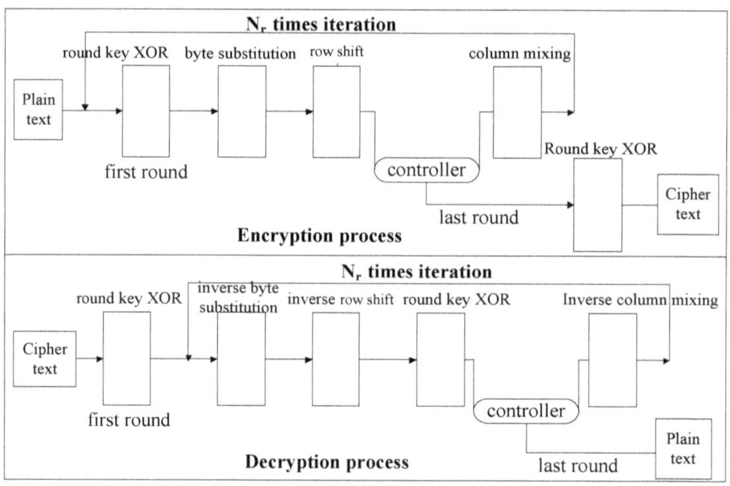

Fig. 1. AES encryption and decryption processes.

In order to facilitate the IP core as a custom component for the Nios II system integration, the memory mapping space is designed in the IP core. In the Nios II system, the Avalon bus width is 32 bits. All write operations are based on the write transmission to the slave port in the IP core. 32 bits data is transmitted in a clock cycles, encryption and decryption data blocks need multiple clock cycles to complete the input, please reference documentation. Therefore, the key to design the IP core are the interface and its address mapping and the finite state.

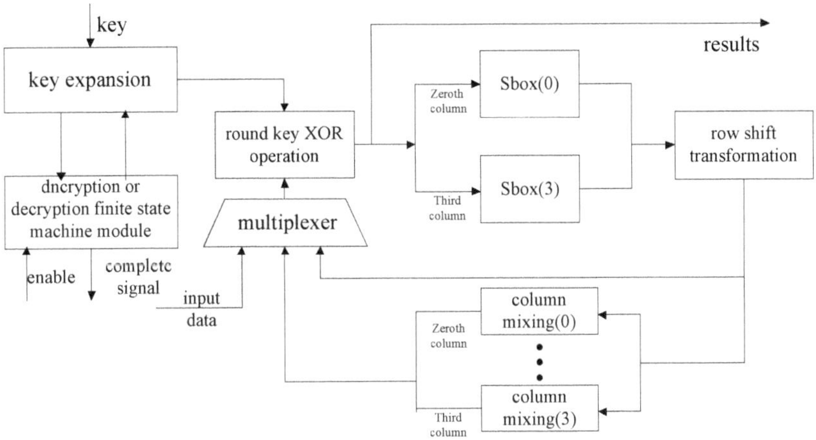

Fig. 2. The structure of IP core

3.1. *The interface and its address mapping*

All signals are synchronous in the IP core, and the rising edge of the clock is valid. All the interface signals are designed according to the Avalon bus specification, which can make the IP core directly integrated into the Nios II system as a custom component. The encryption and decryption data and all control signals are transmitted through the Avalon 32 bits data bus in the IP core. At the same time, reasonable address mapping is needed to design. This design only uses zeroth bytes, and keeps the first to third bytes for future reference. The mapping of individual data bit is shown in Table 2[6].

Table 2. Control register address mapping

Address (Decimal)	Name	Meaning
31-8		reserved
7	KEYVALID	Key valid
6	IRQENA	Interrupt enable
5-2		reserved
1	DEC	Decryption
0	ENC	Encryption

3.2. *The finite state machine module*

The IP core firstly needs to initialize a key. Secondly, the key expansion module will be started to produce the keys which used in each round operation. Thirdly, input the data block which is needed to operate. Fourthly, under the control of the finite state machine complete each round encryption and decryption. Finally, finish the all operation, as shown in Figure 3.

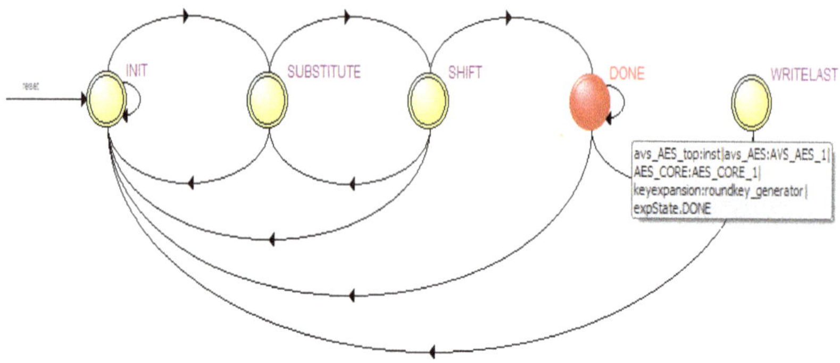

Fig. 3. The finite state machine of encryption and decryption

4. System Integration

Network controller will store data in the data buffer pool. The block state machine read data in a 128-bit block from the buffer pool. In the FPGA, the AES encryption and decryption module consists of a preprocessing unit. And it selects the data needed to deal with. Get the data and input them to the AES component under the control of the Nios II processor. Then complete the data encryption or decryption[7]. Finally, store the results in the buffer pool and transfer them to the network by the network controller, as shown in figure 4. The designed AES module is packaged into Avalon device. And as a custom component, it and the network controller are connected to the Avalon bus. All the used Avalon devices are integrated into the encryption and decryption system [8].

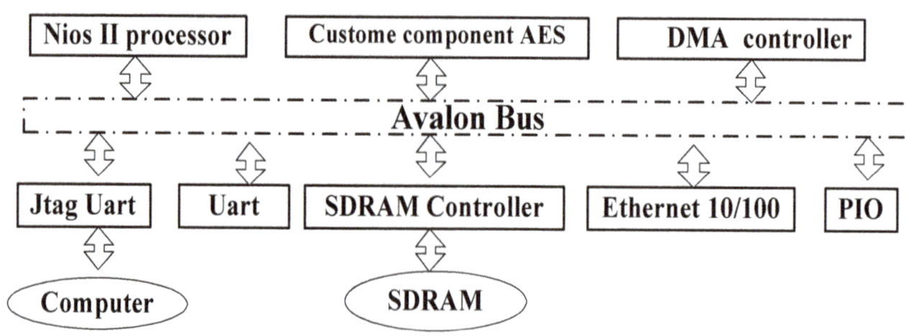

Fig. 4. The hardware structure of the system

5. Timing Simulation of AES Components

In the system address space 00000-00111, store the key. When the address (11111)=00000080, the key is valid. In the address space 01000-01011, store the data. When the address (11111)=000000C1, encrypt the data and the results are stored in the address space 10000-10011. When the address (11111)=000000C2, decrypt the data. And read the decrypted data from the address space 10000-10011, it is the same as before encrypt. Figures 5 to 8 show them [9].

Fig. 5. The keys are stored

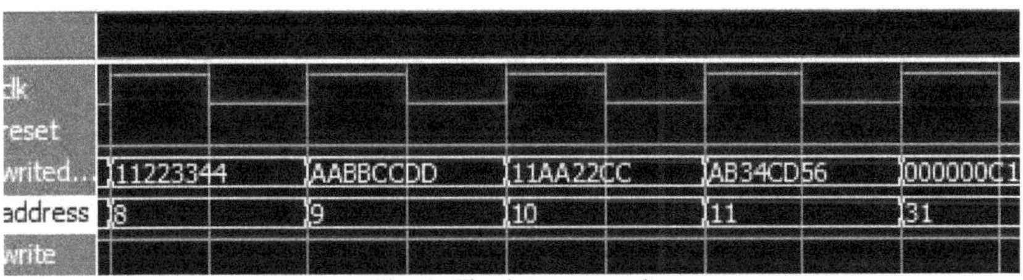

Fig. 6. The data are stored

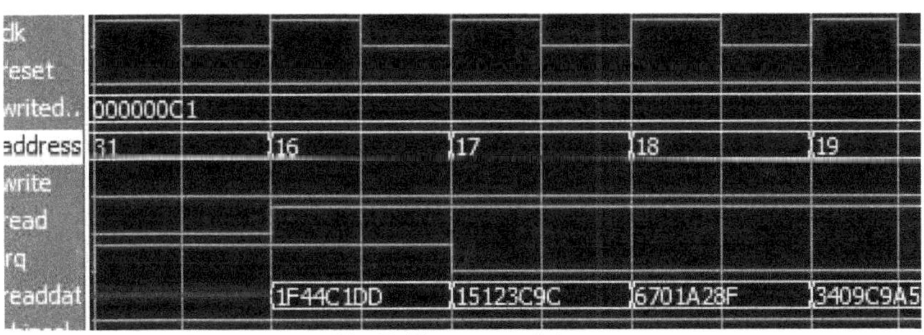

Fig. 7. The encrypted data

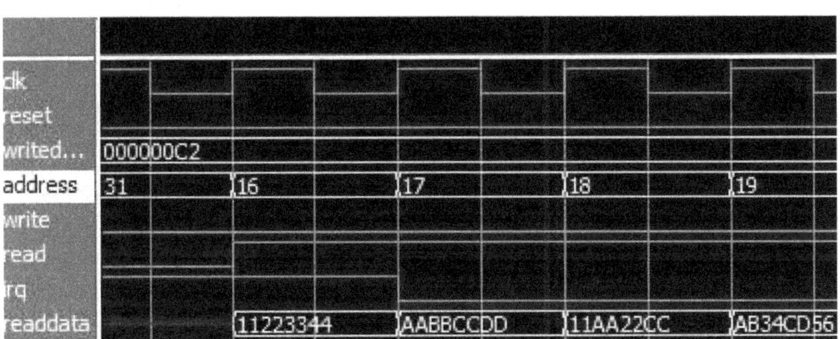

Fig. 8. The decrypted data

6. Summary

This paper designs an AES Encryption and Decryption system applied to embedded processor. 32 bits Avalon data bus specification which not only making the design data with friendly interface but also making the IP core use software control more flexible. The whole design has a simple hardware structure and high running speed advantage, can be widely used in network and files security system.

References

1. Huang Chunyu, Han Chunlei. *A design of FPGA-based AES algorithm mobile hard disk encryption system*[J]. Journal of Changchun University of Science and Technology, Natural Science Edition. 2014 Fourth Issue.
2. He Qingping, Feng Jinpeng, Xu Bairu. *FPGA optimized implementation of AES-128 algorithm*[J]. China Science and Technology Information. 2014 Tenth Issue.
3. Zhang Huixia, Zhao Jinaping, Li Xiaoli, Lu na. *FPGA implementation and simulation of AES algorithm*[J]. Communications Technology. 2013 Ninth Issue.
4. Han Jinsheng, Lin Jiajun, Zhou Wenjin, Ye Jianwu. *Design of FPGA-based AES core*[D]. Computer Engineering and Science. 2013 Third Issue.
5. Yin Weifeng. *Resource optimization design of AES encryption and FPGA implementation*[J]. Computer and Modernization. 2012 Eleventh Issue.
6. Liang Xu, Ling Chaodong, Zhang Lihong. *FPGA-based implementation of AES algorithm.* Communications Technology. 2011 Twelfth Issue.
7. Zhang Bin, Yan Huizhong. *FPGA implementation of Advanced Encryption Standard AES algorithm.* Computer and Telecom. 2010 Ninth Issue.
8. Qi Zhongbing, Li Bin, Wang Zifu. *An optimized implementation of AES encryption algorithm based on FPGA* [J]. Science and Technology Innovation Herald. 2010 Twenty-sixth Issue.
9. Li Lang, Zou Wei, Li Renfa, Li Kenli. *FPGA optimized implementation of AES-128 algorithm using Verilog HDL language*[J]. Journal of Chongqing University, Natural Science Edition. 2014 Sixth Issue.

Application of faceted classification to organizing tourism product information on the web

Wei-Dong Yu[†]
Chengde Petroleum College,
Chengde, Hebei067000, China
[†]*E-mail: suren94@163.com*
www.cdpc.edu.cn

The rapid growth of the tourism market has led to a dramatic increase in both the number and diversity of tourism commodities, or tourism products. Organization of the corresponding product information has thus become extremely important. Many websites tend to combine faceted classification schemes with their websites' tourism product information organization. This paper first introduces facet classification and its advantages, before attempting to analyze how it is applied to websites' tourism product information organization. These are done by listing three perspectives separately, namely product information description, classification navigation systems construction, and product information visualization display.

Keywords: Facet Classification; Tourism Management; Information Organization; Knowledge Organization.

1. Introduction

With the rapid development of computer technologies and networking, web information organization has become an inevitable public resource. Although we can obtain information via Internet conveniently and swiftly, it is not easy for us to get accurate and proper information – just like we cannot find knowledge fitting us in ocean of information[1]. It is universally acknowledged that only by being controlled and organized can information become resource[2], which is exactly why web information organization occurred. Actually, due to prosperity of tourism all over the world, it is popular to display tourism commodity online. Therefore, it is significant to study on how to better organize tourism commodity information online.

To organize information, there are generally two ways: classification and subject indexing method. In terms of classification, two sub-methods are often mentioned: Enumerative classification and facet classification. This paper mainly focuses on how to apply facet classification, a traditional classification, into modern web tourism commodity information.

Facet classification divides subject concept into many simple concepts, according to whose aspects and angles tables are formed. By combining different simple concepts together, a complex concept is represented. This is what facet classification operates.

Facet classification has at least these advantages[3]: Firstly, it can reveal perplex subjects in information resources thoroughly by combining with different simple concepts. Secondly, the order of combining can be changed according to various needs, which benefits multi-gram retrieval. Thirdly, new concepts can be added at any time, enhancing the ability to represent popular knowledge. Fourthly, class tables are easy to regulate, supplement, and revise due to its small space.

Here we list several typical application of facet classification on web tourism commodity information organization. Note that these three aspects are ordered from basic information description to deep display, i.e. the surface to the center.

2. Commodity Information Description

Commodity information, a typical and significant information resource, can also be regarded as a complex concept. According to above, when describing complex concepts, facet classification has much more advantages.

On the one hand, facet classification can identify and reveal complex subjects[3], which is actually multi-dimensional, by combining several simple concepts. On different facets, users can choose to combine several of them, which helps information retrieval on varieties of dimensions and ways[1]. When describing tourism commodity information, such method facilitates more subject concepts into classification system[4], resulting in higher recall ratio.

On the other hand, besides deciding whether to select certain facets, users can also determine the order of combination[5]. Take "3-day trip to Hong Kong" for example. A user can click "Hong Kong trips" firstly, and then click "3-day" button; he can also click "3-day trip" to select all available tourism line on website, and then click "Hong Kong" button in destination information. From this instance, we can see that facet classification actually plays a very important role in tourism commodity information organization, one of whose reason is that every facet in classification system stands on the same level (layer) without any priority part.

3. Classification Navigation Systems Construction

3.1. *Combination between facet classification and enumerative classification*

In terms of classification navigation systems, nowadays websites combine facet classification and enumerative classification to form classification systems[6], the essence of which is revision of enumerative classification by adding facet elements in it[3]. Specifically speaking, tourism commodity e-commerce websites firstly classify all commodities basically and generally with enumerative classification. These categories are called top categories, meaning that they are the top level (layer) of categories. The second layer of categories are called "sub-top categories". After entering into top categories or sub-top categories (remember that at that time one classification has been finished), it appears combination between two types of classification methods, i.e. one facet is also displayed as one layer.

For example, when searching "tourism insurance" -> "personal accident insurance"[1] on Jingdong[7], one of the most significant e-commerce websites in China, the website will occur several facets, including price, insurance company, type, insured person. What needed to be explained is that these facets can be seen different lower classes from different angles and aspects. See, this is a typical combination between two classification methods.

3.2. *Order of use of facet classification and enumerative classification*

Though being used in the same circumstance, order of use of facet classification and enumerative classification is needed to be considered. Generally speaking, e-commerce website often use enumerative classification at first and then try to use facet classification. Let's see Xiecheng, an also-important website on tourism commodities[8]. To begin with, it lists top categories, including tourism trips, tourism insurance, travel goods by enumerative classification, then sets different facets in each module (i.e. top category).

Why facet classification are often used after enumerative classification? Actually the reason attributes to huge difference between diverse top categories. Imagine if e-commerce websites used facet classification, it would be very difficult to determine exact name of each facet. For instance, facets of "insurance" are often "insurance person" or "insurance company", while facets

---

[1] Note that this example is proper because tourism insurance is also an important kind of tourism commodity.

of "trips" are often "destination", "travel time", or even "travel season". That is to say, that website sets facets after 1-2 times classification can practically cater the needs of information retrieval – which as a result improves the speed of searching. Note that although websites usually use facet classification after enumerative classification, using 1 or 2 time should depend on the scale of website as well as characteristic of tourism commodities on it.

3.3. *(Sub-)top categories in enumerative classification and facets*

According to above, we did not regard top categories as facets. Actually, however, we can see them as special facets – the reason why it is special is that at one time top categories show only one facet. For example, on Taobao, main website of Alibaba company, there are only 16 top categories, like clothes, shoes and bags, sports, etc[9]. These top categories actually can be regarded as facets that set according to their "use" or "purpose". What's more, lower class categories have multiple facets, yet top categories or sub-top categories has only one. Needless to say, this is closely related to diversity of different facets mentioned in 3.2 part.

3.4. *Multi-direction classification navigation systems*

As is stated above, it is a premier advantage of web tourism commodities information organization to retrieve in different dimensions and ways. Hypertext technology leads multi-direction searching, which means that if a commodity G can be categorized into both category A and B, users can search G either in A or B. Here, category G is called alternative category in library and information science (LIS)[10]. E-commerce websites prefer to use alternative categories to show commodities when organizing their information by facet classification. In order to better help users to search commodities, information classification systems have to provide many retrieval directions and alternative categories.

For example, digital cameras can be regarded as household electrical appliances, digital products, or camera equipment; MP4 can be regarded as household electrical appliances or digital products; Game disc can be regarded as software or game and related products[7]. Therefore, redisplaying categories and labels of commodities provides more information entrances to users, bringing much more convenience[12].

4. Commodity Information Visualization Display

Facet classification can save view space, helping users find what they need easily by combining different facets[13]. On the one hand, facets visualization of web commodity information anatomizes complex concepts into several simple concepts, which simplifies users' thinking procedures. That will be sure to enhance recall ratio and accuracy ratio[3]. On the other hand, users can determine whether facets are displayed and whether **ALL** facets are displayed. This, to some extent, also helps to save view space, which optimized user experience. For example, when searching "Sports ->Clothes -> Tennis" in Jingdong, many facets like band, price, and type, are included on the website. And there are some bands under "band" facet. In addition, "More" button are set for users to click[7].

5. Conclusion

This paper analyzes three aspects of application of facet classification on web tourism commodity information organization, including commodity information description, classification navigation systems construction, and commodity information visualization display. Among them, the second part is more detailed. Due to prosperity of tourism commodities, information organization of tourism commodities is called for urgently. Since facet classification can anatomize perplex concepts, i.e. commodities which need to be displayed in different levels and perspective, into several facets and represent them by integrating various facets, it caters the needs of trends which is 3-multi (multi-entrance, multi-dimension, multi-direction) information retrieval of web tourism commodity. With the rapid development of information and digitalization, tourism commodity information organization will become more important, and we all believe that facet classification will be sure to be applied more widely and comprehensively.

References

1. Z. Ma & Z. Huang. *Web information resource organization.* (Peking University Press, 2007). (In Chinese).
2. X. Zhang. *Introduction to facet analysis and facet classification*, Library Field, 1985(4): 36–39. (In Chinese).
3. Z. Ma. *Information Organization.* (Tsinghua University Press, 2008). (In Chinese).
4. G. Cao. *Design thinking on web information classification systems.* Modern Information, 2004(12): 56–57. (In Chinese).
5. Z. Zhang, H. Cheng, & L. Li. *Research on theories and application of web information facet classification systems.* Journal of Information, 2009(08): 62–66. (In Chinese).

6. Z. Son. Research on application of facet metadata on website information retrieval systems. Nanjing Agriculture University, Master Thesis, 2010. (In Chinese).
7. Jingdong. http://www.jd.com.
8. Xiecheng. http://www.ctrip.com.
9. Taobao. http://www.taobao.com.
10. H. Zhang & C. Li. *New product classification in web marketing*. Enterprises Management, 2011(6): 88–89. (In Chinese).
11. X. Chen. Application of facet classification on web resource organization. Library and Information Works, 2002(1): 59–61. (In Chinese).
12. D. Wu. Designing thinking of web information classification systems. Library and Information Knowledge, 2002(05): 37–39. (In Chinese).
13. X. Yu. Research on visualization commodity information. Journal of Hubei University of Economics, 2007(4): 89–91.

Sampling properties of complex networks: the average degree

Dai-Jun Wei[†], Ning-Kui Wang, Jing Wang and Fu-Yu Liu
School of Science, Hubei University for Nationality,
Key Laboratory of Biological Resources Protection and Utilization of Hubei Province
Enshi, Hubei, China
[†]*E-mail: eswdj@swu.edu.cn, eswdj@163.com*
www.hbmy.edu.cn

The average degree is a fundamental concept of complex networks. Some properties of the general aggregate are deduced from the properties of the sample using random sampling. In this paper, the relationship of average degree between the subnetwork and networks is investigated. Different relationships between the subnetworks and networks for some real complex networks are described by numerical tests. The results show that the average degree of subnetwork is not directly proportional to the value of sampling rate for the scientific collaboration network and the neural network. However, the average degree of subnetwork is directly proportional to the value of sampling rate for the American football games network. The findings are useful in revealing some of the properties of unknown complex networks.

Keywords: Complex networks; Random sampling; Average degree.

1. Introduction

Complex networks has become a hot topic in vary fields such as sociology, linguistics, economics and physics, etc[1–6]. In complex systems, individual described as a point and the relationship between the individuals is described as a connection, which is noted by edge, then the complex systems can be described as a complex networks. The average degree is one of the most important concepts of complex networks. The values of average degree is a key indicator for identifying random complex networks and regulation networks [7–8].

For complex network, due to data quantity of the real complexity network is too big, it is very important that by studying the properties of sampling sub-network of real complex networks to study the properties of the nature of real complex networks. Then, statistical properties of the sample subnetworks has received more and more research attention from researchers from engineering and scientific fields[9–13]. In Ref.[14], random networks was considered by the random sampling method, the results shown that the degree distribution of subnetworks was same as the original random networks. However, subnetworks of scale-free networks was not the scale-free networks[15]. For real complex networks, J. Han predicted some properties of the protein networks by studying subnetworks of the protein networks[16]. Some social networks were considered in Ref. [17] using Bootstrap sampling method.

This paper is focus on several real complex networks in real life. The degree distribution and average degree of subnetworks are considered under different sampling proportion by random sampling method.

2. Preliminaries

2.1. *The average degree of complex networks*

For given complex networks $G=(N,V)$ and $N=(1,2,\cdots n)$, $V=(1,2,\cdots m)$, where n is the total number of nodes, and m is the total number of edges. The degree of node i is defined as in Eq.(1).

$$k_i = \sum_j^n x_{ij} \tag{1}$$

where x_{ij} represents the connection between node i and node j. The value of x_{ij} is defined as 1 if node i is connected to node j, and 0 otherwise. Then, the average degree of node is given by Eq.(2)

$$\bar{k} = \frac{\sum_{i=1}^n k_i}{n} \tag{2}$$

$\bar{k}$ is represents density between two nodes. The value of $\bar{k}$ is more higher, density is more bigger. The rule network is the most simple networks, which has the same number of neighbor for any a lattice. The classical random networks ER model is given by Erdos and Renyi. Two nodes connected with probability p. The average degree of ER networks is $2(n-1)p$.

2.2. *Some basic concepts of sampling*

The overall and sample are the most basic and most commonly used concept in sampling. Overall is made up of the elements of study object, and elements which component the overall refer to overall unit or population unit. The sample is a collection consisting of part of the pump according to the sampling rules. Samples included in the overall unit number are called sample size.

The sampling is in accordance with the principles of probability theory and mathematical statistics, to select samples from the study population according to the principle of random, through the sample survey accessing to data, in order to make judgments on population characteristics. The random sampling belongs probability sample.

3. The Average Degree of Real Complex Networks

For given some real complex networks, we have sampled on the number of nodes of original networks with the sample proportion by 10%, 20%, 10%, 40%, 50%, 60%, 70%, 80%, 90%, and generated the corresponding size of the sample network. For same proportion, we have five times for sampling, then the results of average are calculated.

3.1. *The scientific collaboration network*

The scientific collaboration network with 1461 nodes and 2742 edges, the edge represents co-author of the paper(excluding single authored papers)[18]. The average degree of subnetwork of the scientific collaboration network is shown in Table 1 by different percents.

Table 1. The values of nodes, edges and average degree of subnetworks of the scientific collaboration based on different proportions.

Percentage	Number of edges	Number of nodes	The average degree
10%	39	52	1.50
20%	152	61	4.98
30%	224	254	1.76
40%	449	431	2.08
50%	597	566	2.11
60%	947	773	2.45
70%	1288	929	2.77
80%	1773	1099	3.23
90%	2202	1294	3.40
Original data	2742	1461	3.75

Form Table 1, number of nodes and edges is increased percentage of sampling. However, the average degree has a singular point, which is 4.98 when the value of percentage equal 20%. These data are used to fit by MATLAB, and the result is shown in Figure 1. The fitting linear model is as Eq.(3)

Linear model Poly8:

$$f(x) = p1*x^8 + p2*x^7 + p3*x^6 + p4*x^5 + p5*x^4 + p6*x^3 + p7*x^2 + p8*x + p9 \tag{3}$$

Coefficients:

p1 = -9.914e+04

p2 = 4.178e+05

p3 = -7.451e+05

p4 = 7.31e+05

p5 = -4.291e+05

p6 = 1.53e+05

p7 = -3.199e+04

p8 = 3507

p9 = -146.1

Goodness of fit:

SSE: 1.45e-20

R-square: 1

Adjusted R-square: NaN

RMSE: NaN

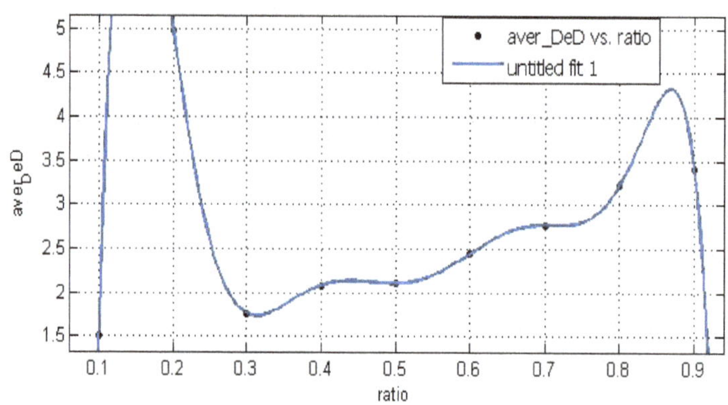

Figure 1. The scientific collaborate network: The relationship between the average degree and proportion. The blue curve represents the parameter value of sampling network.

3.2. *The neural network*

The neural network of the nematode C. Elegans Compiled by Duncan Watts and Steven Strogatz from original experimental data by White et al[19]. The neural network has 297 nodes and 2345 edges. The average degree of subnetwork of thenerual network is shown in Table 2 by different percents.

Table 2. The values of nodes, edges and average degree of subnetworks of the neural network based on different proportions.

Percentage	Number of edges	Number of nodes	The average degree
10%	57	28	4.07
20%	98	51	3.84
30%	186	76	4.89
40%	334	109	6.13
50%	754	158	9.54
60%	1030	183	9.54
70%	1241	206	12.05
80%	1452	235	12.36
90%	1834	262	14
Original data	2345	297	15.79

Form Table 2, number of nodes and edges is increased percentage of sampling. However, the average degree has a singular point too, which is 3.84 when the value of percentage equal 20%. These data are used to fit by MATLAB, and the result is shown in Figure 2. The fitting linear model is as Eq.(4)

Linear model Poly8:

$$f(x) = p1*x^8 + p2*x^7 + p3*x^6 + p4*x^5 + p5*x^4 + p6*x^3 + p7*x^2 + p8*x + p9 \quad (4)$$

Coefficients:

p1 = 1.424e+05
p2 = -5.822e+05
p3 = 9.995e+05
p4 = -9.346e+05
p5 = 5.161e+05
p6 = -1.707e+05
p7 = 3.268e+04
p8 = -3260
p9 = 130.8

Goodness of fit:

SSE: 3.874e-20
R-square: 1
Adjusted R-square: NaN
RMSE: NaN

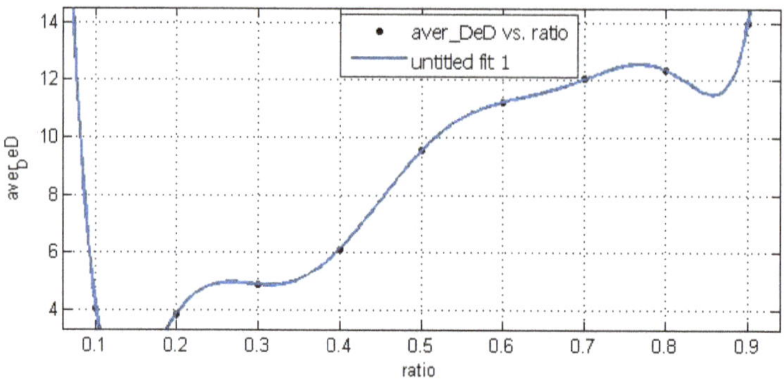

Figure 2. The neural network: The relationship between the average degree and proportion. The blue curve represents the parameter value of sampling network.

3.3. *The American football games network*

The file football.gml contains the network of American football games between Division IA colleges during regular season fall 2000, as compiled by M. Girvan and M. Newman. The network has 613 edges and 115 nodes. Analogously, the average degree of subnetwork of the America football games networks is shown in Table 3 and the fitted curve is given in Figure 3.

Table 3: The values of nodes, edges and average degree of subnetworks of the America football games network based on different proportions.

Percents	Number of edges	Number of nodes	The average degree
10%	4	7	1.14
20%	22	20	2.2
30%	54	34	3.18
40%	80	41	3.9
50%	172	61	5.64
60%	269	76	7.08
70%	322	83	7.76
80%	360	89	8.09
90%	445	98	9.08
Original data	613	115	10.66

The fitting linear model is as Eq.(5)

Linear model Poly1:

$f(x) = p1*x + p2$

Coefficients (with 95% confidence bounds):

p1 = 10.29 (9.048, 11.54)

p2 = 0.1935 (-0.508, 0.8951)

Goodness of fit:
SSE: 1.168
R-square: 0.982
Adjusted R-square: 0.9794
RMSE: 0.4084

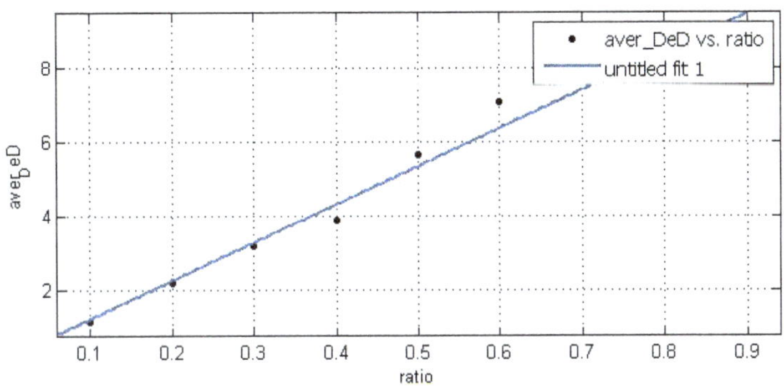

Figure 3. The America football games network: The relationship between the average degree and proportion. The blue curve represents the parameter value of sampling network.

Form Table 3 and Figure 3, number of nodes , edges and the average degree are increased percentage of sampling in the America football games network.

4. Conclusions

In this paper, three real complex networks are considered. The relationship between the average degree of subnetwork and sampling rate are obtained.

1)The relationship between the average degree of subnetwork and sampling rate is different for vary real complex networks. In the scientific collaboration network and the neural network, the value of average degree is not completely increased with sampling rate. However, the value of average degree is completely increased by sampling rate for the America football games network.

2)For the scientific collaboration network and the neural network, outliers of the average degree is obtained when the sampling rate is always 20%.

When the sampling rate increases, the node number also will increase. At the same time, the edges also increases with the sample proportion increasing. However, the rate of growth of nodes and edges may be different. So the changing rule of the average degree may be different. it may be associated with the type of network, also may be associated with the research methods. we are looking forward to researching deeper in further.

Acknowledgments

The work is partially supported by National Natural Science Foundation of China (Grant No. 61364030), the funding project of National Bureau of Statistical of China (Grant No. 2014LY128), the Doctoral Scientific Research Foundation of Hubei University for Nationalities (Grant No. 4148010), the Training Programs of Innovation and Entrepreneurship for Undergraduates of Hubei Province (Grant No. 201410517018).

References

1. M. E. Newman, *SIAM Review* **45**, (2003).
2. Q. Song, J. Cao and F. Liu, *Physics Letters A* ***374,*** (2010).
3. L. Lu, Y. C. Zhang, C. H. Yeung and T. Zhou, *Plos One* **6**,(2011)
4. D. Wei, X. Deng, X. Zhang, Y. Deng and S. Mahadevan, *Physica A: Statistical Mechanics and its Applications*, **392**, (2013).
5. M. Prec and P. Grigolini, *Chaos, Solitons & Fractals*, **56**, (2013).
6. D. Wei, Q. Liu, H. Zhang, Y. Hu, Y. Deng and S. Mahadevan, *Scientific Reports*, (2013).
7. L. Freeman, *Social networks*, **1**,(1979).
8. T. Opshl, F. Agneessens and J. Skvoretz, *Social networks*, **32**,(2010).
9. Y. He, X. Zhu and D. He, *J. Mod. Phys. B*, **18**, (2004)
10. K. Sznajd-weron and R. Weron, *International Journal of Modern Physics C*, **13**, (2002).
11. W. Li and X. Cai, Physical Review E, 69,(2004).
12. S. H. Lee, P. J. Kim and H. Jeong, Physical Review E, 73,(2006)
13. A. Clauset and C. Moorel, Physical Review Letter, 94, (2005).
14. P. H. S. Michael, Physical Review E, (2005)
15. P. H. S. Michael, PANS, 102,(2005).
16. J. J. Han, D. Dupuy, N. Bertin, M. E. Cusick and M. Vedal, Nature Biotechnology, 23, (2005).
17. J. Adibi, P. R. Cohen, C. T. Morrison, Proceedings of the ACM special interest group on knowledge discovery and data mining, (2004).
18. M. E. J. Newman, Physical Review E, 64,(2001).
19. D. J. Watts and S.H. Strogatz, Nature, 393,(1998).

Design of mathematical model of dairy safety and corresponding warning system based on C#

Yu Sun, Jin-Liang Liu and Hai-Ling Zhang, Xiao-Yi Huang,
Shanghai Ocean University,
Pudong New District, Shanghai, P.R.China;
E-mail: 499409708@qq.com, 578090748@qq.com

Wen-Fang Mei
Shanghai Yuang Food Quality Inspection Co. Ltd

Chao-Jin Wang[†]
Corresponding author
College of Food Science and Technology, Shanghai Ocean University,
No.999, Huchenghuan Rd, Nanhui New City, Shanghai, P.R. China;
[†]E-mail: cjwang@shou.edu.cn

In this paper, a mathematical analysis is made via C# programming language based on a common computer system through mathematical analysis knowledge and standards of dairy product indexes. Relevant programming language is also compiled and a mathematical analysis model is established. In addition, a comprehensive detection, analysis and warning system of danger coefficients about dairy product indexes is designed, which issues security warnings for dairy products.

Keywords: Dairy Product; Security Warning; C#.

Dairy industry develops rapidly in China during recent years. According to data published by National Bureau of Statistics, the milk (fresh and raw milk) output of China was 37.25 million tons in 2014, increasing by 1.94 million tons when compared with 35.31 million tons in 2013, with a growth rate of 5.5%. Large-scale dairy herds monitored by China's cow industry technology system increased by 4% in 2014 when compared with 2013, and the output of fresh and raw milk rose by 10%. Moreover, there are richer and richer varieties of dairy products, consumers become more and more rational in consumption, and market competition turns white-hot. Therefore, a higher requirement is raised for quality security of dairy products.

Thereby, it is very important for the either present or the future and for either consumers' health or development and construction of dairy product enterprises to increase the dairy product quality and enhance the security control force of dairy products.

1. Material and Methods

1.1. *Realization of dairy product security warning*

By utilizing relevant data about the existing dairy products and combining with the functions of C# program, relevant codes are compiled and a dairy product quality security warning program is established. This program compares important indexes like protein content, psychrophilic bacteria content, fat content, aflatoxin content, and melamine content in dairy products with national standards. Then multiple curves are drawn and unqualified samples are found out rapidly and intuitively, so as to conduct security warning for dairy products.

1.2. *Data selection*

The expert program requires many test data, so this paper collects test data of dairy products on some northern and southern markets as well as some data gained from experiments from 2011 to 2013. In this project, several data that have important influences on dairy product quality are selected for analysis, and the data include: protein content, fat content, aflatoxin content, melamine content, and psychrophilic bacteria content. Besides, the contrast value in the program is set as national standard value. By taking the detection of ultra-temperature sterilized milk as an example, the standard values of various data in the program are set as follows: protein content, 2.9g/100g[1]; fat content, 3.1g/100g[1]; melamine content, 2.5mg/kg[2]; aflatoxin content, 0.5μg/kg[3]; psychrophilic bacteria content , 1000cfu/ml[4].

1.3. *Implementation instance*

Part of protein testing code of Dairy safety warning is as follows:

```
static class Program
{ [STAThread]
  static void Main()
  {
    Application.EnableVisualStyles();
    Application.SetCompatibleTextRenderingDefault(false);
    Application.Run(new Form1());
  }
}
```

```
public class MilkData
{
   private double Protein;
 public MilkData()
   {
      Protein = 0;
}
   public MilkData(DataRow dr)
   {
      Protein = Convert.ToDouble(dr[0]);
   }

   ~MilkData()
   {
   }
   public double Pro
   {
      get
      {return Protein;
      }
      sct
      {
         Protein = value;
      }
   }
public bool Judgment()
   {
      if(Protein < 2.8)
      {
         return false;
      }
public string StrJudgment()
   {
      if (Judgment() == true)
      {
         return "QUALIFIED";
      }
      else
      {
```

```
            return " UNQUALIFIED ";
          }
        }
      }
      public class AllMilk
      {
        private List<MilkData> all = new List<MilkData>();
        private int count;
        public int Count
        {
          get
          {
            return count;
          }
        }
        private int accordCou;
        public int AccCou
        {
          get
          {
            return accordCou;
          }
        }
        DataSet LoadDataFromExcel(string filePath)
        { try
          { string strConn;
            strConn = "Provider=Microsoft.ACE.OLEDB.12.0;Data Source=" +
filePath + ";Extended Properties='Excel 12.0;HDR=Yes;IMEX=1;'";
            OleDbConnection OleConn = new OleDbConnection(strConn);
            OleConn.Open();
            String sql = "SELECT * FROM  [Sheet1$]"; OleDbDataAdapter
OleDaExcel = new OleDbDataAdapter(sql, OleConn);
            DataSet OleDsExcle = new DataSet();
            OleDaExcel.Fill(OleDsExcle, "Sheet1");
            OleConn.Close();
            return OleDsExcle;
          }
          catch (Exception err)
          {
```

```
            MessageBox.Show("data into Excel err! err reason: " + err.Message,
"tips",MessageBoxButtons.OK, MessageBoxIcon.Information);
            return null; }
        }
          public AllMilk()
        {
          accordCou = 0;
          count = 0;
        }
        public AllMilk(DataSet ds)
        { accordCou = 0;
          count = ds.Tables[0].Rows.Count;
          for (int i = 0; i < ds.Tables[0].Rows.Count; i++)
          { MilkData md = new MilkData(ds.Tables[0].Rows[i]);
            if (md.Judgment())
            {  accordCou++;
            }
            all.Add(md); }
        }
        public AllMilk(string Path)
        {
          accordCou = 0;
          DataSet ds = LoadDataFromExcel(Path);
          count = ds.Tables[0].Rows.Count;
          for (int i = 0; i < ds.Tables[0].Rows.Count; i++)
          {
            MilkData md = new MilkData(ds.Tables[0].Rows[i]);
            if (md.Judgment())
            {
              accordCou++;
            }
            all.Add(md);
          }
        }
        ~AllMilk()
        {
        }
        public void GetAllMilk(string Path)
        {
```

```
        accordCou = 0;
        DataSet ds = LoadDataFromExcel(Path);
        count = ds.Tables[0].Rows.Count;
        for (int i = 0; i < ds.Tables[0].Rows.Count; i++)
        {
            MilkData md = new MilkData(ds.Tables[0].Rows[i]);
            if (md.Judgment())
            {
                accordCou++;
            }
            all.Add(md);
        }
    }
    public void InitMilk(DataSet ds)
    {
        count = ds.Tables[0].Rows.Count;
        for (int i = 0; i < ds.Tables[0].Rows.Count; i++)
        {
            MilkData md = new MilkData(ds.Tables[0].Rows[i]);
            if (md.Judgment())
            {
                accordCou++;
            }
            all.Add(md);
        }
    }
    public void Clear()
    {
        all.Clear();
        count = 0;
        accordCou = 0;
    }
    private DataTable RetAllPro()
    {
        DataTable dt = new DataTable();
        dt.Columns.Add("Amount");
        dt.Columns.Add("Protein");
        dt.Columns.Add("Protein testing value");
        for (int i = 0; i < all.Count; i++)
```

```
        {
          DataRow dr = dt.NewRow();
          dr["Amount "] = i + 1;
          dr["Protein "] = all[i].Pro;
          dr["Protein testing value "] = 3.2;
          dt.Rows.Add(dr); }
        return dt; }public bool
DataToChart(System.Windows.Forms.DataVisualization.Charting.Chart
ca,string dataname)
      {if (Count > 0) {string Company = "";
          DataTable dt = default(DataTable);
          switch (dataname)
          { case "Protein":
                dt = RetAllPro();
                Company = "g/100g";
                break;    default:
                MessageBox.Show("Your option does not exit! ", "Error!");
                return false; }
          ca.DataSource = dt;
          System.Windows.Forms.DataVisualization.Charting.Series  s  =  new
System.Windows.Forms.DataVisualization.Charting.Series();
          s.ChartType                                                      =
System.Windows.Forms.DataVisualization.Charting.SeriesChartType.Line;
          s.LegendText = dataname + "testing value" + Company;
          s.YValueMembers = dataname + " testing value ";
          ca.Series.Add(s); ca.Series[0].YValueMembers = dataname;
          ca.Series[0].XValueMember = "amount";
          ca.Series[0].LegendText = dataname+Company;
          ca.DataBind(); ca.Visible = true;
          return true; }
        else{
          MessageBox.Show("Unselect Data! ", "Error!");
          return false;
        }
      }
      public bool DataToDataGridView(DataGridView dgv)
      { if (count > 0) {
          dgv.Rows.Clear();
          dgv.Columns.Clear();
```

```
        dgv.Columns.Add("Pro","Protein(g/100g)");
        for (int i = 0; i < all.Count; i++)
        { dgv.Rows.Add(all[i].Pro, all[i].Fat, all[i].Afl, all[i].Mel, all[i].Psy,
all[i].StrJudgment());}                dgv.AutoSizeColumnsMode                =
DataGridViewAutoSizeColumnsMode.AllCells;
        return true; }
      else
      {
        return false;
      }
    }
```

The data are processed before the program is operated for warning. The first line of data is protein, the second line is fat, the third line is aflatoxin, the fourth line is melamine, and the fifth line is psychrophilic bacteria. Then data files are selected for judgment according to the hints of the program.

2. Figures

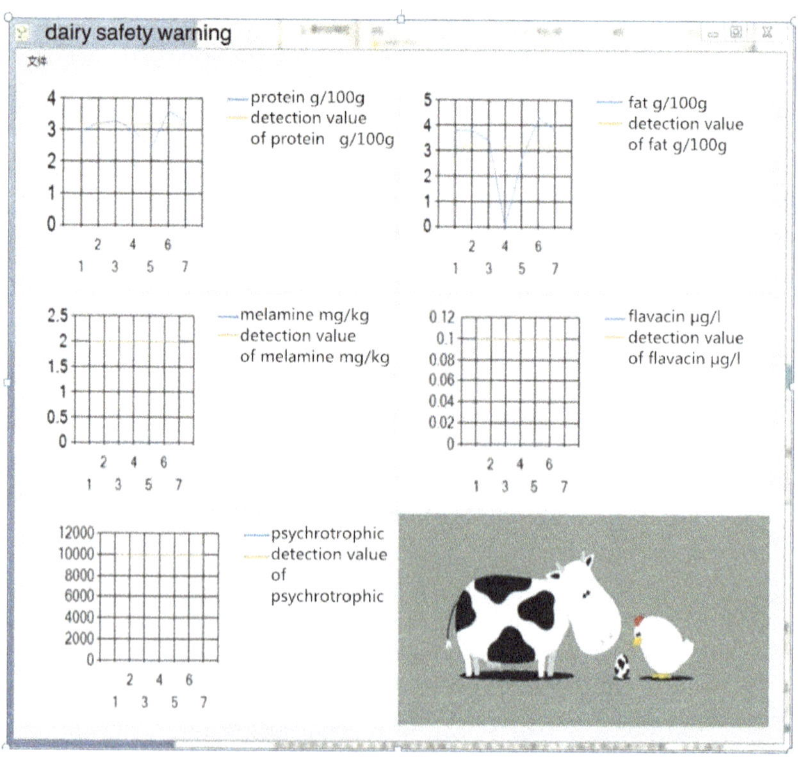

Graph 1. Result of the program

3. Result and Discussion

By reading a large number of national standards about dairy products and relevant literatures, and combining with the quality security situations on the current dairy product market and the actual demands of the market and consumers, this project designs a dairy product quality security warning program via common PC system and C# programming language. By comparing several important indexes that have great influences on dairy product quality like protein content, psychrophilic bacteria content, fat content, aflatoxin content and melamine content with national standards, multiple curves are drawn and unqualified samples are found out rapidly and intuitively. Unqualified dairy products are picked out and security warning is given. In this way, the purpose of giving security warning for dairy product market is realized, and unqualified dairy products are prevented from entering the market. Thus the security of the market and consumers can be guaranteed. In addition, the software has a low requirement for hardware, so it is applicable to the dairy product trading market of most regions. With convenient operation, it will greatly improve the detection efficiency and can be generalized in a large scale throughout the country.

Acknowledgments

Wen-fang Mei, Shanghai Yuang Food Quality Inspection Co. Ltd.

References

1. GB 25190-2010 National Food Safety Standard: Sterilized Milk.
2. Announcement on the Limit of Melamine in Food (Announcement of Ministry of Health 2011, No. 10).
3. GB 2761-2011 National Food Safety Standard: Limit of Fungaltoxin in Food.
4. GB 19301-2010 National Food Safety Standard: Raw Milk.

Effect of chitosan on quality attributes and cell wall enzymatic activity of blueberry fruits*

Jia Deng†, Zheng-Jun Shi‡ and Hui-Ming Liu§

Key laboratory for Forest Resources Conservation and Use in the Southwest Mountains of China, Ministry of Education, Southwest Forestry University, Kunming, 650224, PR, China

E-mail: †dengjia1983@163.com, ‡shizhengjun1979@163.com, §swfuedu@163.com

Blueberry fruits have a very short shelf-life period due to rapid loss of weight, senescence, and pathogen infection, which quickly causes the fruit to deteriorate. This study aimed to evaluate the effect of chitosan coating treatments on the quality attributes and the softening of blueberry fruits. Chitosan coatings at concentrations of 0.5 and 1.0% were applied to fresh blueberries before being stored at 4°C. Changes in quality characteristics and cell wall enzymatic activity were evaluated. Chitosan-coated blueberries lost less weight and had an increased firmness than the control group. The chitosan coating at 1.0% decreased the activity of cell wall hydrolase activities. These findings suggest that the use of 1.0% chitosan coating at cold temperatures is useful for slowing deterioration, maintaining quality and extending the shelf-life of blueberry fruits.

Keywords: Blueberry; Chitosan; Cell wall enzymes; Shelf-life.

1. Introduction

Blueberry fruit have a very short shelf-life due to their susceptibility to mechanical injury, excessive texture softening, physiological disorders and infection caused by several pathogens that can rapidly reduce the quality of fruit, and that make economy losing [1]. Therefore it is necessary to develop strategies to increase their storage life.

During ripening, the cell wall polysaccharides are extensively modified, mostly by the action of a large number of cell wall hydrolase enzymes, resulting in solubilisation, depolymerisation and rearrangements of their associations, which eventually affect cell wall strength and lead to fruit softening [2,3]. Polygalacturonase (PG) and Pectinesterase (PE) are two major enzymes that act on the pectin fraction of the cell wall. PE catalyzes pectin demethylation, making the walls susceptible to further degradation by PG [4]. Cellulase (Cx)

* Work partially supported by grant of the Yunnan Educational Scientific Research Fund Project.

leads to the loosing of cell wall by degrading cellulose which is the skeleton material of cell wall.

Blueberries are traditionally treated with different fungicides to control postharvest decay, but they leave residues that have potential risks to humans and the environment [5]. Globally consumers demand more natural, environmentally friendly food production, with high quality and an extended shelf-life, without any chemical preservatives. Therefore, effective measures for increasing shelf-life have to be explored in markets, while maintaining the nutritional quality of fruit [6].

Chitosan has been studied for extending shelf life of some fresh berry fruits. Chitosan can create a modified atmosphere, which delays produce senescence. It can be potentially be used for minimizing fruit respiration rate and may serve as a barrier to water vapor, reducing moisture loss and delaying fruit dehydration [7]. In addition, chitosan has shown delayed decay of the fruit, possibly due to a direct or indirect defense response of the fruit to chitosan [8].

The objective of the study was to evaluate the effect of chitosan postharvest treatments on fruit quality attributes, cell wall hydrolase activities, and softening inhibition of blueberry fruit.

2. Materials and Methods

2.1 *Materials*

Blueberries of the 'Brightwell' were grown in the experimental orchard of Southwest Forestry University, Kunming, Yunnan, in China. The fruits were collected at their commercial harvest maturity according to ground color. After harvest, the fruits were immediately transferred to the horticulture laboratory of Southwest Forestry University.

2.1.1. *Sample preparation*

After harvest, size uniformity and an absence of defects fruits were randomly divided into three lots, each about for 500 g, the fruits were treated with 0.5% chitosan, 1.0% chitosan and untreated (control). The fruits were packaged in plastic cases (30×30cm) with lots of holes (5 mm diameter), and then stored at 4°C for 25d. Fruits quality attributes and cell wall hydrolase were evaluated at 5 day intervals storage.

2.1.2. *Quality attributes*

Weight loss was determined with three replicates (twenty fruits per replicate) per treatment and was recorded initially and each time during storage. Weight loss was calculated as: weigh loss = $(W_i-W_f)/W_i \times 100\%$, W_i being the initial sample weigh and W_f the final sample weight. Results were reported as percentage weight loss.

Fruit firmness was measured using a fruit pressure tester equipped with a 3.5 mm diameter flat probe. Maximum force for tip insertion to a depth of 0.5 cm was recorded. Ten fruits were measured for each treatment and storage time. Results were expressed in Newton (N).

2.1.3. *Determination of hydrolyzed enzyme activities*

Cell wall hydrolase were extracted according to the method described by Deng [9]. The frozen tissue (1.0 g) was homogenized in 8 mL of Tris–HCl (50 mM, pH 7.0) and then centrifuged for 10 min at 10,000×g. The supernatant was used to assay for enzyme activity. All of these steps were performed at 4 °C.

Polygalacturonase (PG) activity was measured by the method of Deng [9]. The reaction mixture contained 5 mL of 20 mM sodium acetate buffer (pH 4.0), 2 mL of 1% (w/v) polygalacturonic acid, 1 mL of the crude enzyme. One unit of enzyme was defined as the amount of enzyme required to liberate 1 μg of galacturonic acid released g^{-1} fresh weight (FW) min^{-1}.

Cellulase (Cx) activity was measured by the method the same way as PG. The reaction mixture contained 1 mL of crude enzyme, 2 mL of 1% (w/v) carboxymethyl cellulose, and 5 mL 0.1 M sodiumacetate buffer (pH 5.0). one unit being defined as the amount of the enzyme that catalysed the formation of 1 μg of glucose released g^{-1} FW min^{-1}.

Pectinesterase (PE) activity was measured by the method with 2 mL of crude extract was mixed with 2 mL of 1% (w/v) citrus pectin and titrated with 0.05 M NaOH to maintain pH 7.4. One unit of enzyme activity was calculated as 1 μmol NaOH consumed g^{-1} FW min^{-1}.

3. Results and Discussion

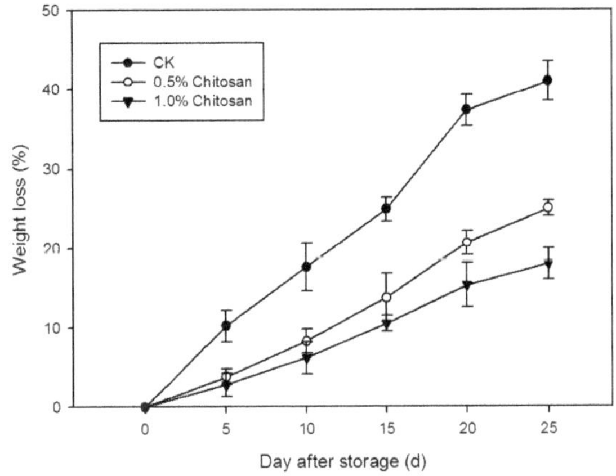

Fig. 1 Change in weight loss of blueberry fruit treated with chitosan coating application during 25 days storage at 4°C. Values are the means of replicates ± SE.

The effect of chitosan coating on the weight loss of blueberry is shown in Fig. 1. The weight loss increased during the shelf-life in all coated as well as control samples, with a slow increase observed with coated fruit. The 1.0% chitosan coating was the most effective treatment in maintaining lower levels of weight loss at 25 days of storage. It is commonly believed that migration of water from fruit to the environment is the major cause of weight loss of fruit during storage [7]. Our study was consistent with previous studies showing a reduction in weight loss was due to the effects of chitosan coatings, which served as barrier to water vapor thus reducing weight loss.

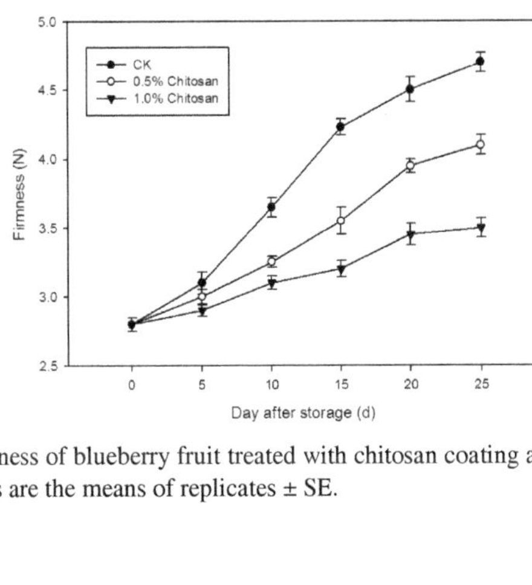

Fig. 2 Change in firmness of blueberry fruit treated with chitosan coating application during 25 days storage at 4°C. Values are the means of replicates ± SE.

The results recorded for the firmness of chitosan coating and control blueberry fruits are shown in Fig. 2, which showing a significant increase in firmness during a prolong storage period, and the greatest firm occurred in control samples, probability due to water loss [10]. Control fruit exhibited dramatic increase in fruit firmness from 2.8 N at 0 d to 4.7 N at 25 d, while chitosan coating treated fruits showed lower value 4.1 N and 3.5N respectively, at 25 d at low temperature storage. It is commonly believed that migration of water from fruit to the environment is the major cause of weight loss of fruit during storage. Results confirmed that chitosan coating had a significant effect on reducing weight loss, delaying hardening especially for later storage.

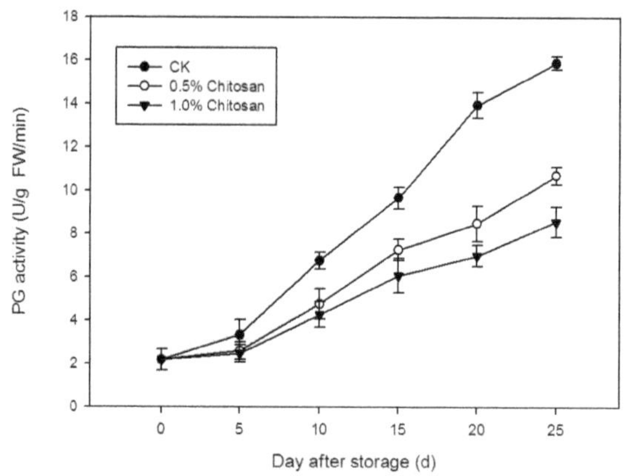

Fig. 3 Change in PG activity of blueberry fruit treated with chitosan coating application during 25 days storage at 4°C. Values are the means of replicates ± SE.

Fruit softening was closely related to an increase in the activity of PG, and the partial inactivation of PG could result in reduced pectin solubilization and slower softening [6]. Significant differences in the activity of PG were observed among the coated and control samples (Fig. 3). The results showed that at the beginning of the storage there was less PG activity, and the activity tended to increase significantly during the storage period in both the control and coated blueberry fruit. Application of chitosan coating fruit showed dramatic reductions in activity of PG throughout the storage period, especially the treatment of 1.0% chitosan which had the maximum influence in terms of restriction of PG activity compared to 0.5% chitosan. The results demonstrated that the chitosan coating markedly inhibited PG activity. Chitosan is a negatively charged polysaccharide and is likely to bind directly to the pectins which form the chitosan-pectin and thereby effectively prevent the access of pectinolytic enzymes such as PG to the substrate of the cell wall and thus help maintain fruit firmness [8].

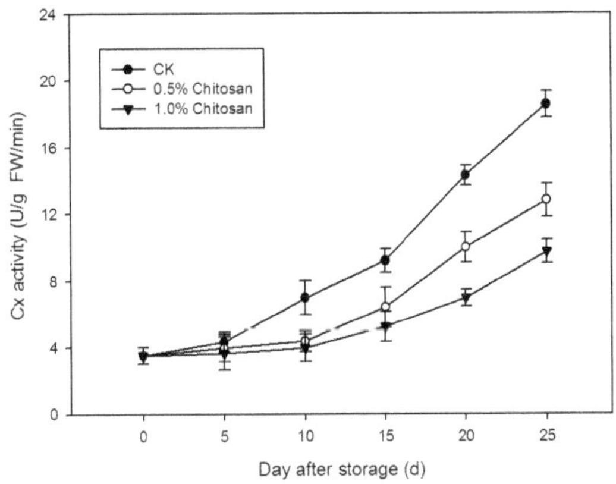

Fig. 4 Change in Cx activity of blueberry fruit treated with chitosan coating application during 25 days storage at 4°C. Values are the means of replicates ± SE.

As shown in Fig. 4 the activity of Cx in the coated blueberry was significantly lower than the control sample throughout the shelf-life. The activity of Cx was found to be low in the fruit treated with chitosan during 25 days of storage. The results revealed that the Cx activity increased during storage in blueberry fruits, but the activity levels were less in the coated samples, and among the tested coatings, 1.0% chitosan was superior in inhibiting Cx activity. During ripening, Cx plays an important role in fruit softening, and the results from this study are agree with previous studies [11]. As has been reported, that chitosan has the ability to inactivate or inhibit activity of several enzymes which cause deterioration of fruit [12].

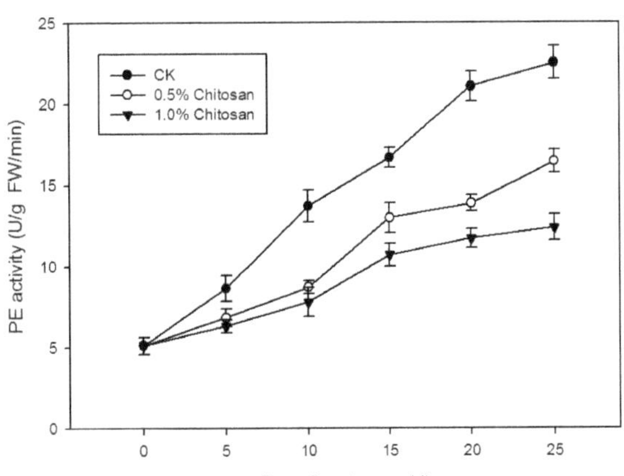

Fig. 5 Change in PE activity of blueberry fruit treated with chitosan coating application during 25 days storage at 4°C. Values are the means of replicates ± SE.

PE is another important enzyme related to the softening of fresh fruit. As shown in Fig. 5, PE activity in coated as well as control blueberry had significantly increased during storage. Application of chitosan coating on showed noticeable reductions in the activity of PE during storage. At the end of the storage period, in comparison with control, the 1.0% chitosan coating showed the maximum influence in terms of restriction of PE activity. Pectins are the main constituents of the middle lamella and primary cell wall of the fruit, and are hydrolyzed by PE to generate demethylated pectins that can be more easily hydrolyzed by PG and thus causing the depolymerization of pectins [11].

4. Summary

Using application chitosan inhibited loss of weight, loss of firmness, cell wall enzyme activities, and prolonged shelf-life of blueberry fruit at cold temperature.

In conclusion, this research demonstrates that chitosan coating application at harvest provide positive results for postharvest texture quality of blueberry during storage by reducing hydrolyzed enzyme activities, and delaying softening.

5. Acknowledgment

This study was supported by Yunnan Educational scientific research fund project (2014Y331), and Higher Educational Institutions scientific research fund project (111406). The corresponding author of this article was Zhengjun Shi.

References

1. J. Hancock, P. Callow, S. Serce, E. Hanson and R. Beaudry, Effect of cultivar, controlled atmosphere storage, and fruit ripeness on the long-term storage of highbush blueberries, HortTechnology. 18.2 (2008) 199–205.
2. D. A. Brummell, V. Dal Cin, C. H. Crisosto, J. M. Labavitch, Cell wall metabolism during maturation, ripening and senescence of peach fruit, J. Exp. Bot. 55.405 (2004) 2029–2039.
3. L. F Goulao, D. J Cosgrove, C. M Oliveira, Cloning, characterisation and expression analyses of cDNA clones encoding cell wall-modifying enzymes isolated from ripe apples, Postharvest Biol. Tec. 48.1 (2008) 37–51.
4. D. A Brummell, M. H Harpster, Cell wall metabolism in fruit softening and quality and its manipulation in transgenic plants, Plant Mol Biol. 47 (2001) 311–340.

5. H. Li, T. Yu, Effect of chitosan on incidence of brown rot, quality and physiological attributes of postharvest peach fruit, J. Sci. Food. Agri. 81.2 (2000) 269–274.
6. N. B Gol, P. R Patel, P. T. V Ramana Rao, Improvement of quality and shelf-life of strawberries with edible coatings enriched with chitosan, Postharvest Biol. Tec. 85 (2013) 185–195.
7. J. Duan, R. Wu, B. Strik, Y. Zhao, Effect of edible coatings on the quality of fresh blueberries (Duke and Elliott) under commercial storage conditions. Postharvest Biol. Tec. 59.1 (2011) 71–79.
8. G. A. González-Aguilar, E. Valenzuela-Soto, J. Lizardi-Mendoza, F. Goycoolea, M. A. Martínez-Téllez, M. A. Villegas-Ochoa, I. N. Monroy-García, J. F. Ayala-Zavala, Effect of chitosan coating in preventing deterioration and preserving the quality of fresh-cut papaya 'Maradol', J. Sci. Food. Agri. 89.1(2009) 15–23.
9. J. Deng, Z. J. Shi, X. Z. Li, H. M. Liu, Effects of cold storage and 1-methylcyclopropene treatments on ripening and cell wall degrading in rabbiteye blueberry (*Vaccinium ashei*) fruit, Food Sci. Technol Int. 2013: 1082013213483611.
10. P. M. Allan-Wojtas, C. F. Forney, S. E. Carbyn, K. U. K. G. Nicholas, Microstructural indicators of quality-related characteristics of blueberries-An integrated approach, LWT- Food Sci. Technol. 34.1 (2001) 23–32.
11. R. Zhou, Y. Li, L. Yan, J. Xie, Effect of edible coatings on enzymes, cell membrane integrity, and cell-wall constituents in relation to brittleness and firmness of Huanghua pears (*Pyrus pyrifolia Nakai*, cv. Huanghua) during storage, Food Chem. 124.2 (2011) 569–575.
12. M. V. Bhaskar-Reddy, P. Angers, F. Castaigne, J. Arul, Chitosan effects on black mold rot and pathogenic factors produced by Alternaria alternate in postharvest tomatoes, J. Am. Soc. Hortic. Sci. 125.6 (2000) 742–747.

Author Index